Maurizio Tiepolo, Enrico Ponte, Elena Cristofori (Eds.)

Planning to Cope with Tropical and Subtropical Climate Change

Maurizio Tiepolo, Enrico Ponte,
Elena Cristofori (Eds.)

Planning to Cope with Tropical and Subtropical Climate Change

Managing Editor: Jan Barabach

Language Editor: Cindy Bongard

DE GRUYTER
OPEN

Published by De Gruyter Open Ltd, Warsaw/Berlin
Part of Walter de Gruyter GmbH, Berlin/Boston

ISBN: 978-3-11-048078-8
e-ISBN: 978-3-11-048079-5

Managing Editor: Jan Barabach
Language Editor: Cindy Bongard

www.degruyteropen.com

Cover illustration: Nsanje district, Malawi, 17 january 2015. During the flood
(Source of the picture: Department of Disaster Management Affairs, Malawi)

Contents

Acknowledgments

This book contains an edited selection of papers presented at the second international conference "Urban Impact of Climate Change in Africa: Planning with Scant Information" which took place at the Politecnico of Turin on 13th November 2013. The Conference was jointly organized by Maurizio Tiepolo, Alessandro Pezzoli (DIST – Politecnico and University of Turin) and Silvia Macchi (DICEA – Sapienza University of Rome) with the support of the Province of Turin.

The event closed the National Interest Research Program (PRIN) 2009SX8YBBH "Assessing, planning and managing locally the territory and the environment in Sub-Saharan Africa" realised with the scientific coordination of Maurizio Tiepolo.

The editors would like to thank all the participants in the Conference particularly those who presented papers and shared their experience with other speakers and participants.

At the Province of Turin, we would hereby like to thank Councillor Marco D'Acri, Paolo Foietta, Furio Dutto and Simonetta Alberico for making the event possible. We would also like to thank Sarah Braccio (chair of the early research achievements), Madhav Giri (secretariat) and Professor Raffaele Paloscia (DIDA – University of Florence) for his participation and the final intervention.

All the papers were adapted to the book. To this end, some integrative information was collected for which we thank Ying Yang (climate plans in large Chinese cities), Morena Zucchelli (Tabarre plans), Eleonora Nadal and Elena Ferro (photo interpretation of Niamey), Katiellou Gaptia Lawan, Abourhamane Daouda, Abdou Adam (climate and hydrological data on Niamey) and Manuel Simoncelli (surveys on Paraguay).

We would also like to thank Tiziana Vicario (COOPI) for having authorised the use of the report we carried out for the Chaco Rapére project (ECHO AM/BUD/2013/91001) and Vieri Tarchiani (Ibimet-CNR) for having authorised the use of the report we drawn up for the ANADIA-Niger Project (Italian development cooperation).

We are also indebted to Alessandro Pezzoli, Alessandro Toffoli and the anonymous referees on whose comments we were able to transform our papers into chapters of the book.

Finally, we would like to thank Luigi Genta Traduzioni for the translation of some chapters into English.

Maurizio Tiepolo[1] and Elena Cristofori[2]

1 Planning the Adaptation to Climate Change in Cities: an Introduction

1.1 Climate Change in Subtropical and Tropical Cities

Very heavy rains, the early start and end of the rain season, river floods, sea level rise (SLR), high temperatures and drought have been reported in subtropical and tropical cities for many years now. These phenomena regard a "change in the state of the climate that can be identified (e.g., by using statistical tests) by changes in the mean and/or the variability of its properties, and that persists for an extended period, typically decades or longer" (IPCC WGII, AR5 2014: 5) known as climate change (CC).

In large subtropical cities the impact of CC is important, due to the higher population densities and a higher concentration of goods and services relative to elsewhere. In these large tropical cities the impact is devastating due to size, the concentration of industrial and tertiary activities of an entire country localized within them, and because often there are not the resources to prepare for catastrophe or to rebuild.

The availability of long term data on daily rainfall, recorded by weather stations or by satellites, and of daily discharge of larger rivers has led to the characterization of CC in cities mainly in terms of pluvial and river floods and, sometimes, in terms of SLR. Heat waves, dust storms, and alteration of aquifers are less known components of CC due to the lack of complete and sufficiently long term databases to evaluate the change. They are less evident hazards but nonetheless considered as having serious impact, especially on human health (respiratory and gastrointestinal diseases, death).

Moreover, it is by now recognised that the urban form (land cover, texture and building density) has an impact on the microclimate. This is changing in all tropical and subtropical contexts. Consequently, we can expect it to influence local CC. Apart from some general studies (Alcoforado and Matzarakis 2010, Stone *et al.* 2010), the consequences of urban form on CC vulnerability are still only investigated in few cases (Tiepolo and Braccio 2015).

Climatic analysis methodologies, which are now widely represented in the literature, are not practiced often in the local climate plans adopted for subtropical and tropical cities.

---

1 Maurizio Tiepolo, associate professor of Urban and regional planning at DIST-Politecnico and University of Turin, is author of 1.2 and 1.3 paragraphs, maurizio.tiepolo@polito.it

2 Elena Cristofori, PhD student at Politecnico di Torino, is author of paragraph 1.1, elena.cristofori@polito.it.

Although some cities have been careful to reduce GHG emissions for over a decade now in order to help reduce global warming and, therefore, CC, it has only been in the last three years that "climate plans" have sharply increased in subtropical and tropical cities.

1.2 Planning the Adaptation of Cities to Climate Change

Adaptation to CC is "the process of adjustment to actual or expected climate and its effects" (IPCC 2014: 1). Adaptation can be realised before, during or after a climatic disaster (flood, heat wave, drought, etc.) through individual measures, plans, strategies and policies.

These tools may be directed at reducing GHC emissions that are considered responsible for global warming and, therefore, CC (e.g., mitigation plans), aimed at reducing impact (adaptation plans), and preparing for and managing emergencies in the event of catastrophes (emergency plans) or a mix of the first and second types (mitigation and adaptation plans).

Some cities have chosen, instead, to mainstream individual adaptation measures within the existing tools (master plans, local development plans) without producing other plans specific to CC.

Following the initial reviews on CC adaptation planning in the urban ambit (Perkins *et al.* 2007; Füssell 2007; Moser *et al.* 2008, Wheeler 2008), the literature has multiplied.

Among the books, some deal with cities from very heterogeneous climatic zones (Davoudi *et al.* 2009; OECD 2010; UN-Habitat 2011; Hoornweg *et al.* 2011; Hammer *et al.* 2011; Ford and Berrang Ford 2011; Otto-Zimmermann 2012), while others focus on mega-cities (World Bank 2010; Krellenberger 2014), macro-regions or large countries (Birkmann *et al.* 2009; Tanner *et al.* 2009; Sharma and Tomar 2010; World Bank 2010; Macchi and Tiepolo 2014; Krellenberg *et al.* 2014). For some years now, adaptation guidelines have been published by UN-Habitat and the World Bank (Hoornweg *et al.* 2012; Dodman 2012; World Bank 2013; Ingram *et al.* 2014), and many national guides to adaptation by cities are available (Buendía 2010; LGASA 2012; MDC 2013, to name but a few).

Despite this wide production and the latest, monumental fifth assessment report of the working group II of the IPCC (2014), gaps in understanding adaptation planning are still present in at least four aspects. Firstly, the comparison of similar cities. The adaptation measures depend on the hazard (therefore, the climatic zone), and on the way the city expands (formal/informal) and carries out specific functions (coastal port, internal interexchange point, etc.). A comparison of adaptation planning should be done on similar case studies.

Secondly, the consideration of the city and its surrounding area. Our understanding of hazards and risks in metropolitan belts and in the nearby rural surroundings from which, in the event of natural disaster, (climatic) migrations move towards the city, is still quite poor.

Thirdly, the integration of climate analysis and planning. The coincidental hazards in urban settlements are rarely characterised and do not determine specific adaptation measures.

Fourthly, analysis and planning methodologies that address the needs of local governments. In the case of scant information on hazards and hazard prone areas, local governments with low capacities and budgets often need snapshot methodologies for the identification of hot spots and decision-making instead developing a new information system for detailed risk assessments.

This book contributes to understanding these aspects.

1.3 Structure of the Book

This book is a collection of case studies in subtropical and tropical zonesand considers different types of cities: large (over 1 million population), intermediate (0.1–1 million population), secondary (less than 0.1 million population).

The book is divided into three sections: hazard, adaptation planning, best practices.

Different chapters could have taken another position, as they deal with specific aspects in several sections.

The first section brings together the chapters on the characterisation of the hazard in climate plans (Tiepolo and Cristofori), heavy rainfall and SLR (Sakai *et al.*), atmospheric drought (Pezzoli and Ponte), flood hazard (Bacci), alteration of aquifers due to sea water intrusion (Sappa and Luciani), marine aerosol transport towards shore (Piazzola *et al.*).

The second section includes the chapters on flood early warning (Cristofori *et al.*), flood risk preliminary mapping on the urban and regional scale as a support to decision making (Tiepolo and Braccio), drought risk (Bacci and Tarchiani), community-based adaptation measures in water sector (Biconne), CC vulnerability on the local scale (Giri), adaptation and contingency planning (Ponte) and the mainstreaming of adaptation measures in other tools (Macchi and Ricci; Shemdoe *et al.*).

The third section includes the chapters on some European best practices that are of interest for subtropical and tropical contexts such as hazard monitoring and risk assessment at regional (Prohom and Puig; Franzi *et al.*) and city level (Ronco *et al.*).

Overall, 12 contexts are explored: large cities (Dar es Salaam, Niamey), intermediate cities (Caraguatatuba, Tabarre, Zurich), secondary cities (Mekhé, Pragatinagar, Nawalparasi) and regions (Catalonia, Chaco, Gaza province, Piedmont, Réunion, Tillabéri). With the exception of Zurich, the case studies are divided equally between subtropical and tropical zones according the Koppen-Geiger classification after the

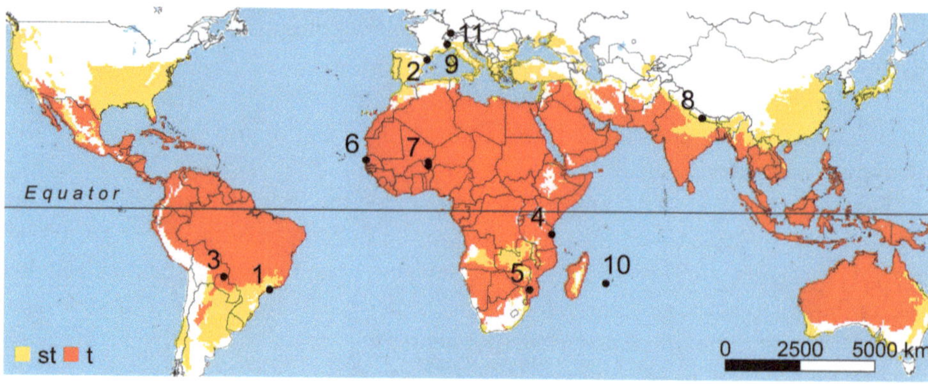

Figure 1.1: Subtropical (st) and tropical (t) zones and the twelve case studies: 1 Caraguatatuba, 2 Catalonia, 3 Paraguayan Chaco, 4 Dar es Salaam, 5 Gaza province, 6 Mekhé, 7 Niamey and Tillabéri region, 8 Pragatinegar and Nawalparasi, 9 Piedmont, 10 Réunion Island, 11 Zurich (by S. Braccio).

categories and subcategories re-unification of Trewartha (Rubel and Kottok 2010, Belda *et al.* 2014) (Figure 1.1).

References

Alcoforado, M.-J., and A. Matzarakis. 2010. Planning with urban climate in different climatic zones. *Geographicalia* 57: 5–39.

Belda, M., E. Holtanová, T. Holenka, and J. Kalvová. 2014. Climate classification revisited: from Köeppen to Trewartha. *Climate Research* 59: 1–13.

Buendía, E.C. 2010. *Guía metodológica para la adaptación a los impactos del cambio climático en las ciudades y opcion de mitigación de emisiones de gases de efecto invernadero*. Lima: Foro ciudades para la vida.

Davoudi, S., J. Crawford, and A. Mehmood. 2009. *Planning for climate change: strategies for mitigation and adaptation*. Earthscan Ltd.

Dodman, D. 2012. *Developing local climate change plans. A guide for cities in Developing countries*. Nairobi: UN-Habitat.

Ford, J., L. Berrang Ford eds. 2011. *Climate change adaptation in developed nations*. Springer.

Fussell, H.-M. 2007. Adaptation planning for climate change: concepts, assessment approaches, and key lessons. *Sustainability Science* 2 (2): 265–275.

Hammer, S.A., S. Mehrotra, C. Rosenzweig, and W.D. Solecki eds. 2011. *Climate change and cities first assessment report of the urban climate change research network*. Cambridge: Cambridge university press.

Hoornweg D., M. Freire, M.J. Lee, P. Bhada Tata, B. Yuen eds. 2012. Cities and climate change. Responding to an urgent agenda, Washington, The World Bank

Ingram, J., C. Hamilton et al. 2014. *Planning for climate change: A strategic, values-based approach for urban planners – Toolkit*. Nairobi: UN-Habitat.

IPCC. 2014. Annex II: Glossary. In *Climate change 2014: Synthesis report. Contribution of working groups I, II and III to the fifth assessmeny report of the Intergovernmental panel on climate change*, IPCC. Geneva: IPCC.

Jha, A.K., R. Block, and J. Lamond. 2012. *Cities and flooding. A guide to integrated urban flood risk management for the 21st century*. Washington: The World Bank.

Krellenberg, K., R. Jordan, J. Rehner, A. Schwarz, B. Infante, K. Barth, and A. Pérez eds. 2014. *Adaptation to climate change in megacities of Latin America. Regional learning network of the research project climate adaptation Santiago (CAS)*. Santiago: United Nations-ECLAC.

LGASA – Local Government Association of South Australia. 2012. Guidelines for developing a climate change adaptation plan and undertaking an integrated climate change vulnerability assessment. Adelaide.

Macchi, S., and M. Tiepolo eds. 2014. *Climate change vulnerability in southern African cities. Building knowledge for adaptation*. Springer. doi: 10.1007/978-3-31900672-7.

Moser, C., and D. Sattertwhite. 2008. Towards pro-poor adaptation to climate change in the urban centers of low- and middle income countries. University of Manchester GURK Working Paper # 1.

MDC-Municipio del Distrito Central. 2013. *Guía metodológica para incorporar la adaptación al cambio climático en la planificación del desarrollo*. Tegucigalpa.

OECD. 2010. Cities and climate change. OECD Publishing.

Otto-Zimmermann, K. ed. 2012. *Resilient cities* 2. Springer.

Perkins, B., D. Ojima, and R. Corell. 2007. *A survey of climate change adaptation planning*. Washington: The Heinz center.

Roggema, R. 2009. *Adaptation to climate change: A spatial challenge*. Springer.

Rubel, F., and M. Kottok. 2010. Observed and projected climate shifts 1901–2100 depicted by world maps of the Köppen-Geiger climate classification. *Meteorol. Z.* 19: 135–141.

Setiadi, N., J. Birkmann, and P. Buckle. 2009. *Disaster risk reduction and climate change adaptation: case studies from south and southeast Asia*. Bonn: United Nations University – Institute of Environment and Human Security.

Stone, B., J.J. Hess, and H. Frumkin. 2010. Urban form and estreme heat events: are sprawling cities more vulnerable to climate change than compact cities? *Environmental Health Perspectives* 118 (10): 1425–1428.

Tanner , T., T. Mitchell, E. Pollack, B. Guenther. 2009. Urban governance for adaptation: assessing climate change resilience in ten asian cities, IDS Working Paper 315

Tiepolo, M., and S. Braccio. 2016. Urban form dynamics in three sahelian cities: consequences on climate change vulnerability and sustainable measures. In *Adaptation planning in a mutable environment. From observed changes to desired futures*, ed. S. Macchi and L. Ricci. Springer.

UN-Habitat. 2011. *Cities and climate change. Global report on human settlements 2011*. UN-Habitat.

Wheeler, S.M. 2010. State and municipal climate change plans. The first generation. *Journal of the American Planning Association* 74 (4): 481–496.

World Bank. 2013. *Guide to climate change adaptation in cities*. Washington: The World Bank.

World Bank. 2010. *Climate risk and adaptation in asian coastal megacities. A synthesis report*. Washington: The World Bank.

Maurizio Tiepolo[3] and Elena Cristofori[4]

2 Climate Change Characterisation and Planning in Large Tropical and Subtropical Cities[5]

Abstract: In recent years, the number of large subtropical and tropical cities with defined climate plans has increased as a result of the initiatives of local governments, multi-bilateral development aid and development banks. Surveys carried out to date on climate planning consider the overall cities, at times by continent, without underscoring those that present planning deficiencies. For instance, we have no idea whether the cities that are most affected by hydro-meteorological and climatic disasters have plans, nor if their climate plans are ready to be implemented. Clarifying these aspects would strengthen the foundation of the current discussion on the United Nations' Sustainable Development Goals 2016–2030. Hence, the objective of this chapter is to ascertain the relevance and quality of climate planning in large subtropical and tropical cities populated by over 1 million inhabitants. Our survey found 344 large cities in the two climate zones concerned, and 82 of these have mitigation, adaptation, resilience or emergency plans, strategies or policies. We verified the relevance of these tools for the climate zones concerned, the type of economy and the frequency of hydro-meteorological and climate-related disasters. The quality of plans was assessed, ensuring that they had taken climate characterisation into account, that every measure was managed by a designated agency or office, and that funds were secured for implementing measures, as well as a monitoring and reporting sytem was defined.

The analysis of collected information underscores considerable differences between large cities in terms of per capita greenhouse gas emissions (which were double in the subtropics relative to the tropics) and exposure to hazards (which were greater in the subtropical zone). Emergency and mitigation plans were the most common, while adaptation plans and resilience strategies were more unusual. The relevance of plans is still weak, given that barely 1/4 of the large cities had a plan. Plans were unquestionably more common in the subtropics, especially in OECD countries and in the BRICS, while they were absent in the Least Developed Countries (LDCs), despite the

---

3 Maurizio Tiepolo is an associate professor of Urban and Regional Planning at the DIST (Inter-University Department of Territorial Sciences, Project and Policies) of the Politecnico and University of Turin and is author of paragraphs 2.1, 2.2, 2.3, 2.4, 2.5, maurizio.tiepolo@polito.it

4 Elena Cristofori is a PhD student at the Turin Polytechnic. She has authored paragraph 2.3.4.1, elena.cristofori@polito.it

5 The authors wish to thank Congling Liu, Kexing Huang, Xinman Hu, Jiaqi Ge, Gan Li, Suijie Liu, Yequi Ma, Shengfang Siu, Yichen Song, Ming Wu, Jing Yan and Weiyi Zhang for identifying contingency and adaptation plans in the largest cities of China, and Sarah Braccio for locating large cities in the climate zones.

presence of large cities that have been repeatedly affected by hydro-meteorological and climate-related disasters. Planning quality was good for 30% of cities only. In the remaining 70% of cities, climate characterisation was briefly defined; the planning process was fully funded by multi-bilateral development aid; measures were without a clear manager; cost, funds and monitoring of measures were not specified. Thus the indication being that local plans were still scarcely action-oriented. Hence, the fact that two sustainable development goals (# 11 and 13) address human settlements and climate change, respectively, especially through assistance to LCDs, seems justified.

Key words: Climate change, Climate-related disasters, Emergency plan, Mitigation plan, Adaptation action plan, Resilience strategy, Subtropics, Tropics, Large cities

2.1 Introduction

Climate change (CC) "refers to a change in the state of the climate that can be identified (e.g., by using statistical tests) by changes in the mean and/or by the variability of its properties, and that persists for an extended period, typically decades or longer" (IPCC 2014: 5). One of the most dominant drivers for CC is the concentration of greenhouse gases (GHG) in the atmosphere that alters the energy balance of the climate system. Carbon dioxide (CO_2) is the primary GHG responsible for CC, and it is mainly produced by burning fossil fuels and deforestation. Large cities (with over one million inhabitants) are responsible for GHG emissions, since they are heavily reliant on fossil fuel consumption to produce energy. Urban areas account for between 49% and 76% of CO_2 emissions from global final energy use. (Marcotullio *et al.* 2013; Seto *et al.* 2014).

The climate in large cities is also closely related to spatial configuration, density, topography and land cover, all factors that can modify temperature, humidity, precipitation as well as wind and air quality (Alcoforado *et al.* 2010). When extreme winter temperatures, droughts, floods, heat waves, landslides, storms and wild fires affect large cities, their impact is devastating if the cities are not well prepared, due to population density and the concentration of economic activities (McClean 2010; UN-Habitat 2009, 2011).

However, large cities also concentrate resources to reduce the impact of CC by adopting appropriate measures. Actually, a growing number of large cities is defining plans to reduce emissions that cause global warming in order to adapt to CC and withstand the consequences of any natural disasters. This mobilisation arises from the initiatives of several multilateral bodies (multilateral development banks, UN-Habitat, UNESCO, UNISDR, WHO, WMO), bilateral aid, associations and movements of local governments (ICLEI, C40, Covenant of Mayors), and commitments made by individual countries in the framework of international agreements (UNFCCC, HFA). These commitments are often converted into national laws that enforce local climate planning.

Past surveys on climate planning (Vergara 2005; CAI 2012; Carmin, Nadkami and Rhie 2012; Fraser 2012; CDP 2014) consider the overall cities, at times analysing them by continent, without distinguishing between towns and mega cities, tropical and temperate or boreal settlements, least developed countries and wealthiest economies. This fails to provide information about the extent of planning in cities that are most exposed to hydro-meteorological and climate-related disasters or in the largest ones, or the quality of the plans. Since the first generation of plans presented poor quality (Preston *et al.* 2011; Tang *et al.* 2010), this issue seems important for the current discussion on Sustainable Development Goals 2016–2030, where access to safe housing, the reduction of affected people and decrease in economic losses caused by disasters (UN 2014) are accepted as targets.

Hence, the objective of this chapter is to ascertain the relevance of climate plans and their quality.

We first defined two climate zones that contain half the urban population of the World and record the major hydro-meteorological and climate-related disasters, namely the subtropics and the tropics. The former includes large cities of countries that have joined the Organisation for Economic Co-operation and Development (OECD), such as Australia, Chile, France, Greece, Israel, Italy, Japan, Mexico, Portugal, South Korea, Spain, Turkey, United States, and the five BRICS (Brazil, Russia, India, China, South Africa). The tropics include parts of OECD and BRICS countries as well, but especially large cities in Developing Countries and Least Developed Countries (LDCs). This distinction must be taken into account when we consider adaptation measures that, besides responding to specific hazards presented by single climate zones, are funded consistently with the expansion capacity and model of the city, which are typical of the various economies.

Within each climate zone we first defined cities that are populated by over 1 million inhabitants (344) and then, among these, the ones that have enforced climate planning tool (82).

We then defined the types of plans, their relevance (by climate zone, by country and related to the frequency of disasters), and quality (implementation features and implementation control program).

The chapter is organized as follows: (i) the emergence and dissemination of climate planning, the types of plans, the measures, and the quality of climate plans, (ii) the general significance of the results achieved, and (iii) the recommendations.

2.2 Materials and Methods

Climate plans can be explored by contacting local administrations (Carmin, Nadkami and Rhie 2012; Baker *et al.* 2012), examining the plans (Corell *et al.* 2007; Birkmann *et al.* 2010; Preston *et al.* 2011), or by adopting both methods (Wheeler 2008; Bassett *et al.* 2010, GIZ-ICLEI 2012). This chapter is based on the second method. Its original approach, compared to previous surveys, can be found in three aspects, namely (i) narrowing the investigation to a homogeneous group of cities in terms of climate zone (subtropical and tropical) and number of inhabitants (over 1 million), (ii) comparison between cities that have a plan, versus those affected by hydro-meteorological and climate-related disasters during the past decade, (iii) the quality of plans.

The two climate zones were defined with Köeppen-Geiger's classification based on temperatures and rainfall observed over the period 1971–2000 (Rubel and Kottok 2010) on one 0.5 degree latitude/longitude regular grid, as presented on the website http://koeppen-geiger.wu-wien.ac.at/shifts.htm (Figure 2.1). Categories and subcategories were used according to Trewartha's classification (Belda *et al.* 2014), which is adopted by the FAO, the Joint Research Centre of the European Commission and by IPCC (www.fao.org/docrep006/ad6528/ad652e07.htm). The subtropical zone includes the categories dry summer Mediterranean, humid and dry winter, and the tropical one includes the categories wet-tropical rain forest, tropical wet in dry called savanna, subtropical desert, subtropical steppe.

Cities with over one million inhabitants were identified from the last census or projections of national demographic services. The geographical coordinates of every city were considered in order to choose those located within the subtropical (196) or tropical (148) zone.

A review of the literature helped to determine whether these cities had a strategy, a policy, or a plan. And if they did, the tool was downloaded from the city council's website. At times, the plans mentioned by the literature were either offline or not even mentioned in municipal documents that were accessible online. In the latter case, we did not consider the plan. Exclusion arises from the assumption that a plan must be public, especially if it is designed to involve stakeholders and when it appeals to the contribution of a broad range of local actors for its implementation.

Local governments with less than one million inhabitants were not considered, unless the plan concerned a supramunicipal jurisdiction (metropolitan area), such as Hyderabad, India, Lagos, Miami, Montevideo, Naples, Sydney and Turin, all of which had over one million inhabitants. This led to the exclusion of San Francisco, one of the first subtropical cities to define a climate action plan, as well as Melbourne and Perth. Plans that were in the process of being drawn up were not considered.

After the plans were collected, their relevance was assessed by comparing such planned cities against the ones listed in the EM-DAT database of the Centre for Research on the Epidemiology of Disasters (CRED), which records hydrological (general flood, flash flood, coastal flood, landslide), meteorological (tropical cyclone, local storm) and climate-related (drought, extreme winter conditions, heat wave, forest or scrub fires) disasters. The 283 disasters recorded between 2005 and 2014 were probably fewer than the ones that actually affected large subtropical and tropical cities during the past decade, since we did not consider disasters that affected the states and regions to which these cities belong, due to the existence of very vast ter-

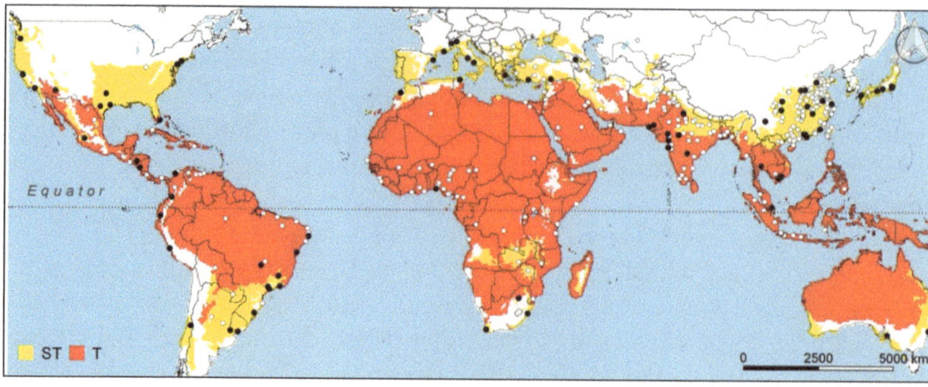

Figure 2.1: Subtropics (ST) and Tropics (T), 2014. Large cities provided by a mitigation and adaptation plan or strategy or an emergency/contingency plan (black dot) and without plan (circlet) (by S. Braccio).

ritories in which the disaster might only have grazed the urban area in question. This is often the case in Brazil, China, India and the USA.

The analysis of cities also took into account two aspects that are at the root of climate planning. First, we considered the amount of per capita GHG emissions, as estimated by emission inventories completed either prior to or concurrently with the climate planning process. Toward this end, we collected a total number of 48 GHG inventories. These inventories were produced with heterogeneous methods, but, until new ones are defined consistently with the COP20 protocol established in Lima (WRI 2014), they remain the only source of information to analyse the CO_2 emissions of such a considerable number of large cities. Secondly, we considered studies on the hydro-meteorological and climatic hazards that threaten cities. This information is also gleaned from climate plans.

2.3 Results and Discussion

2.3.1 Emergence and Dissemination of Climate Planning

Reducing the impact of CC entails (i) reducing the causes thereof, (ii) protecting the population and assets exposed both before and (iii) during a hydro-meteorological or climate-related disaster.

In 70% of the cases examined, the local governments planned these activities as the application of specific national or regional laws, as in the case of the Global Warming Solutions Act of California (2006). In the remaining cases, they do so after signing the US Conference of Mayors' Climate Protection Agreement (2005), the Covenant of Mayors (2008) or other unilateral initiatives to reduce the risk (Table 1.1). Autonomous planning initiatives that are foreign to national or state trends or to the influence of multilateral bodies, such as the ones underscored a few years ago in Quito and Durban (Carmin, Roberts and Anguelovski 2012), seem to be less feasible today.

Table 2.1: Large subtropical and tropical cities, 2015. Legal framework for 44 local climate planning tools.

Large city	Type of plan	Not specified	Supra-national	National/ Federal Law	Local government Act	State/ Regional Law	Local Law
Adelaide	A					•	
Ahmedabad		•					
Buenos Aires	MA		•				
Bangkok	M		•				
Barcelona	M		•				
Belo Horizonte	M		•				•
Cape Town	M			•			
Cartagena	A	•					
Casablanca	A	•					
Chongquing	E/C			•			
Durban	S						
Ganztiep	M			•			
Guayaquil	A	•					
Ho Chi Minh	A	•					
Houston	M	•					
Indore	S	•					
Karachi	S	•					
Kobe	M			2011			
Lagos	P			1999			
Los Angeles	MA	•					
Marseille	MA			2010			
Miami	M	•					
Milan city	M		•				
Milan metro	E/C					2004	
Montevideo	MA			•			
Mumbai	E/C			2005			
Nanjing	E/C			2014			

continued **Table 2.1:** Large subtropical and tropical cities, 2015. Legal framework for 44 local climate planning tools.

Large city	Type of plan	Not specified	Supra-national	National/ Federal Law	Local government Act	State/ Regional Law	Local Law
Nagoya	OP			•			
Naples	E/C					•	
Philadelphia	MA	•		2000			
Phoenix	M					•	
Portland	M				2007		
Porto Alegre	E/C						•
Recife	S			•			
Rio de Janeiro	S	•					
Rome city	M		•				
Rome metro	E/C				1998		
Salvador de Bahia	E/C			2012			2013
San Antonio	E/C		2000				
San José	S					•	
Santiago de Cali	E/C			1989			
São Paulo	MA	•					
Shenzhen	MA			•	•		
Singapore	M		•				
Surat	S		•				
Tegucigalpa	E/C			1993	•		
Tbilisi	M		•				
Tokyo	M	•					
Tunis	A	•					

*Kyoto protocol, Convenant of Mayors, Asian Cities Climate Change Resilience Network, US Conference of Mayors, USA HMA 2000.

A mitigation action is "a human intervention to reduce the sources or enhance the sinks of greenhouse gases" (IPCC 2014: 19). Mitigation planning is the most common climate tool in subtropical and tropical cities, as it is the outcome of long-term actions implemented by local government organisations, such as ICLEI (from 1991, over 1,000 members), C40 (from 2006, 75 members) and by movements like the Covenant of Mayors (from 2008, 5,881 participants), dedicated funds of Mulitlateral Development Banks (MDB) and of certain foundations (1,100 M US$ at the ADB, 140 million US$ from the Rockefeller Foundation and other partners). It also issues from dissemination of knowledge, and from pilot programs launched by the IPPC and other multilateral and bilateral bodies. The first mitigation plans for subtropical large cities were defined in the USA (Los Angeles, Philadelphia 2007), and this trend was rapidly adopted in Asia, South Africa, Latin America and Europe. They focus on reducing emissions in the sectors of transportation, commerce and industry, the residential sector as well as the field of waste collection and disposal by introducing changes that are at times very costly.

Adaptation planning focuses on "the adjustment in natural or human systems in response to actual or expected climatic stimuli or their effects, which moderates harm or exploits beneficial opportunities" (UNISDR 2009). Plans include non-structural measures (e.g. early warnings, flood drills) and structural measures (flood barriers, stormwater drainage, and resettlement of inhabitants from flood-prone areas). In the latter case, if local adaptation plans are not expressly envisaged by the general environment protection law (Uruguay) or by a specific national law (Philippines 2007), they are implicit in the Local Government Act of various countries (Australia 1999, South Africa 2000, Peru) when council functions are specified and often extended to "protect its area from natural hazards and to mitigate the effects of such hazards." In other cases (Argentina), the individual cities implement the specific national commitments made at the United Nations Framework Conference for Climate Change (UNFCCC). Adaptation planning is funded by dedicated MDB programmes, but is still scarcely practised as it has not been able to be implemented on the same global mobilisation levels as GHG mitigation.

Emergency or contingency planning is a "management process that analyses specific potential events or emerging situations that might threaten society or the environment, and establishes arrangements in advance to enable timely, effective and appropriate responses to such events and situations" (UNISDR 2009: 7). Once again, a national law envisages the disaster prevention device (Chile, China, Colombia, India, Philippines, etc.) or civil defence (Brazil), and subsequently the creation of municipal emergency committees to draw up dedicated plans. In other cases (Texas), the contingency plan is drawn up in compliance with rules defined by state commissions. Emergency planning is a very common practice. Several cities have committed to implement it by adhering to various declarations, the last of which is the Aqaba Declaration (2013).

Resilience strategies are the ultimate planning tool. They focus on strengthening the "ability of a community exposed to hazards to resist, absorb, accommodate and recover from the effects of a hazard in a timely and efficient manner, including through the preservation and restoration of its essential basic infrastructures and functions" (UNISDR 2009: 24). Since 2010, UNISDR has supported local resilience strategies, especially partnered in Asia by the Rockefeller foundation. The target of

Table 2.2: Large subtropical and tropical cities, 2005–2014. Incidence of hydro-meteorological and climate-related disasters, and establishing climate planning (source: EM-DAT and Survey by M. Tiepolo).

Large cities (> 1 M inhabitants)	N°	%
Affected by natural disasters 2005–2014	122	42
Idem, with plan	45	
Not affected by natural disasters 2005–2014	170	58
Idem, with plan	11	
Total affected	292	100
Idem, with plan	50	

these strategies is to strengthen planning, organisational and management skills in view of a disaster, rather than implementing structural measures.

At least 82 large cities have a climate planning tool today (Figure 1.1). Five of them have both a mitigation/adaptation or resilience plan as well as an emergency plan. Two thirds of the cities are subtropical. Sixty-six cities belong to the most developed economies (OECD) or to rapidly developing ones (BRICS), and only sixteen are situated in Developing Countries. Seventeen cities are Chinese, eight are in the United States and eight are European.

During the past five years, climate plans in large subtropical and tropical cities have increased by 150% compared to 2010; to be precise, 48 new plans have been added to the 33 existing at the time. Another 10 plans are being drawn up, basically in the subtropics, particularly in OECD and BRICS countries, namely Abu Dhabi, Algers, Bangkok, Curitiba, Fortaleza, Kathmandu, Porto Alégre, Rio de Janeiro, San Antonio (TX), San José (CA). Large LCD cities are still excluded.

However, 262 millionaire cities (76% of the total) still lack planning tools to face CC, even though they are not spared from hydro-meteorological and climate-related disasters. According to the EM-DAT database, 43 subtropical and tropical cities that were repeatedly affected by disasters during the past decade still lack a plan, including Quezon city-Manila (7 disasters during the past decade), Dakar, Havana, Luanda (5), Brazzaville, Managua, Maputo, Ndjamena, Niamey, Santo Domingo (4).

2.3.2 Tool Types and Their Characteristics

Large subtropical and tropical cities have strategies (10%), policies (1%) and plans (89%).

The strategy "refers to a general plan of action for addressing the impacts of CC, including climate variability and extremes. It may include a mix of policies and measures, selected to meet the overarching objective of reducing the vulnerability" (Lim *et al.* 2004: 250). The mitigation strategy starts from the emissions baseline, which is at times calculated by sector (transportation, commerce, industry, residence, waste), and is organised in initiatives, each of which lists the principal actions, without either localising them or specifying the expected results in terms of reducing emissions and, even less so, costs and funding modes (Tokyo) (Table 2.3). Some Brazilian cities have defined a mitigation policy (Belo Horizonte, Recife), that indicates the tools a city must have (plans, programmes), the initiatives to be pursued (renewable energies, compact city, strategic environmental assessment of plans, protection of biomasses that can reduce GHGs), and establishes the general and specific goal (e.g. percentage of emissions to be reduced over a period of time). The policy indicates mitigation and adaptation measures without localising them, namely transportation, energy, waste, health, construction works, land use, high risk buildings, permanent protection areas, maintenance of permeable surfaces, and forestation. The policy then indicates the implementation tools (annual report on emissions, tax reductions if renewable energies are used, reduction of GHG emissions/absorption, contributions to private reserves of natural heritage, training programmes, devices to reduce the GH effect, CO_2 market, civil defence, early warning system for extreme events, and CC municipal committee).

Table 2.3: Large subtropical and tropical cities. Localisation and impact of measures envisaged by 33 climate plans.

Large city	Zone	Plan type	Baseline	Goals	Options priorities	Measures		
						Localization	Impact mitigation potential	Scheduling
Adelaide	ST	A				occasional		
Antalya	ST	M	•	•		occasional	•	•
Bangkok	T	M				occasional	•	
Barcelona	ST	M						
Belo Horizonte	T	M	•	•			•	
Buenos Aires	ST	MA					•	
Cape Town	ST	M					occasional	
Cartagena	T	A						
Casablanca	ST	A				•	•	•
Durban	ST	A						

continued **Table 2.3:** Large subtropical and tropical cities. Localisation and impact of measures envisaged by 33 climate plans.

Large city	Zone	Plan type	Baseline	Goals	Options priorities	Measures		
						Localization	Impact mitigation potential	Scheduling
Gaziantep	ST	M					generic	
Guayaquil	T	A						
Houston	ST	M					•	
Los Angeles	ST	M						
Marseille	ST	MA						
Miami	T	M					•	
Milan city	ST	M	•	•			•	
Montevideo	ST	MA						
Naples	ST	M	•					
New York City	ST	A				•		
Philadelphia	ST	M					•	
Phoenix	T	M					•	
Portland	ST	M						
Rio de Janeiro	T	R						
Rome city	ST	M	•	•			•	
San José	ST	M					•	
Saitama	ST	M						
São Paolo	ST	MA		•				
Santiago Chile	ST	A						
Semarang	T	A						
Tbilisi	ST	M					•	
Tunis	ST	A				•		•
Zapopan	ST							
Total			5	5	0	6	15	3

In the best cases, plans localise the measures, describe them in detail, estimate the cost, define the lead city agency and funding sources, thus specifying priorities and the monitoring, evaluation and reporting program.

2.3.2.1 Mitigation

This group collects 25% of the planning tools traced. The prevalence of mitigation confirms the observations of Revi, Satterthwaite *et al.* (2014). Mitigation plans in subtropical zones are conceived in the USA, and usually comprise the emission inventory, GHG reduction goals and measures. The most detailed plans estimate the expected reduction of emissions, the risks resulting from CC, the cost, funding sources and the timing of every measure. The capability assessment, which considers the technical and political ability to implement the measures, is rarely conducted.

The GHG emission inventory of 48 subtropical and tropical large cities underscores the fact that per capita emissions indicate a mean value of 6.2 T (Table 2.4). But the value is more than two-fold in the subtropics (8 T) relative to the tropics (3.5 T). The first sector for emissions is transportation and the second is industry in both zones, while the residential sector (the considerable seasonal temperature variation demands heating and cooling) ranks third in the subtropics, and commerce ranks third in the tropics (where heating is not required and often households can't afford cooling costs). There are no direct relations between the demographic size of the city and per capita GHG emissions.

Table 2.4: Large subtropical and tropical cities. Total GHG emissions according to 48 plans detailed by sector (39 plans).

Climatic zone	GHG emissions*							GHG**	P**	GHG/pc
	Transport	Commerce	Industry	Residential	Waste	Other	Total	Total		
	%	%	%	%	%		MT	MT	Millions T	
Subtropical	29	13	24	22	3	3	594	1275	160	8.0
Tropical	30	6	28	6	4	3	176	348	100	3.5
ST & T	29	11	24	18	3	3	770	1623	260	6.2

*39 plans, **48 plans

2.3.2.2 Adaptation

Adaptation plans (11%) usually describe the consequences of CC in terms of expected hazards (temperatures, rainfall, etc.). Then they define the expected impact (vulnerability), which is rarely analysed in terms of risk (hazard probability* expected damage). Areas exposed to the greatest impact are often highlighted on maps to facilitate the definition of measures to be implemented. Adaptation plans then define the objectives and intervention axes. Usually, the measures are described with charts that also specify priority, temporal distribution and who must implement them. Adaptation plans rarely specify performance indicators and the monitoring and reporting

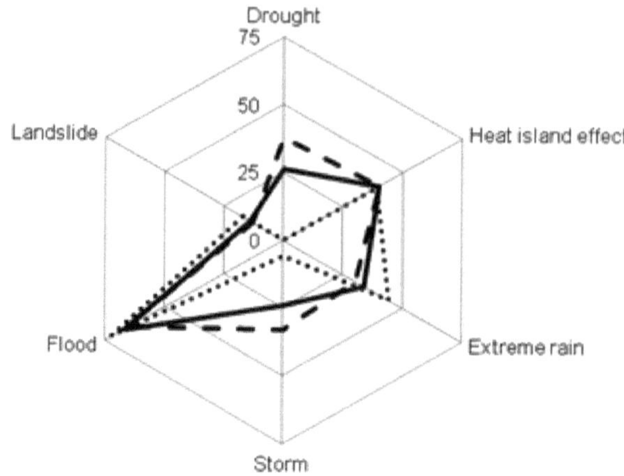

Figure 2.2: Large subtropical and tropical cities. Frequency (%) of main hazards mentioned in plans.

program (Table 2.3). The adaptation plans of large subtropical and tropical cities differ in terms of the hazards they must face. A study of 23 climate plans underscores the fact that the increased incidence of floods (generated by river or by sea level rise) is the hazard that is most cited as a consequence of climate change (61% of plans mentions them), followed by heat waves (52%), extreme rainfalls (30%), drought (30%) and storms (22%) and, with a lower incidence, landslides, snow and fire. But subtropical cities are exposed to a larger number of hazards than tropical ones. Secondly, subtropical cities feature drought, storms (hurricanes, cyclones) and extreme rain, while tropical cities are not affected by drought and storms are less frequent (Figure 2.2, Table 2.5).

2.3.2.3 Resilience

Resilience strategies (6%) are especially common in Asia, and only concern large tropical cities for the time being. Aside from structural measures, they define many actions based on accumulation of information (databases on hazards, hazard prone areas identification), management (establishing a CC coordination office), training and awareness sectors. The most accurate resilience strategies identify vulnerabilities, list the actions required to reduce them, define implementation phases and the relevant financial mechanism.

2.3.2.4 Emergency/Contingency

This group (40%) is especially common in Europe, Latin America and China but not in Africa, with the exception of South Africa. Sixty-one percent of plans traced were drawn up over the past three years. There are two types of plans, ones that specify

Table 2.5: Large subtropical and tropical cities. Main hazards according to 23 climate plans.

Large city	Zone	Plan	Hydrological		Meteorological		Climatic				
			Flood	Land slide	Intense rain	Storm	Drought	Cold	Snow	Heat wave	Fire
Adelaide	ST	A			•					•	•
Bangkok	T	M	•								
Barcelona	ST	M								•	
Buenos Aires	ST	MA	•		•					•	
Cape Town	ST	M	•		•						
Cartagena	T	A	•				•				
Casablanca	ST	A									
Durban	ST	A	•			•				•	
Los Angeles	ST	M	•				•			•	
Marseille	ST	MA			•		•			•	
Miami	T	M	•								
Milan city	ST	M									
Montevideo	ST	MA			•	•	•			•	
Naples	ST	M	•	•							
New York City			•		•	•	•			•	
Philadelphia	ST	M	•			•			•		
Phoenix	T	M								•	
Rio de Janeiro	T	R									
Rome city	ST	M	•								
Saitama	ST	M								•	
Santiago Chile	ST	A	•							•	
São Paolo	ST	MA	•		•	•	•			•	
Tunis	ST	A	•				•				
Total			14	1	7	5	7	0	1	12	1

the arrangements a local government must make in order to become operative in case of a disaster (Cape Town, Tshwane), and operative ones (Milan, Naples, Rome, San Antonio, Santiago de Cali, Turin), which describe the sequence of established operations the various actors are called upon to implement in order to respond to the emergency. These plans are, at times, limited to only one hazard (drought in Campinas and Dallas, floods in Jinan, Luoyang and Zhangjiagang) (Table 2.6). Though they do not solely refer to CC-related disasters, they overlap mitigation/adaptation plans with respect to the definition of zones that are either exposed or at risk. In cities that have both an emergency plan and an adaptation plan, the two tools refer to different bodies, specifically civil defence or fire brigade in the first case, city council/environmental sector in the second one. The plan is implemented after an early warning. The warning should be strategically communicated to areas that are exposed to the hazard, especially in case of floods. This only occurs in 30% of cases. However, only 18% of plans define the threshold of early warning, barely 20% define hazard prone areas and only in half the cases do they make use of maps.

The emergency plans of large cities in tropical zones better define the pre-emergency phase by describing the hazard prone areas (40%) and the early warning system (60%) more often. Conversely, those of large subtropical cities better detail the emergency phase by reporting the guidelines for potentially affected populations (30%), defining escape routes (20%) and emergency refuges (30%) that are virtually absent in tropical zones (Table 2.6).

Table 2.6: Large subtropical and tropical cities. Choice of 31 emergency/contingency plans.

Large city	**Zone**	**Flood**	**Land/Mud slide**	**Heavy rain**	**Storm**	**Wind**	**Drought**	**Cold**	**Snow**	**Heat**	**Fire**	**Thunderstorm**	**Tsunami**	**EWS**	**EW level**	**Hazard prone areas**	**Losses**	**Guidelines to people business**	**Escape routes**	**Emergency refuge location**
Ahmedabad	T									•				•						
Bazhong	ST								•											
Campinas	ST						•				•			•						
Cape Town	ST																			
Dallas city							•							**						-
Delhi	T	•												•						
Fukuoka	ST	•	•	•		•														
Hong Kong	T	•	•		•							•	•	•						
Hyderabad, PAK	T	•		•										•	T	•				•
Jinan	ST	•	•				•													
Lima	T	•	•													map				
Luoyang	ST	•													P					

continued **Table 2.6:** Large subtropical and tropical cities. Choice of 31 emergency/contingency plans.

Large city	Zone	Flood	Land/Mud slide	Heavy rain	Storm	Wind	Drought	Cold	Snow	Heat	Fire	Thunderstorm	Tsunami	EWS	EW level	Hazard prone areas	Losses	Guidelines to people business	Escape routes	Emergency refuge location
Milan	ST	•			•	•		•	•						SW					map
Mumbai	T	•														list				list
Nanchang	ST	•			•	•	•									map				list
Nanjing	ST	•	•	•			•		•		•			•		map				map
Naples	ST	•	•	•											P	map/ list			list	list
Panzhihua	ST	•					•								D				map	map
Philadelphia	ST	•			•	•	•			•						map	•	•	•	
Porto Alegre	ST																			
Quanzhou	ST	•		•	•					•						•				
Rio de Janeiro	T			•										•		•				
Rome	ST	•												•	R	map		•	map	map
Salvador de Bahia	T			•										•						
San Antonio	ST						•											•		
Santiago de Cali	T	•	•								•					map				
Tegucigalpa	T	•	•											•		list				
Tshwane	ST	•																		
Turin metro	ST	•		•										•						
Xiamen	ST				•															
Zhengzhou	ST				•			•		•										
Total		20	8	8	7	4	8	2	3	4	3	1	1	11		12	1	3	4	8

2.3.2.5 Other Plans

Master plans, sustainability or “green” plans, and local development plans that include adaptation measures converge in this group (8%). Mainstreaming adaptation within existing tools (or integration of adaptation concepts) (Levina and Tirpak

2006:15) is deemed to be the key to successful adaptation (Revi, Satterthwaite *et al.* 2014: 45; Bassett *et al.* 2010). This approach is adopted in Brazil where large municipalities act independently with their own policies and plans, which also occurs in India (Sharma *et al.* 2010) and other countries (Preston *et al.* 2011).

The main advantages of this approach should theoretically include the five items described below.

1. Low exposure to hazards after the inclusion of adaptation in land use planning (prohibition of building on land that is exposed to floods, creation of green areas to reduce the impact of heat waves) (Hyderabad).
2. Increased possibility of implementing new measures by including adaptation in the tool the local government is already carrying-out.
3. Charge the cost of mitigation to the private sector where urban extension occurs prevalently by subdivision plans (OECD countries) via including mitigation measures among the requisites a developer must meet to be authorised to develop the land (San José, CA).
4. Ensure cost reduction by optimising actions that are already either envisaged or funded in other sectors, as in the case of Millennium Development Goals for sewage devices and the adduction of drinking water.
5. Reduce the contribution of the large cities to CC when regulations on the use of construction materials (that affect albedo), building density (urban heat island) and natural ventilation are favourably modified (Alcoforado *et al.* 2010).

In practice, the cases of mainstreaming adaptation into physical planning tools for large cities present several limitations compared to climate plans. Specifically, these include lack of characterisation of the hazard, prevalence of actions that only concern land use (set back to be complied with during construction works, and land use allowed), lack of priority and scheduling, no reference to the expected impact of works, rare implementation of monitoring (Table 2.7).

2.3.3 Measures

Comparing plans by considering the measures envisaged can be misleading. When climate planning is carried out, each city does not start from the same baseline. For example, separate glass, metal, paper, and food waste and recycling programs or the use of LED bulbs for street lighting can be innovative measures used in one city, while they are so consolidated in another that they need not even be mentioned among the measures established by the plan. Hence, the absence of certain measures does not always indicate lack of detail or visionary planning. That said, we shall now discuss measures that are most frequently adopted.

The 29 climate plans of large subtropical and tropical cities traced (7 in Latin America, 7 in the USA, 7 in Middle East and Asia, 4 in Europe and 4 in Africa), one third of

Table 2.7: Large subtropical and tropical cities. Quality of climate measures in other plans.

Large city	Plan	Hazard						Measure							
		Flood	Rain	Storm	Drought	Heat	Fire	Baseline	Prone area	Localization	Priority	Cost	Impact	Scheduling	Monitoring
Athens	Mo											Partial			
Goiania	S	•													•
Hyderabad, IND	M	•	•		•	•				Map					
Managua	S	•						•	•		•	•		•	
Nagoya	B					•		•							
Sydney	W				•									•	
Wuhan	M	•					•								•
Total										1	1	1	0	2	2

B-Biodiversity, Mo-Mobility, MP-Master, S-Sustainability, W-Water

which were defined after 2012, concern the main jurisdiction only, with the sole exception of Miami, Montevideo and Portland, where all the metropolitan area was considered. Planning using a metropolitan scale stems from the need to harmonise the measures of many jurisdictions and authorities (water, etc.), whose consent is required (Revi, Satterthwaite *et al.* 2014: 44). Plans for metropolitan areas include building awareness, studies and assessments to ensure that mitigation/adaptation measures become rooted in each jurisdiction, and fundraising initiatives.

Municipal plans focus instead on direct impact, especially on the municipal facilities (offices, transportation, employees), and on sectors in which the Municipality has regulatory authority (private construction works, road systems, waste, education, etc.).

The 29 plans analysed contain over 100 mitigation measures and 50 adaptation measures, which we collected by area (Table 2.8). The most frequently proposed mitigation measures include low-carbon transportation (21), building adjustments to reduce the total energy use (19) and local government operations (use of LED bulbs for traffic signals, street and parks lighting, 16 plans). The above are closely followed by waste reuse/recycling to recover gases emitted by landfills and wastewater treatment plants (15), and information and communication measures (15).

Table 2.8: Large subtropical and tropical cities. Measures in 28 climate plans by sector. The numbers indicate the number of measures proposed in every sector.

Measure sector	Adelaide 2011	Antalya 2013	Baires 2009	Bangkok 2007	Barcelona 2010	Cape Town 2011	Cartagena 2013	Casablanca 2011	Durban 2010	Ganziantep 2011	Guayaquil 2012	Houston 2008	Los Angeles 2007	Marseille 2012	Miami 2008	Milan 2009	Montevideo 2012	New York City 2013	Philadelphia 2007	Phoenix 2009	Portland 2009	Rome 2010	Saitama 2012	Santiago Chile 2012	São Paulo 2011	Tbilisi 2011	Tunis 2011	Zapopan 2013	All plans
Mitigation																													
Energy, efficiency	1	8	3	3	2	2				2		3		2	2	24	1			2	8	16	1	1	1	1		2	20
Building, green	2	8	3		14	4		1		3		2		1	4	9	1		10		9	2	4	3		10	1		19
Water, reducing consumption	2	1				1		1		2		2	1												2		1	1	10
Fuel, clean		3	1	2		1						1	4		2					2					1				9
Transport	1	5	3	5	18	2				1	2	2	1	7	4	12			7	2	6	8	2		5	8		1	21
Waste, reuse/ recycling	1	3	3	3	3					2	2		1	5		1			5	2	13				3			2	15
Food, organic, bio														1							6				2				3
Agriculture														1		2								1				1	4
Land use		3		1		1	1					1	3	1					2				2		1				10
Information/ communication	1	3	1			1	1	1			1		1		1				4		6		4	2	1			2	15
Funding projects														3															1
Monitoring														2															1
Adaptation																													
Knowledge and studies			2			2	4	13	3	1	1		1	4			5	6						2	2		6		14
Preparedness																		13											1
Built environment	1								3		1		2				3	14							1			1	8
Water conservation	1							2					1		2									1	1		1	1	8
Drainage			1				1	4	1															1			4		6
Utilities																		47											1

continued **Table 2.8:** Large subtropical and tropical cities. Measures in 28 climate plans by sector. The numbers indicate the number of measures proposed in every sector.

Measure sector	Adelaide 2011	Antalya 2013	Baires 2009	Bangkok 2007	Barcelona 2010	Cape Town 2011	Cartagena 2013	Casablanca 2011	Durban 2010	Ganziantep 2011	Guayaquil 2012	Houston 2008	Los Angeles 2007	Marseille 2012	Miami 2008	Milan 2009	Montevideo 2012	New York City 2013	Philadelphia 2007	Phoenix 2009	Portland 2009	Rome 2010	Saitama 2012	Santiago Chile 2012	São Paulo 2011	Tbilisi 2011	Tunis 2011	Zapopan 2013	All plans
Coastal protection						1	1		1								2	37											5
Green extension/managem.	1		1	1			1				2		3		1			16						2	2			3	11
Health									2									38											2
Insurance																	1	16											2
Total measures	10	34	18	13	37	15	9	22	10	11	9	11	18	27	16	28	13	195	28	8	48	24	26	14	22	19	13	15	-

To reduce energy consumption, actions must also target the physical structure of the city by enhancing compactness, concentrating activities on certain points or axes to reduce the per capita daily vehicle-miles traveled, and by making the mass transportation system economically sustainable. But investments must also focus on work organisation (telecommuting, flexible work hours). These measures, which were conceived by a city to be elite until a few years ago, are now practised at all latitudes (land use, 10 plans). However, significant changes in the physical and functional structure of the city requires years, if not decades, to be implemented. Meanwhile workers and activities become suburbanised, and the working mode undergoes changes (having two jobs is a common practice for many residents), thus removing the effects of the reformed physical structure on per capita daily vehicle-miles traveled.

Adaptation measures depend on the hazard. Studies have the highest incidence of use (14), followed by green extension (11), which was conceived to limit run-off and urban heat island. Water conservation and built environment follow closely (each with 8), and also stormwater drainage maintenance (6). (Table 2.8)

It must be said that only in two cases was insurance deemed an adaptation measure. There are two possible explanations for this. Firstly, the insurance sector is already highly developed and does not deserve to be mentioned as a measure in cities governed by OECD economies. Secondly, the opposite situation is found in the less developed economies, but there are a few private enterprises that wish to protect their own interests.

Resilience measures are primarily structural (55%), and especially concern the water sector. Non-structural measures are substantial, regardless (45%), and concern community awareness, training, early warning, studies, assessments and plans.

The analysis of 33 climate plans (Table 2.3) emphasizes the fact that half of them neither specify the impact of mitigation (MT CO_2/year) or of adaptation, nor the costs resulting from lack of mitigation and adaptation. Hence, it is hard to assess the benefit that can be obtained by implementing the single measures net of mitigation/adaptation costs. This is why preliminary risk mapping is required to highlight the hot spots that need priority intervention (Tiepolo and Braccio 2015, chap. 11).

Moreover, there are substantial differences between subtropical and tropical cities regarding measures. To understand this, we must not forget that we prevalently find large cities of Developing Countries in tropical zones. Firstly, the climate plans of subtropical cities contain three-fold the measures of tropical cities. They are complex plans with many arrows to strike the targets. Secondly, in large subtropical cities measures focus on modernising building assets (ecological roofs, passive cooling, energy efficiency) and infrastructures to reduce the consumption of power and water (Adelaide, Casablanca, Santiago, Tunis). Instead, in large tropical cities measures focus on increasing consumption, taking into account the lack of sanitation, rainstorm drainage, adduction of drinking water, and removal of solid waste in many districts (Cartagena, Guayaquil).

2.3.4 Quality of Climate Planning

Several parameters, such as internal consistency between objectives, priorities and measures (Baker *et al.* 2012) can be considered to assess the quality of plans. In our case, we are interested in assessing the implementation features of planning tools. Specifically, if they are able, as currently formulated, to guide the implementation of measures that reduce the impact of CC. The focus will then be on six indicators: (i) climate characterisation, (ii) funding for planning, (iii) cost of measures, (iv) responsibility for each measure, (v) funding sources, (vi) implementation monitoring and reporting.

2.3.4.1 Climate Characterisation

Effective climate characterisation allows for improved definition of the measures. Mitigation plans do not analyse the climate but rather estimate GHG emissions according the sectors that generate them, specifically transportation, commercial, industrial, residential and waste. This helps to define emission-reducing measures. The major part of plans describes the possible future scenarios, taking into account the increased energy demand (transportation), population increase, changes in water consumption (water, solid waste) and changes in environmental parameters. Instead, adaptation plans and, at times, resilience strategies characterise the climate. They consider the main environmental parameters (temperature, precipitation, sea level, wind, sea waves), and define climate events and extremes variability, which helps to identify possible measures to reduce vulnerability. About 90% of adaptation plans provide estimates of long-term future trends (2030, 2050, 2100) for each of the environmental parameters analysed. These estimates are obtained by using both global and regional climate models. Different models are usually used, up to 21 in the case of Barcelona and 9 in the case of Surat. Plans that report the results of climate analysis mention the following long-term trends (Table 2.9):

- temperature increase from 0.5–1.5°C by 2030 and up to 4–5°C by 2090;
- increase in heat waves both in terms of number and duration (two-fold and even three-fold);
- drop in the number of rainy days without a significant trend in annual accumulated precipitation;
- increased incidence and intensity of extreme rainfall events (both likelihood of flash floods and/or more intense monsoons);
- increase in sea level rise (15 to 79 cm by 2030–2050).

Not even emergency/contingency plans characterise the climate because they are implemented after an alert is issued by a specialised centre. In some cases (Hyderabad, Luoyang, Milan, Panzhihua, Rome, Turin), the plan specifies the threshold that triggers the various types of alert, which can be mm of rain within a period of time (surface flood), cm reached at the water level gauge (river flood), cm of snow (avalanches), number of consecutive days without rainfall (drought), or temperatures above the maximum seasonal average (heat waves). The other plans do not characterise the climate.

Hence, the plans present two limitations in terms of climate analysis. Firstly, they do not define the probability of concurrent occurrence of different types of hazards, for instance coincidental floods due to a sea level rise, surface flood or river flood. Past disasters indicate that such a circumstance generates catastrophic consequences (e.g. Venice 1966).

Secondly, they do not define the geographic origin of the floods that affect large cities, as already observed regarding sub-Saharan cities (Tiepolo 2014a). Some river floods are determined by changed patterns of rainfall in areas that are distant from cities, at times even in other countries, and can determine non-seasonal catastrophic floods. These analyses, especially if they concern border zones, should start by defining the watersheds of the tributaries of the river that crosses the large city concerned. Daily rainfall recorded by ground-based weather stations situated in other countries generally cannot be easily accessed by the large city concerned. However, the analysis can be conducted by the national weather service by using data on daily precipitations as estimated by satellites, even for neighbouring countries, by exploiting the dedicated regional information systems (e.g. EMSAT MSPE for West Africa). Knowing the full extent of the watershed responsible for river floods in large cities is essential to schedule and coordinate adaptation measures.

Table 2.9: Large subtropical and tropical cities. Climate characterisation in 12 adaptation plans.

Large city	Sea level rise	Pecipitations	River flow Peak discharge increase	Temperature			UAI	Bush fire weather increase	Air quality	Return period	Models	Climate projections
	cm	Trend or mm/Y	%	Min °C	Max °C	Extrm %	°C or %	%			n.	
Adelaide	–					35		20				2030
Barcelona		reduction		2.2	5.1				PM10 NO_2		21	2080–2099
Buenos Aires	60				0.5							2020–2029
Cartagena	15–20	Ext. increase										2040–2050
Casablanca	20	reduction		0.8	1.3					•		2030
Durban	rise				rise							2045–2065
Ho Chi Minh	•			1								2050–2100
Indore	–	increase	50–80	2			2–4				1	2021–2100
Johannesburg		+18–27%		2.5	2.3							2046–2100
Los Angeles							double		78–85%			2100
New York C	28–79	+4–13%					100–200	–				2020–2050
Santiago Chile		reduction		1–2	1–2							2045–2065
Surat	–	200–450			4						9	2070–2100

2.3.4.2 Financing Climate Planning

The process for preparing the plan in OECD countries is funded by municipal resources in 80% of cases, in BRICS in 40% and in Developing Countries, 33%. Hence, climate planning is almost always funded by development aid in the tropics (Table 2.10), which means that it is a gift. "Gifted" plans are increasing in Developing Countries in recent years. This practice weakens planning because it undermines local appropriation of tools. If we also consider plans that, at times, are not drawn up in compliance with a law that prescribes them, this explains why they often have poor operative features.

Table 2.10: Large subtropical and tropical cities. Origin of resources used to draw up 37 climate plans.

Country group	National	Municipal	Multi-bilateral development aid	Mixed	Total
OECD	1	15	1	1	18
BRICS		2	3	0	5
Other		5	9	0	14
Total	1	22	13	1	37

2.3.4.3 Assessing the Cost of Measures

Fifty-six percent of the plans specify the cost of measures, but unfortunately, only too often said calculation is only carried out for some of them (Buenos Aires, Casablanca, Gaziantep, Milan, Montevideo). Amounts vary considerably when the cost is estimated, depending on the nature of the envisaged measures, on the dimensions of the city, and on the calculation method (annual or global). The per capita cost is the best way to compare plans. It can vary from 3 (Phoenix) to 2,378 US$ (New York City) (Table 2.11). Climate plans for the cities in tropical Developing Countries do not report the costs of measures; hence, they run the risk of not being implemented.

2.3.4.4 Identifying Responsibility for Each Measure

A measure without a clear owner is destined to fail. 52% of the plans identify the municipal office responsible for the implementation of each measure (Table 2.11). Among these plans, only one is for tropical cities.

2.3.4.5 Funding for Climate Measures

39% of plans specify the funding source envisaged for the measures. But often the sources of funds are only roughly indicated (Cape Town, Casablanca, Montevideo,

Table 2.11: Large subtropical and tropical cities. Quality indicators of 27 climate plans.

Large city	Quality indicator					
	Climate characterization	Municipal funding of plan	Measure cost	Measure implementation responsible	Funding sources	Monitoring	Total
Adelaide	1	1				1	3
Antalya		1	1	1			3
Bangkok							0
Barcelona	1	1	1		1		4
Buenos Aires	1	1	1				3
Cape Town		1			1		2
Caratgena	1						1
Casablanca			1	1			2
Durban	1	1		1			3
Gaziantep		1	1	1		1	4
Guayaquil		1					1
Houston		1		1			2
Johannesburg	1	1		1			3
Los Angeles	1	1					2
Marseille		1	1			1	3
Miami		1	1	1	1	1	5
Milan city		1	1	1		1	4
Montevideo		1	1		1		3
New York City	1	1	1	1	1	1	6
Philadelphia		1		1			2
Phoenix		1	1		1		3
Portland		1	1		1	1	4
Rome city		1	1	1	1	1	5
Santiago Chile	1	1		1	1		4
São Paulo		1				1	2
Tbilisi		1	1	1			3
Tunis			1	1	1		3

*Permit fees, tolls, traffic fines. *minimum

Tunis) (Table 2.11). Conversely, some plans present detailed costs and specify how they will be covered (Miami, New York, Phoenix). When the costs are not known, it is impossible to assess the efficiency of adaptation (Adger *et al.* 2005). Funding sources are clearly broader in subtropical zones where local governments not only depend on their own resources (fees, fines, tolls), but also on credit, municipal bonds, project financing, and on the contribution of regional/state and federal/national governments. Plans for large cities in tropical Developing Countries do not specify the funding sources of the measures, which is another factor that undermines their implementation.

Implementation of measures can be assigned to private parties in seven ways.

The first is to issue building authorisations only on condition that the developer will implement mitigation measures, some of which were typical of the visionaries of bioclimatic architecture in the 1970s (Mazria 1979; Vale and Vale 1980): from screening buildings with deciduous trees planted near building roofs, to fenestration limited at 40% of the façade surface, composting sites, solar panels, cool roofs, renewable energy systems (solar, wind turbine), the majority of windows facing South, and recycled water systems (San José, CA). The second is the carbon-based adaptation fee. The tax is based on the "polluter pays principle", and serves to fund the adaptation measures and encourage polluting companies to reduce emissions (Kaoshiung). The third is the issuing of catastrophe bonds (Florida, Mexico), which have been recently supported by the World Bank through MultiCat (ICLEI 2011). The fourth is a property and casualty insurance surcharge as proposed by New York City. The fifth is implementation of a mitigation banking programme (New York City) (Zirschky *et al.* 1995). This system consists of authorising construction works on wetlands after developers purchase mitigation credits to pay a third party that implements protection/restoration of environmentally important coastal wetlands. The sixth is involvement of private enterprises through project financing (Rome). Finally, the literature reports a funding mode we found no trace of in the plans considered, namely microfinancing. This option addresses individuals, and cannot be used to implement community works, such as drainage canals or dams. But it would allow for making houses climate-proof (Revi, Satterthwaite *et al.* 2014) and supporting small entrepreneurs who usually have no access to either insurance or formal credit.

The ones mentioned by way of example do not constitute universally effective solutions. Three out of four large cities do not expand in compliance with building/urban regulations, but rather through informal settlements, and potential payers of the carbon-based adaptation fee are few. In such cases we must not forget that the large cities of developing countries, including LDCs, regardless contain valuable assets, such as vacant land with infrastructures. Improving local fiscal systems, activating value capture, and taxing vacant lands (McCarney 2012) can provide sufficient resources to implement measures that reduce the impact of CC in hot spots (Braccio *et al.* 2014).

2.3.4.6 Monitoring and Reporting

The monitoring and reporting program allows for checking progress in the implementation of the plan, assessing it, and communicating the results to both decision-makers and local actors. The literature reports that plans are rarely monitored (Revi, Sattertwhaite *et al* 2014: 59). Our survey on plans, one third of which were very recent, defines a less pessimistic picture. The monitoring device (of measures or of GHG emissions) is envisaged by 10 plans (33%), and 4 plans out of 27 envisage reporting (Table 2.3). However, the program is often only stated and not described in detail (Milan). And even in the latter case, it presents several omissions compared to the conventional Monitoring and evaluation criteria (Tiepolo 2012), precisely the crucial ones concerning the indicators to be used for monitoring, the methods and costs of information collection, monitoring phases, and assessment methods for the impact of the plan, all of which are the target of the reporting process. Once again, climate plans for large cities in tropical Developing Countries stand out for the lack of monitoring and reporting.

2.4 Conclusions

The objective of this chapter was to ascertain the relevance and quality of climate planning. We limited the analysis to large cities in the subtropics and tropics, zones that, regardless, contain half the urban population of the World.

Our survey presents three innovative features compared to previous ones, because (i) it considers the cities by climate zone and dimension, (ii) it analyses the relevance of plans for the number of cities in the segment concerned, taking into account the hydro-meteorological and climate-related disasters and, finally, (iii) it assesses planning quality.

We found that barely a quarter of the large cities in the subtropics and tropics has a climate plan. The percentage is low, if we consider the extensive, greater than ten-year long commitment of multilateral bodies, central and local governments to face CC. Emergency planning is the most common type. It is followed by mitigation plans and, far below in terms of frequency, by adaptation plans and resilience strategies. Other types of plans (sustainability, master plans) are a minority.

The analysis by climate zone revealed aspects that were not highlighted by previous surveys. The tropics differ from the subtropics in the type of hydro-meteorological and climate disasters (no drought, more extreme rains), quantity of per capita GHG emissions (less), local government skills (poor planning), and expansion mode of the cities (informal). Planning measures, especially mitigation ones, change from zone to zone because they consider these differences. Climate zone characterisation also depends on the standard of economic development that prevails in the site. The subtropics include OECD and BRICS nations, while Developing Countries and, especially, Least Developed Countries, prevail in the tropics.

The status of planning outlined so far is steadily evolving. During the past five years, the number of large cities that have a climate plan has increased by 150% compared to 2010. Unfortunately only one fourth of the new plans concern cities in Developing Countries. In particular, none of the 20 large cities in LDCs has a plan, though this group has been repeatedly affected by hydro-meteorological and climate-related disasters. Development aid, which by definition funds half the plans, does not intervene in these cities, not even to co-fund the planning process, which would be more sustainable than the gift if the scope were to guarantee appropriation of the plans.

Some supranational initiatives, such as the Covenant of Mayors, have greatly contributed to disseminate climate planning, standardize its features, and improve its quality. However, 70% of plans attend 3 or less indicators out of the six chosen to gauge quality.

These results are useful for the discussion on Sustainable Development Goals. In the first place, they reveal that the substantial participation of large cities in the associations that promote climate planning does not correspond to an equal number of climate plans. Secondly, the results of our analysis confirm that the trend to support LDCs in reducing the number of affected cities and damages caused by hydrological disasters meets a genuine need that international communities have been unable to meet thus far.

There remains the issue of extending mitigation/adaptation/resilience planning to 75% of large cities that still lack it, and of studying improvement modes for existing plans. Some experts propose learning from the cities that adapted first. The considerable differences between cities classified by climate zone and economy of origin suggest that extreme caution is required before this proposal can be accepted. Promoting climate planning in the sub-Saharan region, based on the Covenant of Mayors model, might instead be the solution, especially if the process of drawing up plans is considered a prerequisite to access multi-bilateral co-funds for the implementation of climate measures.

2.5 Recommendations

The literature and plans we have referred to are deficient, if not silent, regarding eight factors, and we recommend that experts explore these points:

1. state of climate planning in middle-sized cities. The number of subtropical and tropical cities between 0.1 and 1 million inhabitants is over 2,200. The climate planning survey presented in this chapter should be replicated for this important class of cities, which have documented various climate plans since 2003.
2. analysis of GHG emissions relative to city size. This analysis offers a wealth of implications for physical planners. It is important to verify whether the absence of direct relations between per capita GHG emissions and city size ascertained for large cities also applies to middle-sized cities: it might enrich the ongoing debate

on optimal city size with an environmental consideration. In the first place, the analysis of GHG inventories must be extended to medium-sized cities and, later, data must be reanalysed when GHG inventories based on the new post-COP20 standards become available.

We advise local governments and their partners to strengthen or define five points.

1. Urban characterisation of CC. The probability of a coincidental hazard and the geographical origin of events that determine catastrophic river floods can enhance the implementation of adaptation measures. Information on peak temperature, humidity, and direction of dominant winds are useful for planners to establish the direction of road network, building density, and location of green spaces in order to better adapt to the climate.
2. Local planning skills. It is an ingredient without which mitigation/adaptation cannot succeed. There are three essential steps. First, arrange the tools (GIS for risk assessment) and data on previous disasters. Second, acquire working skills in multidisciplinary teams. Meteorological services must collaborate with physical planners, hydraulic engineers, NGOs and CBOs, and with producers of scientific information and knowledge (Gagnon-Lebrun *et al.* 2006), and all these must cooperate with municipal planning bodies, professional planners and decision-makers (Bassett *et al.* 2010). Third, adapting best practices to local specifications. These steps characterise local capabilities, which should be assessed during planning and supported by dedicated measures if one actually wishes to produce local operative planning.
3. Plans for the metropolitan area. Climate planning is almost always limited to the main jurisdiction. Preliminary risk maps are required, and measures must be planned on a metropolitan scale.
4. Priority intervention zones. They are often not defined in the plans, and they cannot be defined with means, times and information as in EU28 after the definition of Directive 2007 on flood risk mapping. Snapshot tools as a means of preliminary risk assessment seem more sustainable and feasible for appropriation by local bodies. However, a definition of risk (Risk = Hazard * Damage) that allows for measuring the single components is required to produce preliminary risk maps.
5. Funding the established measures. A mitigation plan can be funded by requiring compliance with some requisites to issue building authorisations or a 'polluter pays principle' where private individuals have resources and where there is a significant number of industrial companies (OECD, BRICS). Mitigation banking, needs organisation, and law changes. Adaptation can be funded with catastrophe bonds and project financing. But in the other cities, it is essential to reformulate local taxes.

We advise multilateral development agencies to support:

1. climate planning South of Sahara. The example of the Covenant of Mayors, progressively broadened by the EU28 to North Africa and to the Middle-East, should be proposed again to benefit large continental cities. The issuance of funds for planning measures should be subordinated to the adoption of plans on the subject.

References

Adger, W.N., N.W. Arnell, and E.L. Tompkins. 2005. Successful adaptation to climate change across scales. *Global Environment Change* 15: 77–86.

Alcoforado, M.-J., and A. Matzarakis. 2010. Planning with urban climate in different climatic zones. *Geographicalia* 57: 5–39.

Baker, I., A. Peterson, G. Brown, and C. McAlpine. 2012. Local government response to the impacts of climate change: An evaluation of local climate adaptation plans. *Landscape and Urban Planning* 107: 127–136.

Basset, E., V. Shandas. 2010. Innovation and climatic action planning perspectives from municipal plans, *JAPA*, 76 (4): 435–450

Belda, M., E. Holtanová, T. Holenka, J. Kalvová. 2014. Climate classification revisited: from Köeppen to Trewartha, *Climate Research* 59: 1–13.

Birkmann, J., et al. 2010. Adaptive urban governance: new challenges for the second generation of urban adaptation strategies to climate change. *Sustain Sci* 5: 185–206.

Braccio, S., M. Tiepolo. 2014. Atlas des ressources locales : Kébémer, Louga (Sénégal), Niamey (Niger). http://cooptriangulaireins.net. Accessed 2 May 2015.

CAI-Clean air initiative and Cities development initiative for Asia. 2012. *Climate change plans and infrastructure in Asian citis*. Pasig city: CAI and CDIA.

CDP-Carbon Disclosure Project. 2014. *CDP cities 2013. Summary report on 110 global cities*. London: CDP.

Carmin, J.A., N. Nadkami, and C. Rhie. 2012. *Progress and challenges in urban climate adaptation planning. Results of a global survey*. Cambridge, MA: MIT.

Carmin, J.A., D. Roberts, and I. Anguelovski. 2012. Planning climate resilient cities: Early lessons from early adapters. In *Cities and climate change. Responding to an urgent agenda*, ed. D. Hoornway, 430–461. Washington: The World Bank.

Corell, R., D. Ojima, and B. Perkins. 2007. *A survey of climate change adaptation planning*. Washington DC: The Heinz Center.

Fraser, A., and D.V. Lima. 2012. Regional technical assistance initiative on climate adaptation planning in LAC cities. Survey results report.

Gagnon-Lebrun, F., et al. 2006. *Progress on adaptation to climate change in Developed Countries. An analysis of broad trends*. Paris: OECD.

GIZ-ICLEI. 2012. Climate change adaptation in urban India. A status report.

ICLEI. 2011. *Financing the resilient city: A demand driven approach to development, disaster risk reduction and climate adaptation. An ICLEI white paper. ICLEI Global report*. Bonn: ICLEI world secretariat.

IPCC. 2014. Annex II: Glossary. In *Climate change 2014: Synthesis report. Contribution of working groups I, II and III to the fifth assessmeny report of the Intergovernmental panel on climate change*, IPCC. Geneva: IPCC.

Levina, E., and D. Tirpack. 2006. *Adaptation to climate change: key terms*. Paris: OECD and Environment directorate-IEA.

Lim, B., and E. Spanger-Siegfred eds. 2004. *Adaptation policy framework for climate change: developing strategies, policies and measures*. Cambridge: Cambridge University Press.

Marcotullio, P.J., A. Sarzynski, J. Albrecht, and J. Garcia. 2013. The geography of global urban greenhouse gas emissions: an exploratory analysis. *Climatic Change* 121: 621–634.

Mazria, E. 1979. *The passive solar energy book*, Rodale Press.

Mc Carney, R. 2012. City indicators on climate change: Implications for policy leverages and governance. In *Cities and climate change: Responding to an urgent agenda*, ed. D.H. Hoornweg, M. Freire et al., 18–60. Washington: The World Bank.

McClean, D. ed. 2010. *World disaster report 2010*. IFRC International Federation of Red Cross and Red Crescent Societies.

Preston, B.L., R.M. Westaway, and E.J. Yuen. 2011. Climate adaptation planning in practice: an evaluation of adaptation plans for three developed nations. *Mitig Adapt Strateg Glob Change* 16: 407–438.

Revi, A., D. Satterthwaite et al. 2014. Urban areas. In *Assessment report 5*, IPCC-WGII, 1–113.

Rubel, F., and M. Kottok. 2010. Observed and projected climate shifts 1901–2100 depicted by world maps of the Köppen-Geiger climate classification. *Meteorol. Z.* 19: 135–141.

Seto, K.C., S. Dhakal. 2014. Human settlements, infrastructure and spatial planning. In *AR5*, IPCC-WGIII.

Sharma, D., and S. Tomar. 2010. Mainstreaming climate change adaptation in Indian cities. *Environment and Urbanization* 22 (2): 451–465.

Tang, Z., S.D. Brody, C. Quinn, L. Chang, and T. Wei. 2010. Moving from agenda to action: evaluating local climate change actions plans. *Journal of Environmental Planning and Management* 53 (1): 41–62.

Tarchiani V., and M. Tiepolo eds 2016. *Risque et adaptation climatique dans la région Tillabéri, Niger*. Paris: L'Harmattan. Isbn: 978-2-343-08493-0.

Tiepolo, M., and S. Braccio. 2016. Flood risk preliminary mapping in Niamey, Niger. In *Planning to cope with tropical and subtropical climate change*, ed. M. Tiepolo, E. Ponte, and E. Cristofori, 201–220. De Gruyter Open.

Tiepolo, M. 2014. Flood risk reduction and climate change in large cities South of the Sahara. In *Climate change vulnerability in Southern African cities. Building knowledge for adaptation*, ed. S. Macchi and M. Tiepolo, 19–36. Springer. doi: 10.1007/978-3-31900672-7.

Tiepolo, M. ed. 2012. Évaluer l'environnement au Sahel. Premieres réflexions sur la gouvernance locale. Turin-Paris: L'Harmattan. Isbn: 978-2-336-00013-8.

UN-United Nations. 2014. Working group proposal for sustainable development goals. http://undocs.org/A/68/970.

UN-Habitat. 2011. *Cities and climate change. Global report on human settlements 2011*. London-Washington DC: Earthscan.

UN-Habitat. 2009. Climate change. The role of cities. Involvment, influence, implementation. UN-Habitat and UNEP.

UNISDR. 2009. *2009 UNISDR terminology on disaster risk reduction*. Geneva: UNISDR.

Vale, B., and R. Vale. 1980. *The autonomous house*. London: Thames & Hudson Ltd.

Vergara, W. 2005. *Adapting to climate change. Lessons learned, work in progress, and proposed next steps for the World Bank in Latin America*. The World Bank.

Wheeler, S. 2008. State and municipal climate change plans: the first generation. *Journal of the American Planning Association* 74 (4): 481–496.

WRI-World Resources Institute. 2014. Global protocol for community-scale green gas emission inventories. Accounting and reporting standard for cities.

Zirschky, J.H., R. Perciasepe, T.R. Hebert, R.P. Davison, and D.K. Hall. 1995. Federal guidance for the establishment, use and operation of mitigation banks. Federal Register, 60, 228: 58605–58614.

Appendix A

Mitigation, Adaptation, Resilience, Strategy, Policies and Plans

Adelaide city council (2013) Climate change adaptation action plan 2011–2013
Greater Antalya Municipality (2013), Antalya green house inventory and sustainable energy action plan
Ajuntamento de Barcelona (2010) Plan de energía, cambio climatico y calidad del aire de Barcelona (PECQ 2011–2020)
Bangkok metropolitan administration (2007) Action plan on global warming mitigation 2007–2012
Prefeitura de Belo Horizonte (2013) Plano de redução de emissões de gases de efeito estufa PREGEE. Relatório técnico final
Prefeitura Municipal de Belo Horizonte (2011) Política municipal de mitigação dos efeitos da mudança climática
Buenos Aires (-) Cambio climático. Plan de acción Buens Aires 2030
Can Tho City People's Committee, Steering Commitee for Decision 158 (2010) Can Tho city climate change resilience plan
Egis BCEOM International/IAU-IDF/BRGM (2011) Plan d'adaptation et de résilience du Grand Casablanca. Rapport final
Chongqiong Municipality (2007) Energy saving
Invemar, MADS, Alcaldia Mayor de Cartagena de Indias-CDKN (2012) Lineamentos para la adaptacion al cambio climatico de Cartagena de Indias
Ethkwini Municipality (2013) Durban climate change strategy. Draft vision statement
Ethkwini Municipality (-) Climate change adaptation planning for a resilient city
Gaziantep (2011) Climate action plan of Gaziantep Energy and gereenhouse gas emission profile preliminary action plan and implementation strategy. Final report rev 2
VCAPS Consortium (2013) Climate adaptation strategy Ho Chi Minh City. Moving towards the sea with climate change adaptation
Guangzhou (-) Energy saving and reduction of pollution gas plan
Municipalidad de Guayaquil (2013), Decalogo de acciones de Guayaquil frente al cambio climatico, y proyectos de desarollo sostenible
Vietnam Climate Adaptation PartnerShip consortium (2013) Climate adaptation strategy Ho Chi Minh City Moving towards the sea with climate change adaptation
City of Houston (2008) Emission reduction plan
Indore city resilience strategy team (2012) Final report on city resilience strategy Indore
Anwar F (2012) Karachi city climate change adaptation strategy. A roadmap
City of Kobe (2011) Climate and environmental plan
Lagos state government (2012) Climate change policy 2012–2014
Communauté urbaine Marseille provence métropole (2012) Le plan climat de Marseille Provence Métropole
Miami-Dade county (2010) Green print. Our design for a sustainable future
Comune di Milano (2009), Piano d'azione per l'energia sostenibile e il clima
PNUD (2012) Plan climatico de la region metropolitana de Uruguay
City of Nagoya (2008) Nagoya biodiversity strategy
Comune di Napoli (2012) Piano d'azione per l'energia sostenibile
City of New York (2013) A stronger, more resilient New York
City of Philadelphia (2007) Local action plan for climate change
City of Phoenix (2009) Climate action plan for government operations
City of Portland and Multnomah county (2009) Climate action plan 2009

Quingdao Municipality (2012) Energy saving and reduction of pollution gas plan
Cidade do Recife (2014) Política de sustentabilidade e de enfrentamento das mudanças climáticas do Recife e dá outras providências
Prefeitura da cidade do Rio de Janeiro (2015) Rio resiliente Diagnóstico e áreas de foco
Roma capitale (2010) Piano d'azione per l'energia sostenibile della città di Roma
Saitama prefecture (2009) Stop global warming. Saitama navigation 2050. Saitama Prefecture global warming strategy action plan
City of San José (2011) Greenhouse gas reduction strategy for the city of San José
Gobierno regional metropolitano de Santiago (2012) Plan de adaptación al cambio climatico para la región metropolitana de Santiago de Chile
Prefeitura da cidade de São Paulo (2011) Plano de ação para mitigação e adaptação da cidade de São Paulo às mudanças climáticas, comitê municipal de mudanças do clima e eco economia e seus grupos de trabalho
City of Semarang (2010) Semarang's adaptation plan in responding to climate change, ACCCRN
Shenzhen Municipality (2014) Energy saving
Republic of Singapore, National climate change secretariat (2012), National climate change strategy 2012
Surat municipal corporation (2011) Surat city resilience strategy
Tiblisi City (2011) Final Sustainable Energy Action Plan City of Tbilisi for 2011–2020
République tunisienne, Banque mondiale (2011), Adaptation au changement climatique et aux désastres naturels des ville cotieres d'Afrique du Nord. Phase 2: Plan d'adaptation et de resilience – Tunis. Rapport final
Tokyo metropolitan government (2007) Tokyo climate change strategy. A basic policy for the 10-year project for a carbon minus Tokyo
City of Yokohama (2008) Yokohama climate change action plan
Municipio de Zapopan (2013) Plan de acción climatica municipal de Zapopan

Other Plans

South East Europe Mobility management scheme (2010) Mobility management plan for the city of Athens 2010–2014
Prefeitura municipal de Goiânia (2012) Goiânia sustenável. Plan de ação
Hyderabad Metropolitan Development Authority (2013) Metropolitan development plan for 2031
Kobe City (2011) Climate and environmental basic plan
Alcaldia de Managua (2013) Plan de acción Managua sostenible
New South Wales office of water (2010) Sydney 2010 metropolitan water plan, Department of environment climate change and water
City of Wuhan (2010) Wuhan urban master plan

Emergency Plans

Ahmedabad Municipal Corporation (2013) Ahmedabad heat action plan 2013. Guide to estreme heat planning in Ahmedabad, India. Early read version
BaZhong Municipal Peole's government office (2009) BaZhong emergency plan to deal with the disaster off snow and ice
Governo do Estado de São Paulo (2014) Normas de procedimientos do Plano de contingência para o periodo de estiagem da Região metropolitana de Campinas
Cape Town (2012) Municipal disaster risk management plan
Chongqiuong Municipality (2008) Disaster emergency salvation

City of Dallas (2014) Drought contingency plan, adopted by resolution of the Dallas city council on February 26, 2014
Government of National Capital Territory off Delhi (2015) Delhi disaster management plan (2014–15)
Dongguan Municipality (2004) Disaster emergency response
Fukuoka city disaster prevention council (2012) Fukuoka regional climate prevention
Guangzhou Municipality (2013) Disaster of Tsunami and surge emergency response
City of Jinan (2012) Jinan city flood emergency plan
City of Philadelphia (2012) Natural hazard mitigation plan – Final draft
District disaster Management Authority (East) (...) District disaster management plan East Delhi
Emergency Support Unit Security Bureau Government Secretariat (2007) Hong Kong Contingency plan for natural disasters (including those arising from severe weather conditions)
Hyderabad City (2009) Flood emergency rain relief plan 2009
Municipalidad metropolitana de Lima, Region de Callao (2011) Plan de operaciones de emergencia para el área metropolitana de Lima y la región de Callao
Luoyang's Municipal people government office (2013) Luoyang city flood emergency plan
Provincia di Milano (2013) Piano provinciale di emergenza di protezione civile. Rischio idraulico e idrogeologico
Government of Maharashtra (2007), Greater Mumbai disaster management action plan. Risk assessment and response plan vol I, Mumbai
Nangchan City (2011) Nangchan climate and natural disaster relief emergency plan, Office of civil affairs
Nanjing Municipality (2014) Nanjing natural disaster emergency plan
Comune di Napoli (2012) Piano generale di emergenza di protezione civile. Piano comunale di emergenza per il rischio idrogeologico e idraulico
City of Panzhihua (2008), Panzhihua city flood control and drough contingency plans, Office of emergency management
Municipio de Porto Alegre (1998) Plano de defesa civil
Quanzhou City people's government office (2012) Quanzhou natural disaster relief emergency plan (revised)
Prefeitura Rio Defensa civil Rio de Janeiro (2014) PEM Rio Plano de emergência para chuvas fortes da cidade do Rio de Janeiro
Comune di Roma (2008) Piano generale di emergenza di protezione civile
City of San Antonio (2013) Drought operations plan
Prefeitura municipal do Salvador (2013) Defesa civil. Plano de contingência para chuvas 2013
Municipio de Santiago de Cali (2009) Plan local de emergencia y contingencia PLEC
Shanghai Municipality (2009) Disaster emergency response
Shenzhen Municipality (2014) Gas pollution emergency response
Shenzhen Municipal bureau of meteorology (2011) Shenzhen meteorological development 12th five years plan
Tangshan city government (2007) Tangshan forest fire emergency plan
Municipio del Distrito Central, Comité de emergencia municipal (2007) Plan municipal de emergencias, Tegucigalpa, MDC
Provincia di Torino (2013) Schema di piano provinciale di emergenza di protezione civile
City of Tshwane (2007) Disaster risk management plan Level 1
Xiamen (2010) Xiamen anti-typhoon emergency plan
Municipal flood control and drought prevention of Zhangjiagang city (2012) Zhangijagang city flood emergency plan
Zhengzhou municipal people's government office (2011) Zhengzhou city meteorological disaster contingency plan

Appendix B – Subtropics and Tropics, 2015. 82 Large Cities Provided by Climate Planning

Large city (over 1 M P)	Country ISO 3166-1	Zone	P millions	Planning tool	Year
Adelaide city	AUS	ST	1.3	A	2011
Ahmedabad	IND	T	6.3	E	2012
Antalya	TUR	ST	1.0	M	2013
Athens	GRC	ST	3.1	O	2010
Baires municipality	ARG	ST	2.9	AM	2009
Bangkok	THA	T	8.2	M	2007
Barcelona	ESP	ST	1.6	M	2010
Bazhong	CHN	ST	3.3	E	2009
Belo Horizonte	BRA	T	2.5	SP, M	2011, 2013
Campinas	BRA	ST	1.1	E	2014
Can Tho	VNM	T	1.2	R	2010
Cape Town	ZAF	ST	3.5	E	2011/12
Cartagena	COL	T	1.0	A	2013
Casablanca	MAR	ST	2.9	A	2011
Chongquing	CHN	ST	17.2	M	2007/08
Delhi	IND	T	19.0	E	2015
Dallas city	USA	ST	1.3	E	2014
Dongguan	CHN	ST	8.7	E	2004
Durban	ZAF	ST	3.4	S	2010
Fukuoka	JPN	ST	1.5	E	2012
Gaziantep	TUR	ST	1.5	M	2011
Goiania	BRA	T	1.3	O	2012
Guangzhou	CHN	ST	12.7	E	2013
Guayaquil	ECU	T	2.3	A	2012
Ho Chi Minh city	VNM	T	8.2	S	2013
Hong Kong	CHN	ST	7.1	E	2007, 2010
Houston city	USA	ST	2.2	M	2008
Hyderabad metro	IND	T	6.8	O	2013
Hyderabad	PAK	T	1.1	E	2009
Indore	IND	T	1.9	S	2012
Jinan	CHN	ST	6.8	E	2012
Karachi	PAK	T	23.0	S	2012
Kobe	JPN	ST	4.1	O	2011
Lagos	NGA	T	9.0	P	2012
Lima	PER	T	8.5	E	2011
Los Angeles city	USA	ST	3.9	M	2007
Luoyang	CHN	ST	6.5	E	2013
Managua	NIC	T	1.3	O	2013
Marseille	FRA	ST	1.2	M	2012
Miami county	USA	T	2.6	M	2008
Milan city, metro	ITA	ST	1.3, 3.9	M	2009, 2013
Montevideo	URY	ST	1.7	M	2012

Mumbai	IND	ST	12.5	E	2007
Nagoya	JPN	ST	8.9	O	2008
Nanchang	CHN	ST	5.1	E	2011
Nanjing	CHN	ST	8.2	E	2013
Naples city	ITA	ST	1.0	E, M	2012
New York city	USA	ST	8.2	AM	2013
Panzhihua	CHN	ST	1,2	E	2008
Philadelphia city	USA	ST	1.5	M	2007
Phoenix city	USA	T	1.5	M	2009
Portland & Multnomah	USA	ST	1.4	M	2009
Porto Alégre	BRA	ST	1.5	E	1998
Quangzhou	CHN	ST	1.2	E	2012
Recife	BRA	T	1.6	P	2014
Rio de Janeiro	BRA	T	10.0	E, R	2014, 2015
Rome city	ITA	ST	2.6	E, M	2008, 2010
Saitama	JPN	ST	1.2	M	2009
Salvador de Bahia	BRA	T	2.7	E	2013
San Antonio city	USA	ST	1.4	E	2013
San José city	USA	ST	1.0	M	2011
Santiago de Cali	COL	T	2.1	E	2009
Santiago de Chile	CHL	ST	6.1	A	2012
São Paulo municipality	BRA	ST	11.2	AM	2011
Semaarang	IDN	T	1.6	A	2010
Shanghai	CHN	ST	22.3	E	2010/09
Shenzhen	CHN	T	10.4	E	2012
Singapore	SGP	T	3.8	M	2012
Surat	IND	T	4.8	S	2011
Sydney metro area	AUS	ST	4.6	O	2010
Tegucigalpa	HND	T	1.2	E	2007
Tbilisi	GEO	ST	1.2	M	2011
Tokyo	JPN	ST	8.9	S	2007
Tshwane	ZAF	ST	2.9	E	2007
Tunis	TUN	ST	4.0	A	2011
Turin metro area	ITA	ST	2.2	E	2013
Wuhan	CHN	ST	9.8	O	2010
Xiamen	CHN	ST	1.9	E	2010
Yokohama	JPN	ST	...	S	2008
Zapopan	JPN	ST	1.6	AM	2013
Zhangjiagang	Chi	ST	1.2	E	
Zhengzhou	CHN	ST	4.3	E	2011

A–Adaptation plan, E–Emergency/contingency plan, M–Mitigation plan, O–Other plan, P–Policy or Program, R–Resilience plan, S–Strategy, ST--Subtropical, T–Tropical.

Rafael de Oliveira Sakai[6], Diego Lourenço Cartacho[7], Emilia Arasaki[8], Paolo Alfredini[9], Maurizio Rosso[10], Alessandro Pezzoli[11], Wilson Cabral De Souza Junior[12]

3 Extreme Events Assessment Methodology as a Tool for Engineering Adaptation Measures – Case Study of North Coast of São Paulo State (SP), Brazil

Abstract: The North Coastal Region of the State of São Paulo, which comprises the Municipalities of Caraguatatuba, São Sebastião, Ilhabela and Ubatuba, is one of the Brazilian areas most prone to flooding and debris flow deposition, owing to hydrological extreme rainfall events, usually coupled with extreme tidal levels.
The catastrophic scene of the city of Caraguatatuba March 18, 1967, resulting from one of the most serious natural disasters in Brazil, fosters discussions about probabilities of heavy rainfall-caused events and subsequent rise in the sea level in coastal areas. The research is founded on an innovative methodology based on the analysis of past rainfall and tidal station data, complemented with debris flow measurements and coupled with FLO-2D hydrodinamical model. The analysis involved meteorological, hydraulic, geotechnical and statistical knowledge areas applied to the region of the North coastal zone of the State of São Paulo (Brazil).
The obtained results are a good predictor of the probability of occurrence of certain types of heavy rainfall-caused events such as flooding or debris flow, coupled with a corresponding increase in tidal levels.
These practical results are intended to be used for urban planning, designs of macro-drainage, fluvial, maritime projects and debris flow retention structures.

Keywords: Meteorology, Hydrology, Maritime Hydraulics, Rainfall, Tidal Levels, Extreme Events, Natural Disasters, Geomorphology, Debris-Flow, Flooding

---

6 Promon Engenharia, Presidente Juscelino Kubitschek Avenue, 1830, CEP 04543-900, Itaim Bibi, São Paulo city (SP), Brazil, rafael.sakai@promon.com.br

7 Promon Engenharia, Presidente Juscelino Kubitschek Avenue, 1830, CEP 04543-900, Itaim Bibi, São Paulo city (SP), Brazil, diego.cartacho@promon.com.br

8 Department of Arquitetura e urbanismo, Universidade de São Paulo, Brazil, earasaki@usp.br

9 Depto de Engenharia hidraulica e ambiental, Escola Politécnica da Universidade de São Paulo, Brazil, alfredin@usp.br

10 DIATI, Politecnico di Torino, Italy, maurizio.rosso@polito.it

11 DIST-Politecnico di Torino, Italy, alessandro.pezzoli@polito.it

12 Department of Recursos Hídricos, Instituto Tecnológico de Aeronáutica, São José dos Campos (SP), Brazil, wilson.ita@gmail.com

3.1 Introduction

It is well established that the Brazilian Coastal Zone (BCZ) is important to the development of the country, given that most of total internal production is exported by harbors and waterways. To illustrate the relevance of BCZ, Figure 3.1 demonstrates the production of the municipalities, in excess of R$ 1 billion. It is note-worthy that the most important economic regions are situated around the São Paulo State. In spite of the fact that these coastal areas are economically developed, they are exposed to natural forces, such as orographic rainfall, tidal levels, debris flows and flooding. Understanding the relationship with the environment is of critical importance; as a consequence, this research was undertaken in the area of Civil Engineering, involving the interface between Maritime Hydraulics (tidal levels and hyper-concentrated flows), Hydrology (rainfall) and Geotechnics (percolation and landslides) specifically, and applied on the Northern Coast of São Paulo State (Figure 3.2).

Figure 3.1: Production of Brazilian municipalities greater than one billion Reais.

Coastal areas are subject to severe sea action and precipitation. The north coastal region of São Paulo is known for its orographic rainfall, caused by moisture fronts from the Atlantic Ocean; when they collide with the mountain range of Serra do Mar, there is precipitation on the coastal towns.

From the climatic point of view, the most impactful element is rainfall, with areas that have the highest total rainfall in Brazil (with an annual average of over

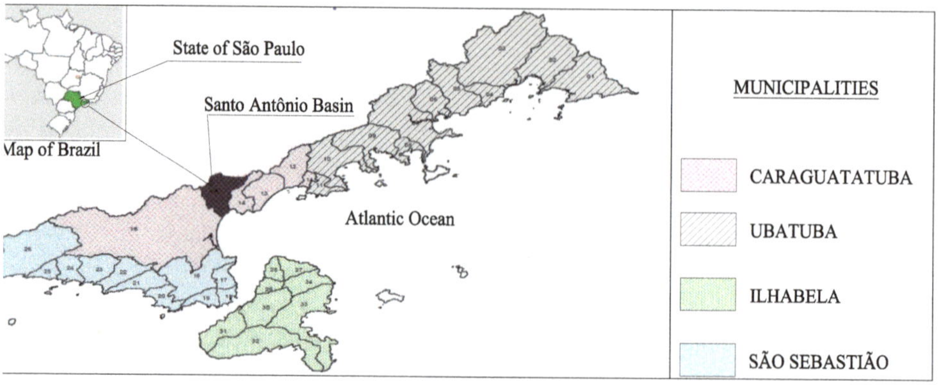

Figure 3.2: Municipalities of the North Coastal Region of São Paulo State (Brazil).

4000 mm/year, 6000 mm/year reached in extreme years) (CPTEC 2009). There is also the presence of "rain shadow islands" provided mainly by the massive island of São Sebastião (Ilhabela), affecting the areas north of the São Sebastião channel and the Caraguatatuba bay region. In these areas, total rainfall is lower (around 1800 mm/year). Due to its climate-influenced position, the region is dominated by tropical masses and presents constant frontal systems (cold fronts), which, together with the morphological structure and the Serra do Mar mountains, account for the events caused by the extreme rainfall (Conti 1975).

There has been a great demand for studies on the subject, mainly due to historical occurrences of disasters in the last century. According to Brigatti and Sant'Anna Neto (2008), the Northern Coast of São Paulo, because of its own natural characteristics and the recent economic dynamics, is regarded as an area where studies are currently aimed at better understanding the increasing effects of natural and anthropogenic factors.

In order to illustrate the magnitude of the 1967 disaster, images of Caraguatatuba were collected immediately after the debris flow (Figure 3.3(a)). Figure 3.3 shows the city of Caraguatatuba in 2012 (Santo Antônio River basin), with a population of approximately 100,000 inhabitants. It is of great concern that another event similar to that of 1967 could cause significant damage.

Some consequences of the 1967 event were:
- 7.56 million tons of mobilized material;
- 436 casualties registered, 400 buildings destroyed, 3,000 displaced people out of 15,000 inhabitants;
- Four to five meters high block deposits were formed along the Santo Antônio River. The largest boulders weighed between 30 t and 100 t;
- widening of the Santo Antônio River: from ten meters up to 20 m and from 60 m to 80 m in some areas.

Well known, the study of the combined effects of the tide, rainfall flooding and debris flow has been one of the most significant areas of study in the new century (Jones

Figure 3.3: Aerial photographs of the Santo Antônio river at different times. (a) scars on mountains after the 1967 rainstorm; (b) bridge over the Santo Antônio river destroyed by the debris-flow in 1967; (c) Santo Antônio river mouth after the 1967 rainstorm (five times greater than usual); (d) Santo Antônio river mouth in 2012.

1998; McInnes et al. 2005; White 2010; Lian et al. 2013). In recent years, a great deal of attention has been given to showing the separate impact of the events; nevertheless, only the combined effects have been successfully studied.

Recently Lian *et al.* (2013) studied the combined effect of rainfall and tidal level of the receiving water body on flood probability and severity in Fuzhou City, which has a complex river network.

The main purpose of this research was to evaluate the events caused by heavy rainfall combined with tidal levels. The obtained study results were applicable to urban planning fluvial and maritime projects and debris flow retention structures in the Santo Antônio River Basin (Figure 3.2) located in the city of Caraguatatuba on the North Coast of São Paulo / Brazil.

In this study, a multivariate probability methodology was developed to study the joint risk probabilities of rainfall and tide, and focused on the area of the São Paulo North Coast where studies and analyses are currently lacking. Furthermore, it provided a combined analysis of historical data of tidal levels and rainfall. Then, basing on the data recorded in the debris-flow catastrophe of 18th March 1967, a simulation with the hydrodynamic model FLO-2D was run to analyze the effects of flooding and debris flow.

Finally, comparing the results obtained by the statistical analysis and the deterministic model, policies for effective land use and management are suggested.

3.2 Methods

Given the evident extreme rainfall events such as flooding and debris flows and random sea – level behavior (applied for meteorological tide), the focus of the discussion will be on whether there is interdependence between these variables and, especially, how the variables are related to climate changes.

The first stage of the study was based on realistic data applied in a statistical methodology of concomitant rainfall and tidal level, producing results that represent the probabilistic scenarios of occurrence which is extremely important to determine drainage system policies. Similarly, these results are applied in deterministic methodologies (mathematical models) as input data, used in the second stage of this study.

The methodology described in the first stage (statistical stage) was based on Hidroconsult (1979), which carried out a similar study for Cubatão city, located on the central coast of the State of São Paulo; moreover, it represents a similar condition of rainfall and topography.

The statistical methodology adopted was founded on the following:

- collection, processing and data validation of tidal levels for the North Coast region of the State of São Paulo;
- collection and processing of rainfall data for the region of the North Coast of the State of São Paulo;
- understanding, development and application of statistical methodology applied to a combined occurrence of rain-tide for the Caraguatatuba region;
- obtaining graphs and tables of probabilities of the occurrence of certain phenomena involving rainfall and tides.

The debris flow and flooding studies, which represent the second stage of the research (deterministic step), related to the extreme rainfall events were performed according to this relationship between both variables (rainfall and tide level).

In this case, the methodology was organized as follows:

- collection, processing and data validation of topographic, geotechnical, rainfall and urban occupation information;
- data input in a digital model (capable of simulating debris flow and flooding);
- calibration and validation of the information input based on the real event which happened in March 1967;
- verification of the results in the area affected by debris flow and flooding, and also their physical parameters such as flow speed and deposition depth in order to aid future urban planning and defense effort. This analysis will also be performed considering the tide level associated with the rainfall event which may trigger these phenomena.

3.2.1 Getting the Database

The database was divided into the following groups: tidal levels, rainfall values and geographic data. Data characteristics and capture are explained below.

3.2.1.1 Tidal Data

The tidal level data come from different sources and institutions, namely:

- IGC tidal station (Cartographic and Geographic Institute) in Ubatuba;
- IOUSP tidal station (Institute of Oceanography, USP – University of São Paulo) in Ubatuba;
- CTH tidal station (Hydraulic Technological Center) on Martin de Sá Beach / Caraguatatuba;
- São Sebastião Harbour tidal station;
- Buoy of CEBIMar (Marine Biology Center, University of São Paulo) in São Sebastião.
- The compilation of information from tidal stations enabled us to generate a large database from 1954 to 2005, with some intermediate gaps. This database comprises over 225,000 hourly values of tidal levels.

3.2.1.2 Rainfall Data

The composition of the rainfall database in the city of Caraguatatuba started with investigation of data available from ANA (Agência Nacional de Águas), which represents a Federal Agency of management of water (ANA 2005).

The E2-046 rainfall station (Caraguatatuba) contains data from 1943 to 2010, totaling 24,603 accumulated daily rainfall values and thus it was taken into consideration in this study.

For the debris flow and flooding simulations, which are events that require greater data resolution (minimum hourly rate), the daily values provided by ANA are not optimal. Hence, a statistical analysis was necessary in order to achieve a reliable conversion between both resolutions, as demonstrated in the paragraph 3.2.3.2.

3.2.1.3 Geographic Data

Geographic data comprises topographic, urban occupation and geotechnical data.

Topographical plans from the Cartographic and Geographic Institute of the State of São Paulo (IGC-SP) at a 1:2000 scale were used to define the topography of the lower region of the Santo Antônio River basin, associated with smaller scale maps obtained from Cruz (1974) for its greater areas (Figure 3.4).

The inhabited areas of the municipality were also mapped in order to better highlight the association of the flow and flooding-affected areas with the risks that the populations face (Figure 3.5).

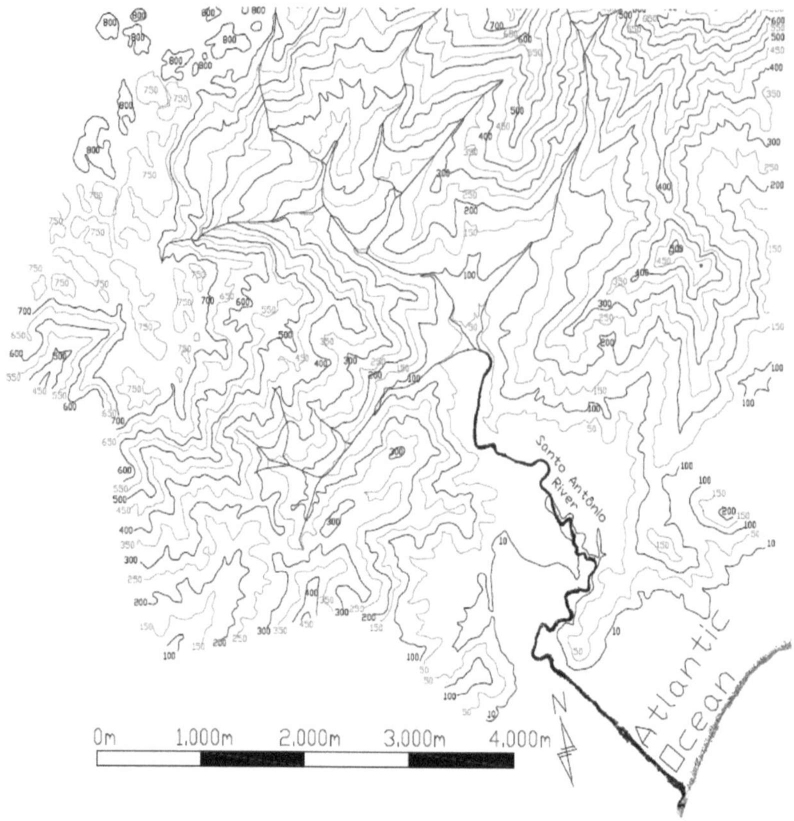

Figure 3.4: Topographical plan for Santo Antônio River Basin (contour lines each 50m).

3.2.2 Analysis of Rainfall and Tidal Data

3.2.2.1 Data Compilation

Referring to the methodology as described in the previous paragraph, the compilation of a tide and rainfall database was performed as initial step.

Tide and rainfall values were organized so as to highlight direct relationships between the daily rainfall in the Santo Antônio River Basin and levels in the tidal stations of the Northern Coast of the State of São Paulo.

Table 3.1 illustrates, as an example, how the data were compiled for the month of January 1954.

At the end of this step, 9.361 days have both rainfall and tides measurements (High tide; Low tide and Mean Sea Level). According to Figure 3.6, tidal behavior of São Paulo North Coast is characterized as semi-diurnal, meaning that in a period of 12.42 hours a complete tide wave occurs (one peak of both High and Low tide); consequently, in a period of approximately 24 hours, two High and two Low Tide levels are recorded.

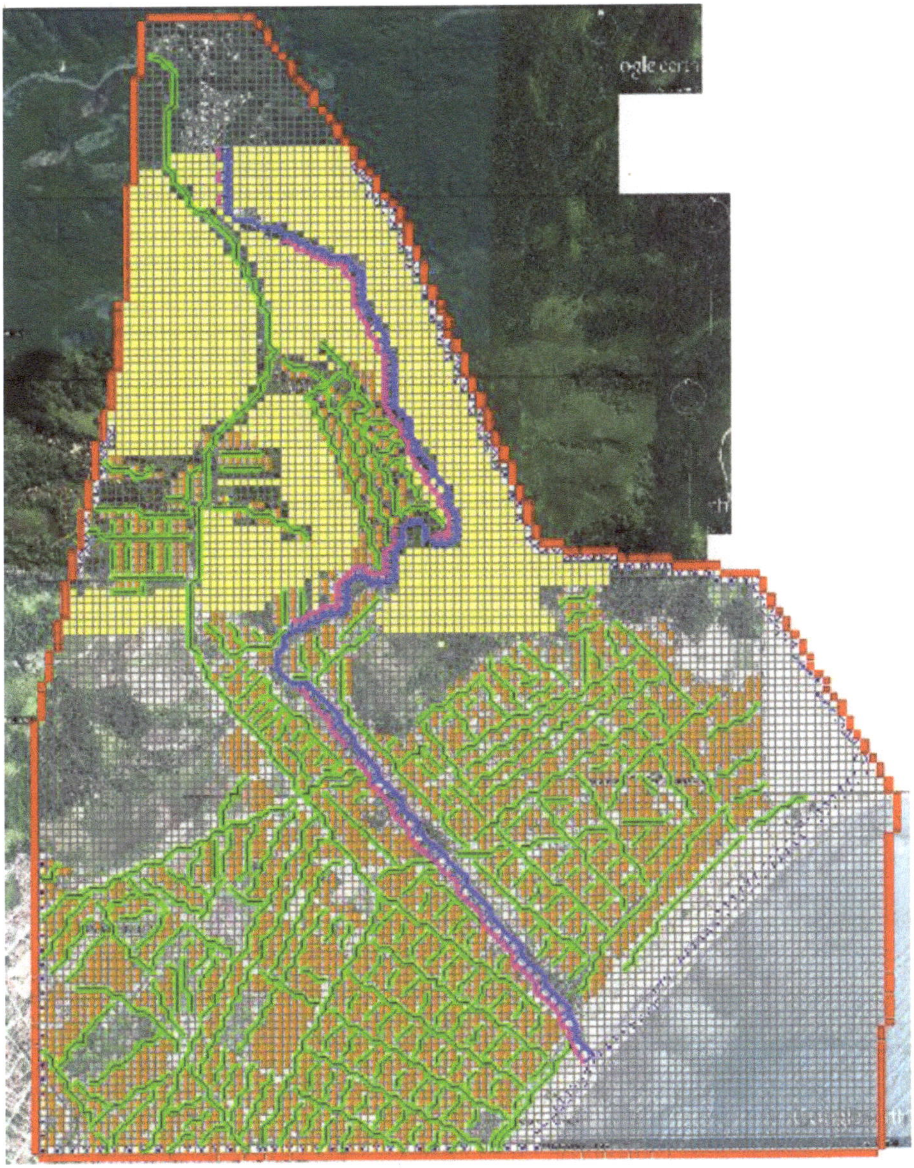

Figure 3.5: Inhabited areas of Santo Antônio River Basin (red – model boundaries, blue – Santo Antônio River's left bank, magenta – Santo Antônio River's right bank, green – roads and streets, orange – buildings, yellow – areas with high resistance to superficial flow).

Based on the methodology proposed by Hidroconsult (1979), only the highest and the lowest peaks in each day are selected for the analysis; therefore, the daily Mean Sea Level (MSL) is the result of arithmetic average of all measurements in a period of 24 hours.

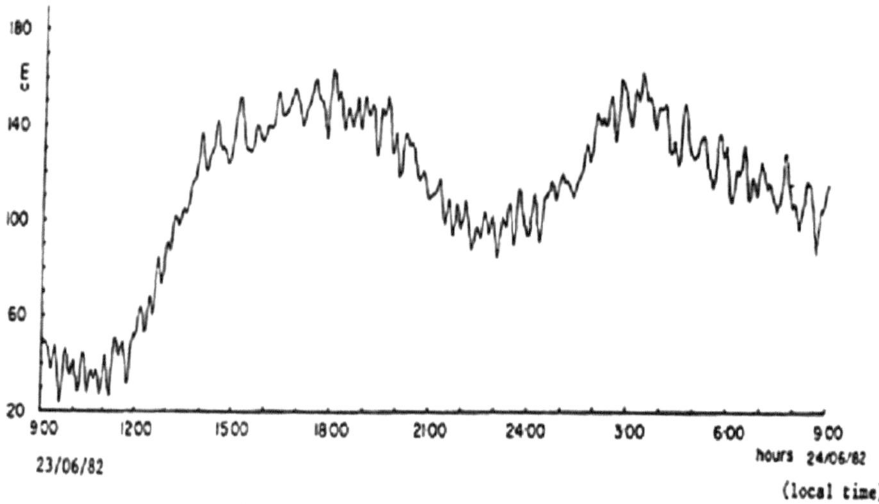

Figure 3.6: Semi-diurnal behavior tide on 23/06/1982 and 24/06/1982.

3.2.2.2 Division of Database Into Rainfall Groups

The next step in the methodology is to divide the database into groups, based on the accumulated daily rainfall: P ≥ 0 mm/day, P ≥ 25 mm/day, P ≥ 50 mm/day, P ≥ 75 mm/day, P ≥ 100 mm/day. The division of accumulated daily rainfall into these groups followed the concept that the range of measurements is between 0 mm/day and 190 mm/day; thus, in order to obtain 5 groups of accumulated daily rainfall, the intervals of each group is 25 mm/day. The application of this methodology in another region, where the rainfall range is larger, can generate large interval as, for example, each 50 mm/day.

The division of the database into groups is an important step in the process, because each rainfall track represents a probabilistic curve of the occurrence of rain-tide phenomena, as demonstrated below; as a consequence, small rainfall ranges implies more probabilistic curves.

The altimetry levels followed the IBGE reference, which represents the Brazilian standardization of surface levels (Brazilian altimeter Datum), according to the Imbituba tidal station located in Santa Catarina State. It is absolutely important to equalize all measurements of tidal levels to the same altimeter reference.

3.2.2.3 Reorganization of the Database Parameterized by Tidal Levels

The data were reorganized, after the separation into rainfall groups, following the guidelines below:

- for each precipitation interval (it was convenient to separate them into different worksheets), the table was rearranged in a decreasing order, using tidal levels

Table 3.1: Example of compilation of daily tidal levels and daily rainfall, for the Santo Antônio Basin, in January 1954.

Date	Data of tidal station (IBGE Reference)			Rainfall (mm/day)
	Lowest Low Tide (cm)	Mean Sea Level (cm)	Highest High Tide (cm)	
01/01/1954	−74.83	−26.96	15.17	0
02/01/1954	−78.83	−34.92	4.17	0
03/01/1954	−62.83	−13.13	20.17	16.7
04/01/1954	−54.83	12.50	60.17	14.5
05/01/1954	−85.83	−8.54	45.17	7.9
06/01/1954	−97.83	−19.50	25.17	3.2
07/01/1954	−75.83	−16.92	25.17	4.2
08/01/1954	−56.83	4.87	45.17	0
09/01/1954	−57.83	−3.83	29.17	0
10/01/1954	−67.83	−20.21	8.17	0
11/01/1954	−64.83	−19.54	4.17	0
12/01/1954	−44.83	−19.79	3.17	0
13/01/1954	−34.83	−11.83	27.17	0
14/01/1954	−31.83	−1.21	30.17	0
15/01/1954	−59.83	−11.08	35.17	0
16/01/1954	−64.83	−14.17	28.17	0
17/01/1954	−71.83	−15.29	28.17	0
18/01/1954	−64.83	0.00	44.17	0
19/01/1954	−77.83	−6.67	38.17	1.2
20/01/1954	−79.83	−12.67	36.17	0
21/01/1954	−62.83	−8.13	41.17	3.3
22/01/1954	−61.83	−7.00	37.17	2.5
23/01/1954	−61.83	−9.92	36.17	0
24/01/1954	−42.83	−6.00	30.17	0
25/01/1954	−39.83	−14.67	13.17	0
26/01/1954	−38.83	−11.25	13.17	0
27/01/1954	−26.83	−7.08	11.17	0
28/01/1954	−40.83	−18.46	−1.83	0
29/01/1954	−51.83	−21.83	2.17	0
30/01/1954	−48.83	−15.46	16.17	0
31/01/1954	−57.83	−6.38	36.17	0

as a parameter (highest tides at the top of the table). This methodology can be applied separately to daily Low tides, daily Mean Sea Level and daily High tides, according to specific applications. As an example, for drainage it is appropriate to use the Mean Sea Level values, for navigation the extreme measurements are appropriate, whereas for flooding analysis, the Highest tidal levels are suitable;
- note that the rainfall values should always refer to the corresponding daily tide values;
- for each precipitation interval, the largest annual tide occurrence (in the case of Low tides, the smallest annual occurrence) should be selected because the Return Period (TR) will be calculated in years. This explains the reason for the use of the maximum annual values in Table 3.2.

As a result of this step, there are several tables (one for each precipitation interval) sorted from the highest to the lowest tides with annual extreme values. It is worth noting that the event (day) must be repeated at different rain's intervals. For instance, it is possible to have the same day with rain over 100 mm ($P > 100$ mm/day) and rain over 75 mm ($P > 75$ mm/day).

3.2.2.4 Calculation of Probability of Combined Events (Rainfall Tidal Level)

At this stage, probabilities of occurrence of certain sea levels are calculated, associated with a rainfall range using a Gumbel mixed-model as suggested by Yue *et al.* (1999). Table 3.2 illustrates this step for the precipitation interval $P \geq 0$ mm/day and mean sea level. For each precipitation interval, a different table was created. The highest sea levels represent lower probabilities of occurrence, compared with lower tides; consequently, the table shows the statistical probability of occurrence of each event (sea level) to be equaled or exceeded by year concomitantly with $P>0$ mm/day. In other words, the application of following values ($P>0$ mm/day and MSL 12.76 cm) represents a probability of 97.14 % of occurrence of this daily event during the year, independently of the other past years. This concept explains the relationship of Return Period in years related with daily event.

3.2.3 Debris Flow and Flooding Analysis

3.2.3.1 Topographic Data Treatment

A data transformation was made from the scanning of paper-formatted documents into the AutoCAD program, and later by tracing the contours of the images. For each of them, the elevation was defined, and the data were then exported into the FLO-2D for subsequent simulation of the event as suggested by Grimaldi *et al.* (2013).

Table 3.2: Calculation of probability of occurrence of combined events (P ≥ 0 mm/day and Mean Sea Level), for the Santo Antônio Basin.

Mean Sea Level (maximum annual) IBGE Reference (cm)	Day	Rainfall (mm)	Probabilities of occurrence P (%)	Order Number
52.00	30/05/1988	26.2	2.86%	1
49.79	07/07/1989	0	5.71%	2
49.28	22/06/1990	0	8.57%	3
48.93	31/07/1980	4.1	11.43%	4
47.79	15/08/1999	0	14.29%	5
45.33	10/05/1956	3.5	17.14%	6
45.11	13/05/1959	1.6	20.00%	7
44.44	20/02/1995	63.5	22.86%	8
44.42	10/02/1966	0	25.71%	9
43.96	23/11/1970	0	28.57%	10
42.56	11/03/1987	17.8	31.43%	11
42.52	11/06/1993	19.4	34.29%	12
41.13	17/07/2000	3.7	37.14%	13
40.66	17/06/1971	6	40.00%	14
40.59	19/12/1994	15.2	42.86%	15
40.20	07/01/1996	21.8	45.71%	16
39.87	12/02/1998	22.3	48.57%	17
39.43	05/07/1991	0	51.43%	18
37.85	16/07/1992	2.1	54.29%	19
37.52	05/05/1963	0	57.14%	20
36.75	22/05/1978	5.3	60.00%	21
36.57	07/04/1979	13.5	62.86%	22
35.69	10/06/1983	3.2	65.71%	23
35.48	09/12/1982	2.3	68.57%	24
34.93	30/09/1981	1.8	71.43%	25
34.28	26/05/1958	8.8	74.29%	26
33.35	29/07/1955	0	77.14%	27
33.04	15/04/1986	0.5	80.00%	28
32.24	13/12/1972	16.5	82.86%	29

continued **Table 3.2:** Calculation of probability of occurrence of combined events (P ≥ 0 mm/day and Mean Sea Level), for the Santo Antônio Basin.

Mean Sea Level (maximum annual) IBGE Reference (cm)	Day	Rainfall (mm)	Probabilities of occurrence P (%)	Order Number
31.37	04/07/1965	0.8	85.71%	30
30.32	26/03/1997	1.1	88.57%	31
30.22	30/03/1964	20.6	91.43%	32
29.62	11/05/1954	1	94.29%	33
12.76	22/12/1977	35.9	97.14%	34

Table 3.3: Daily rainfall observed from three different points of the Santo Antônio river basin in the week preceding the disaster of March 18th, 1967.

Station	Coordinates	Days of March 1967 – Daily Rainfall (mm)						
		12	13	14	15	16	17	18
Rio do Ouro	S 23°38' / W 45°26'	0.8	6.5	10.7	0.3	6.5	50.4	195.5
Caputera	S 23°37' / W 45°26'	7.4	0.1	2.7	63.1	7.2	9.2	240.8
Fazenda São Sebastião	S ? / W ?	10	0	50	0	20	115	420
Daily Average						13.425	72.4	319.075

3.2.3.2 Rainfall Data Treatment

Unlike the previous detailed data treatment (paragraph 3.2.2), the rainfall event simulated here is not based on a random whole series of days, but only on those days of the week which predicted and culminated in the huge debris flow catastrophe of 18th March 1967. The daily rainfall values used are shown in the Table 3.3.

The processing of turning these daily-based values into hourly-based data was carried out by a statistical method that was developed by CETESB and DAEE (government sanitary / electrical energy companies) (DAEE/CETESB 1980). This method is based on transformation coefficients, presented in Table 3.4, which have been defined though statistical analysis of pluviographic rainfall data in relation with total daily rainfall values.

For example, in order to transform a daily rainfall value into an hourly value, it should be multiplied by the factor 24 h/1d, so as to comprehend the highest rainfall in a consecutive period of 24h. After this conversion, the obtained 24h precipitation

Table 3.4: Conversion factors from daily rainfall values to smaller time frames.

Time Relation	24 h/ 1d	12 h/ 24 h	6 h/ 24 h	1 h/ 24 h	30 m/ 1 h	15 m/ 1 h	5 m/ 1 h
Correction Factor	1.14	0.85	0.72	0.42	0.74	0.7	0.34

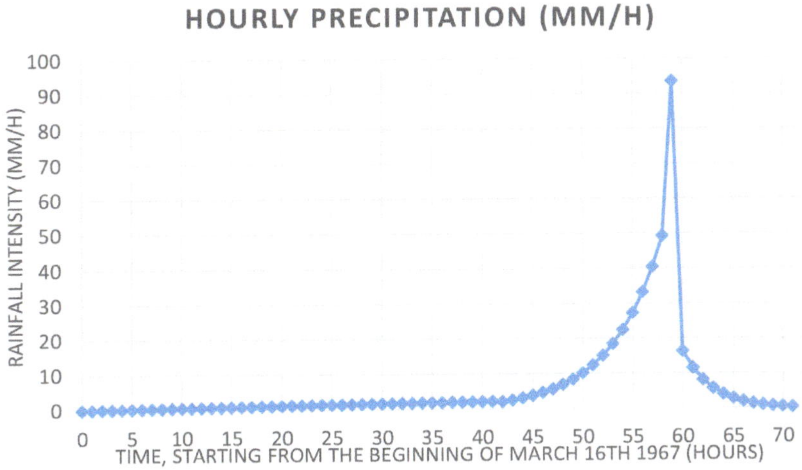

Figure 3.7: Graph of hourly rainfall during the previous days and the day (March 18, 1967) of the catastrophe.

intensity must be multiplied by the 1h/24 h factor in order to estimate the hourly peak rainfall of this day.

Based on the values shown in Table 3.3 and on the conversion factors shown in Table 3.4 as well as on the qualitative description of the event described in Cruz (1974), an hourly hyetograph was created for the specific event, referencing zero as the milestone hour of the beginning of March 16, 1967 (Figure 3.7).

3.2.3.3 Debris Flow and Flooding Modeling

When in possession of sufficient data for modeling the terrain, the computer program FLO-2D (O'Brien 1989) can be used in order to build, calibrate and validate a simulation model. The model used in this program can simulate the spread of a water-sediment mixed flow on terrains of complex topography and roughness, by following the principles of conservation of mass and momentum. The model uses dynamic equations from hydraulics and a finite mesh to predict the progression of a system flow over a grid of elements representing topography and buildings.

For the modeling of a complex hydraulic system, the program uses a set of components and routines, which are calculated for a number of discrete units, called grid cells. Its code contains elements that can simulate rainfall, channel flows, the flow from external sources and along watercourses, infiltration, the effect of dams, bridges, and many other items that can interfere or spread the flood /debris.

The program simulates the channels as a one-dimensional flow through the known geometry of its cross sections while the surrounding areas is a simulated two-dimensional flow, and the inflow exceeds the capacity of the channel bed (full in rough elements). The two-dimensional simulation is performed using numerical integration of the equations of motion and conservation of fluid volume.

In order to input the terrain model, a data transformation was made from the scanning of paper-formatted documents to the AutoCAD program, and later tracing the contours of the images. For each of them the elevation was defined, and the data was then exported to the FLO-2D.

The rainfall hydrogram was input into a HEC-HMS model of the basin and the resulting discharge at the floodplain's mouth was then input into FLO-2D model as a hydrograph.

Other necessary input values like terrain roughness, sediment coefficients and solid concentration were calibrated on the basis of the results of the model comparison with the real event which happened on 18th March 1967.

3.3 Results and Discussion

Some more relevant results from the rainfall and tide level relationships were obtained on the basis of the methodology described in paragraph 3.2.2. Figure 3.8 shows the graphical result from the statistical analysis of the previous paragraph.

From this graph, Table 3.5 summarizes the main useful values for both maritime and fluvial hydraulic projects in the region; in addition, these values can be considered as input in the deterministic model.

For a macro drainage project, for example, a 50-year Return Period is considered. By assuming this hypothesis, it is necessary to adopt a Mean Sea Level of 50.69 cm for any daily rainfall ($P > 0$ mm/day). Note that for the same Return Period (50 years) with accumulated heavier daily rainfall ($P > 100$ mm/day) a lower Mean Sea Level (40.59 cm) must be assumed; hence, an elevated sea level along with heavy rain is less likely to occur. Another important concept, for example, is that when days with the same rainfall characteristics are considered (i.e.: $P > 0$ mm/day), as larger Return Periods (less likely to occur) are selected, higher sea levels must be adopted. According to the explanation in paragraph 3.2.2.3, only the MSL values are shown in Figure 3.7 and Table 3.5, given that these results are most commonly used in engineering applications, which is based on the fact that sea behavior (tide levels) is constantly changing during the day. Moreover, the daily tidal cycle (Figure 3.6) is larger than other natural events (precipitation), meaning that MSL is used properly for the drainage purposes. Notwithstanding the fact that MSL is most commonly applied,

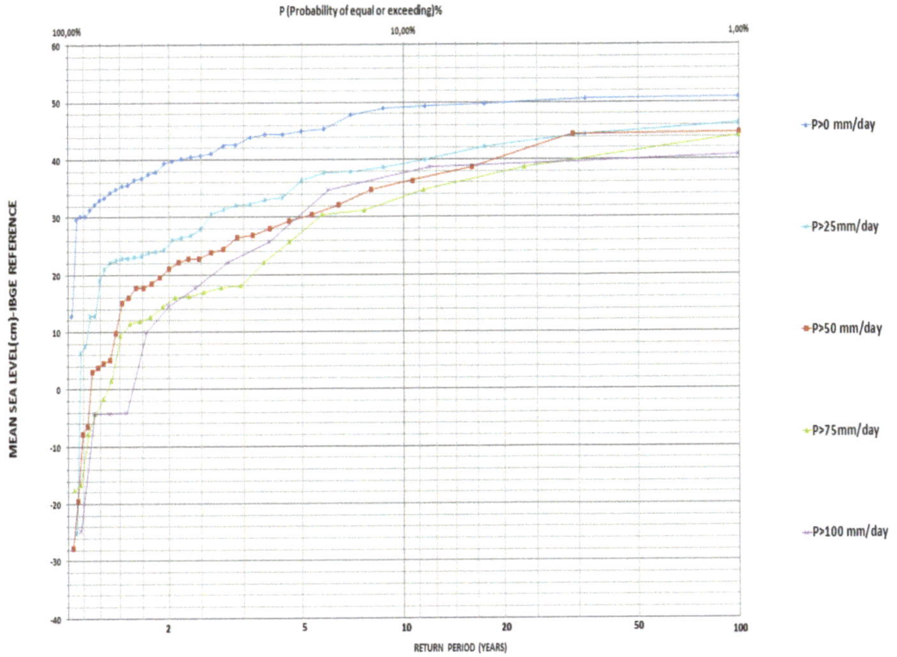

Figure 3.8: Graph of probabilities (Rainfall × Mean Sea Level).

Table 3.5: Results applicable to engineering projects (Rainfall × Mean Sea Level).

Rainfall (mm/day)	Return Period (TR)-Years							
	2	5	10	20	50	75	100	
>0	38.56	46.11	49.18	50.28	50.69	50.75	50.77	Mean Sea Level (cm)-IBGE Reference
>25	25.74	37.86	43.82	45.75	46.27	46.30	46.31	
>50	20.69	30.33	36.77	40.81	43.49	44.11	44.42	
>75	14.89	25.62	33.54	39.13	43.13	44.08	44.57	
>100	11.95	31.64	37.71	39.81	40.59	40.71	40.76	

the extreme analyses (High and Low tides) are important to specific situations, such as flooding and navigation. As an example, in a harbor project, the statistical results using high tide are important to determine the vessel draught in the access channel, jointly with precipitation parameters for drainage and discharges.

By combining the tide level and rainfall analysis, FLO-2D results were obtained for a defined rainfall data input, as described in paragraph 3.2.3, for two situations: debris flow and flooding occurrences.

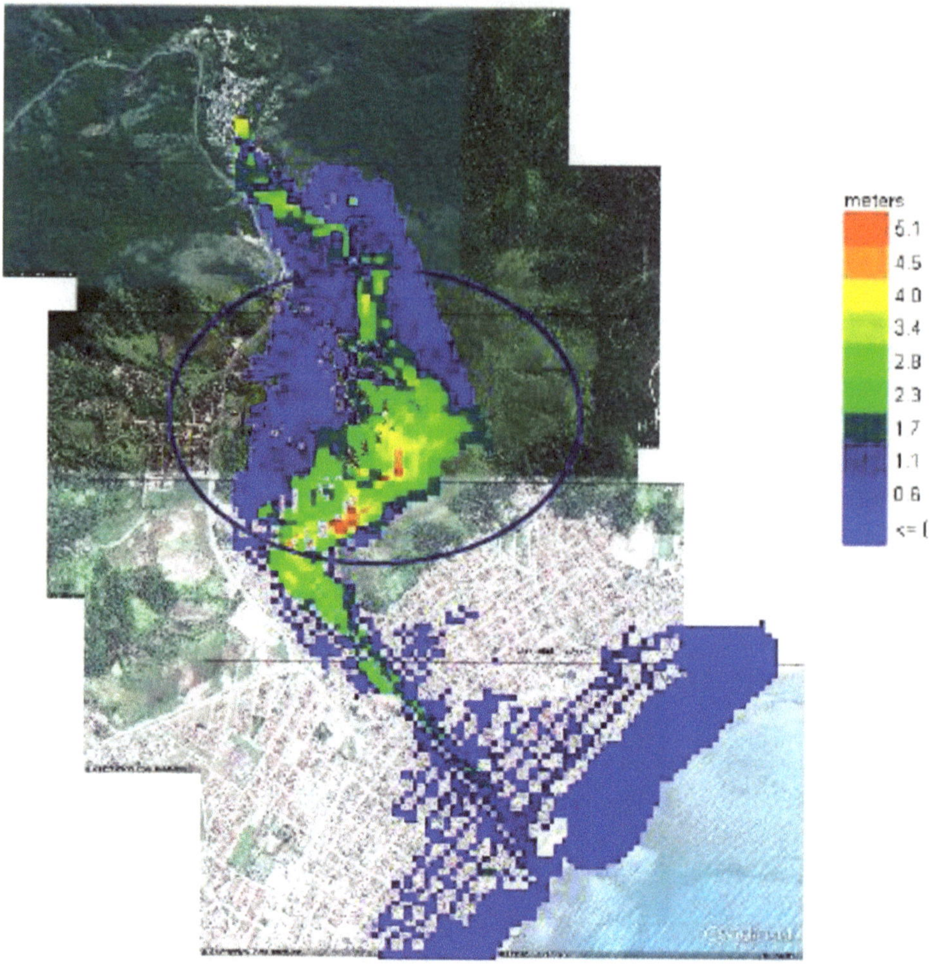

Figure 3.9: Flood depth and affected areas.

The areas affected were defined for these situations and plotted on a map, in order to facilitate understanding of the results.

Finally, by combining these results with the results presented in Figures 3.7, 3.8 and Table 3.5, the real catastrophic scenario of an event of this size is fully characterized (Figure 3.9 and Figure 3.10).

According to Brigatti and Sant'Anna Neto (2008), regarding the occurrence of floods, flooding and debris flow, the north coast has unique characteristics, mainly due to its physical make-up and land use. The land occupation beside the river banks and their mouths (Figure 3.11), together with peculiar atmospheric dynamics and tidal fluctuations, commonly causes serious socio-environmental damage.

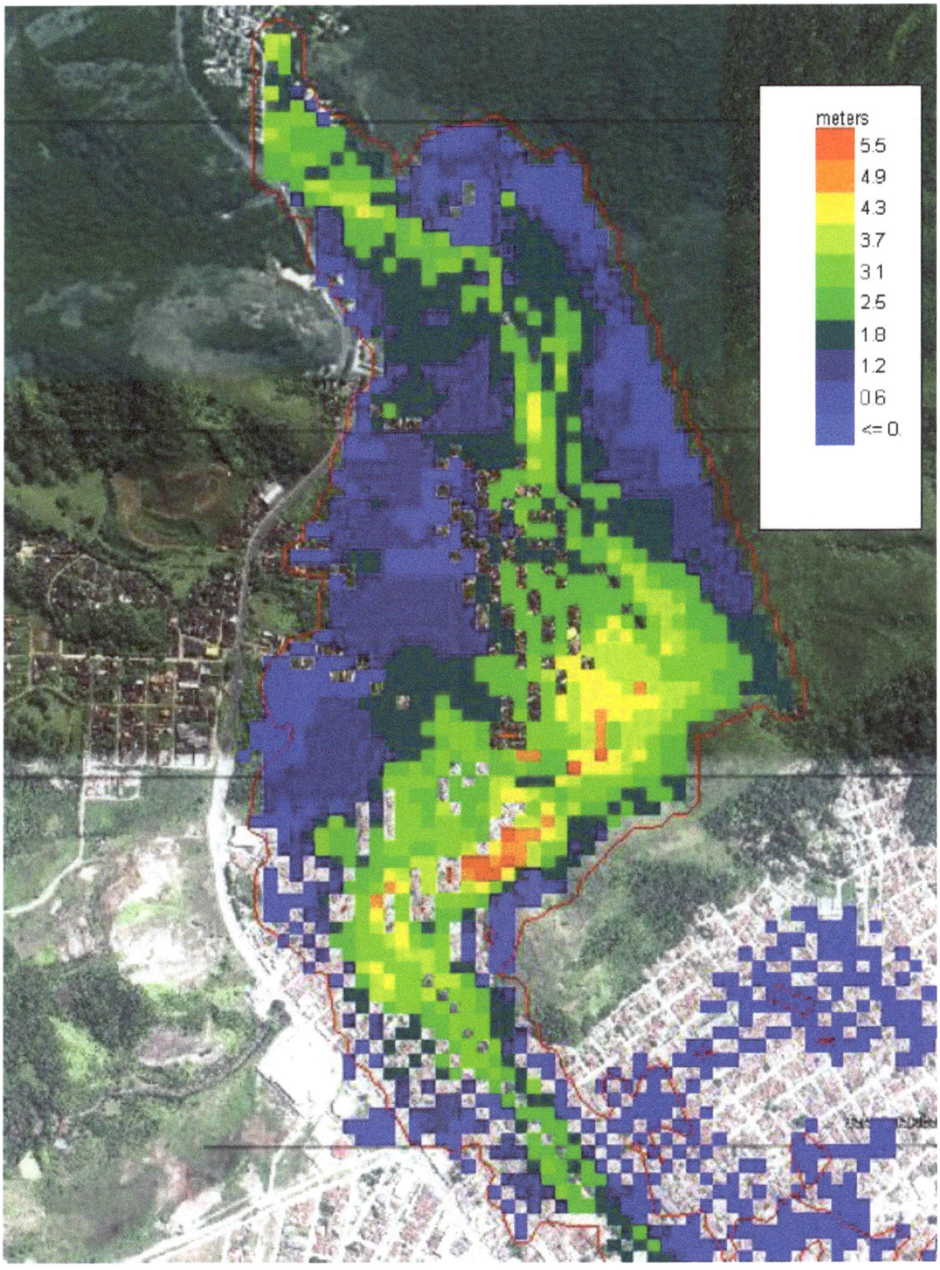

Figure 3.10: Debris-flow depth and affected areas.

Figure 3.11: Occupation by the river flooding area in the northern coast of São Paulo State, Brazil.

The ocean-atmosphere-continent interrelationship is extremely complex and leads to an area of uncertainty. In the specific case of the episodes related to floods, flooding and debris flow, many aspects must be considered, namely:

- the meteorological factors (mainly related to the cold fronts going through the region and the variations of their elements, especially wind, rainfall and atmospheric pressure) (Cartacho 2013);
- the coastal dynamics (its relations with meteorological events, currents and depositional processes that directly influence the rates of discharge of rivers, besides the tidal dynamics, notably related to spring tide episodes) (Cartacho 2013);
- the land use and anthropogenic influences (change in surface flow and absorption along the coast).
- the North Coast region of the State of São Paulo is located in an area with important atmospheric activities. The mountains of Serra do Mar act as a barrier to the atmospheric flow from the ocean, and their presence gives the region a complex configuration in relation to rainfall, as noted by Conti (1975), and the orographic effect greatly participates greatly in this dynamic (Cruz 1974).

The climatic characteristics, coupled with steep slopes, the small extension of the coastal plain, the shapes of the basins of major rivers and the ocean dynamics, characterize a fragile region. This is aggravated by irrational occupation and the construction of numerous roads, with the presence of irregularly occupied areas and poor projects carried out in areas susceptible to extreme episodes (Souza 1998; Fúlfaro *et al.* 1976).

3.4 Conclusions

The coastal regions of Brazil have constantly been subjected to extreme events caused by both heavy rainfall, and sea forces (waves, tides, currents).

The study of natural phenomena must begin with a continuous collection of data. It is understood as essential that any analysis must be based on collection, storage and processing of natural variables data (tidal levels, rainfall heights, waves, currents, etc.), in order to examine phenomena statistically, linking them to the probabilities of occurrence. This research is involved in this initial process.

From statistical studies, the results can be applied to Engineering practices. The coastal projects should consider the lessons learned from past events, both with the direct application of statistical analysis, and by using mathematical models, such as data input for simulations of natural events.

In recent decades, an advanced branch of Engineering has been committed to discussing, from the analysis of databases and mathematical models, whether the projects already built will be affected by climate changes, such as the increase of the Mean Sea Level, more frequent heavy rainfall and debris flow events that destroy cities in the coastal floodplain. If this is the case, repair work will become increasingly more frequent in these regions.

3.5 Acknowledgements

This work was supported by Redelitoral, with the project described in (ITA 2009), funded by CAPES (Coordination for the Improvement of Higher Education Personnel). The authors would also like to thank Promon Engenharia that encourages its employees to participate in academic research and Studio Rosso Ingegneri Associati s.r.l., which provided support in the FLO-2D simulation.

References

CPTEC-Centro de Previsão de Tempo e Estudos Climáticos. 2009. *Estudos da previsibilidade de eventos meteorológicos extremos na Serra do Mar*, Cachoeira Paulista.

Conti, J.B. 1975. *Circulação secundária e efeito orográfico na* gênese das *chuvas na região lesnordeste paulista*. São Paulo: IGEOG/USP.

Brigatti, N., and J. Sant´Anna Neto. 2008. Dinâmica climática e variações do nível do mar na geração de enchentes, inundações e ressacas no litoral norte paulista. *Revista Formação-Especial* 2, 15: 25–36.

Jones, D.A. 1998. Joint probability fluvial-tidal analyses: Structure functions and historical emulation. Institute of Hydrology-Wallingford & Ministry of agriculture, fisheries and food and the natural environment, Project FD04017.

McInnes, K.L., I. Macadam, G.D. Hubbert, D.J. Abbs, and J. Bathols. 2005. The effect of the climate change on storm surges. In *Climate change in eastern Victoria*. Stage 2nd edition: CSIRO.

White, L.W. 2010. Tidal and rainfall flooding evaluation for Cape Henry, Cape Story. Virginia: Parsons & Brinckerhoff.

Lian, J.J., K. Xu, and C. Ma. 2013. Joint impact of rainfall and tidal level on flood risk in a coastal city with a complex river network: A case study of Fuzhou City, China. *Hydrology and Earth System Sciences* 17: 679–689. http://dx.doi.org/10.5194/hess-17-679-2013.

Hidroconsult. 1979. Análise probabilística de ocorrência conjugada de eventos máximos chuva-maré. São Paulo: SOMA-Secretaria de Obras e do Meio Ambiente and DAEE-Departamento Estadual de Águas e Energia.

ANA-Agência Nacional de Aguas. 2005. http://hidroweb.ana.gov.br.

Cruz, O. 1974. A serra do mar e o litoral na area de Caraguatatuba, SP - Contribuição à geomorfologia litorânea tropical. São Paulo: Faculdade de Filosofia, letras e ciências humanas da Universidade de São Paulo. Department of Geography.

Yue, S., T. Ouarda, B. Bobée, P. Legendre, and P. Bruneau. 1999. The Gumble mixed model for flood frequency analysis. *Journal of Hydrology* 226; 88–100. http://dx.doi.org/10.1016/S0022-1694(99)00168-7.

Grimaldi, S., A. Petroselli, E. Arcangeletti, and F. Nardi. 2013. Flood mapping in ungauged basins using fully continuous hydrologic-hydraulic modeling. *Journal of Hydrology* 487: 39–47. http://dx.doi.org/10.1016/j.jhydrol.2013.02.023.

DAEE/CETESB. 1980. Departamento de água e energia elétrica / Companhia de tecnologia de saneamento ambiental. In *Drenagem urbana: Manual de projeto*. São Paulo: DAEE-CETESB.

O'Brien, J.S. 1989. FLO-2D: 2 Dimensional flood routing model software. In *Flood hazard mitigation and planning*. Nutrioso: Federal emergency management agency.

Cartacho, D.L. 2013. Análise probabilística chuva maré para a Bacia do Rio Santo Antônio em Caraguatatuba (SP). São Paulo: Escola Politécnica da Universidade de São Paulo-Departamento de Engenharia hidráulica e ambiental, dissertação (mestrado).

Conti, J.B. 1975. *Circulação secundária e efeito orográfico na* gênese das *chuvas na região Lesnordeste Paulista*. São Paulo: IGEOG/USP.

Souza, C. 1998. Flooding in the São Sebastião region, northern coast of São Paulo state, Brazil. *Anais da Academia Brasileira de Ciências*: 353–366.

Fúlfaro, V.J., et al. 1976. Escorregamentos de Caraguatatuba: expressão atual e registro na coluna sedimentar da planície costeira adjacente. In Congresso brasileiro de geologia de engenharia, vol. 2: 341–350.

ITA. 2009. *Mudanças climáticas globais e impactos na zona costeira: modelos, indicadores, obras civis e fatores de mitigação/adaptação*. São José dos Campos.

Alessandro Pezzoli[13], Enrico Ponte[14]

4 Vulnerability and Resilience to Drought in the Chaco, Paraguay

Abstract: This chapter presents an innovative methodology to identify and characterise the vulnerability and resilience to drought in ten indigenous communities (*comunidad*) in Chaco, Paraguay. The first part studies meteorological drought through the analysis of average daily rainfall over the last 38 years. The second part analyses vulnerability (V) on the *comunidad* scale. The third and final part analyses resilience (R) on the *comunidad* scale. The main sources used are the survey conducted for the Participatory community diagnosis (2014), from which 18 indicators have been chosen to measure exposure (E), sensitivity (S), and adaptive capacity (Ac) according to the equation V = (E * S) / Ac. We ascertain that rainfall in the driest areas of the Paraguayan Chaco varies cyclically in relation to the meteorological phenomena of El Niño and La Niña. The operative consequence of this phenomenon is that in the Chaco, drought may be predicted with enough forewarning to launch an early warning in the driest areas. Furthermore, we highlight that the most vulnerable *comunidades* are also the most resilient. This demonstrates that the various projects undertaken to strengthen resilience have so far benefited the most vulnerable communities and therefore the drought defence has been well directed.

Keywords: Vulnerability, Drought, Climatic analysis, Resilience, Climate change, Indigenous communities, Chaco, Paraguay

4.1 Introduction

This chapter analyses drought experienced by ten indigenous communities (*comunidades*) of the Chaco region in Paraguay, their vulnerability and resilience.

The Paraguayan Chaco (257,000 km^2, 43% of the national territory) covers almost a quarter of the Great American Chaco, a region that extends as far as Argentina and Bolivia (1.1 million km^2) on a sedimentary basin at the foot of the Andes (Pasig 2005; PSAC 1998) (Figure 4.1). The Paraguayan Chaco is located between the Paraguay River

---

13 Alessandro Pezzoli is a lecturer in Geography at DIST-Politecnico and University of Turin. He is author of paragraphs 4.2 and 4.3, alessandro.pezzoli@polito.it.
14 Enrico Ponte, PhD in environment and territory, holds a research grant at DIST-Politecnico and University of Turin. He is the author of paragraphs 4.1, 4.4, 4.5, 4.6, enrico.ponte@polito.it.
The Authors would like to thank Maurizio Tiepolo (DIST-Politecnico and University of Turin) for comments and observations; Manuel Simoncelli and Francesco Anichini (COOPI) for the documentation and support during the mission to the Chaco in April 2014.

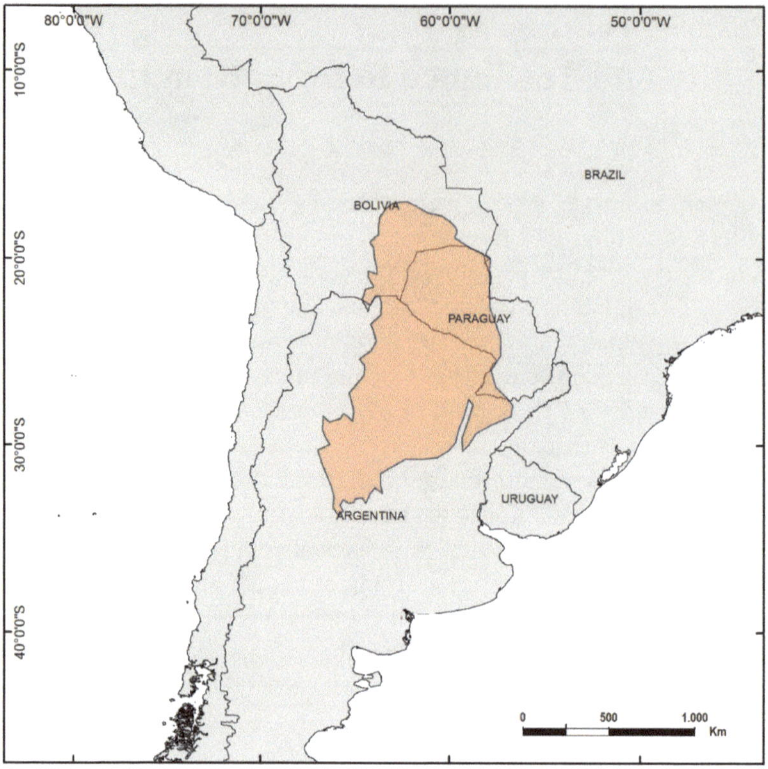

Figure 4.1: The Great American Chaco (pink) in the context of Latin America (elaboration E. Ponte).

to the East and Pilcomayo to the South. From an administrative point of view, it is divided into three regions (Boquerón, Alto Paraguay and Presidente Hayes), which are in turn divided into 15 districts. Only 2.5% of the population of Paraguay lives in the Chaco (density 0.5 inhabitants/km^2) and is made up of indigenous groups (33% of the Great American Chaco population), Mennonite colonies (14% of the Great American Chaco population), and creoles (53%) (PNUMA 2013; Pacheco 2012).

The climate is characterised by two seasons: dry from May to October and rainy from November to April. Drought and flooding characterise the climate of Paraguay, the summer is long, hot and wet, while the winters are short, mild and dry, but frost may occur between June and August (Grassi *et al.* 2005).

The majority of rainfall is convective, produced by storms or series of storms common from spring to autumn (Pasten 2007). Temperatures can reach 45° C and the long periods of drought block the economic development of this region and broaden its marginalisation compared to the economic and financial centres in the south-west of the country. According to the literature (Grassi *et al.* 2005; GTZ 2005; PNUMA 2013; USAID 2007), western Paraguayan Chaco is drier than that to the east. Annual overall rainfall in central Chaco is on average (1960–2010) 1,000 mm, while in western Chaco it is 1,500 mm. Nonetheless, rainfall in the latter has been less than 450 mm in 13 years out of 40.

Nunez *et al.* (2009) have shown how in Paraguay the effects of climate change (CC) caused by global warming consist of rising temperatures, especially during the spring.

In the Chaco, where the indigenous populations survive on agriculture and sheep farming, climate trends have significant impact: the indigenous communities are less prepared to deal with this type of occurrence (Vila 2010; IPMPCC 2011). Drought in the Chaco is a transitory anomaly that is characterised by an insufficient availability of fresh water for plants, animals and humans.

Among water resources studies developed in the sector from the 1970s on, IDAEA (2012) highlighted how the biggest problem for the Chaco is the lack of fresh water. In this vast area it is rare to have underground water, but when it is there, it is so salty as to render it unusable for humans and animals (Pacheco 2012).

The aim of the chapter is to characterise atmospheric drought and assess the vulnerability and resilience of ten indigenous communities of Paraguayan Chaco.

The method used is simple, but innovative, and contributes to demonstrating the relationship between global and local climate phenomena. In effect, we based it on an integrated analysis of the hydrological data and risk evaluation. Based on the Standard Precipitation Index (SPI) as indicated by Khan *et al.* (2008) for the study of drought and the impact of rainfall on agricultural areas, and carrying out an evaluation of those parameters distributed over the territory, we were able to identify the most arid areas and the wettest areas in the territory studied. Finally, the analysis of the thermal sea temperature anomaly models obtained by the NCEP-NOAA oceanic data assimilation system (Behringer and Xue 2004) enabled us to define a connection between the annual variations of the humid and arid areas with El Niño cycles. This evaluation has allowed us to introduce, in the risk evaluation model, a variable of exposure to drought that is, in addition to the population size, also a function of the physical phenomenon. The inclusion of this variable facilitates an effective improvement of the risk evaluation model and resilience. The most obvious application is the possibility of replicating this analysis in similar situations, and adapting it to the information available. Another interesting possibility would include using this analysis to carry out specific impact studies.

The following paragraphs present the innovative methodology of meteorological, vulnerability and resilience analysis, followed by the results obtained. Finally, in the conclusions, we set out some suggestions to develop future strategies.

4.2 Methodology

4.2.1 The Analysis of Vulnerability

Vulnerability may be defined as a "function of the climate variation to which a system is exposed, of its sensitivity and its ability to adapt" (IPCC 2001). Consequently, the components of vulnerability are:

- Exposure (E), which is defined as "The nature and degree to which a system is exposed to significant climatic variations" (IPCC 2001). In our case, the hazard is represented by the meteorological drought.
- Sensitivity (S), which describes "the degree to which a system is affected, either adversely or beneficially, by climate-related stimuli" (IPCC 2001).
- Adaptive capacity (Ac), which defines "the ability of a system to adjust to climate change (including climate variability and extremes) to moderate potential damages, to take advantage of opportunities, or to cope with the consequences" (IPCC 2001), in our case five indicators are used.

The first two components of vulnerability represent potential impact while the last one indicates the measure in which the impacts may be attenuated.

These components are linked in the equation: V = (E * S) / Ac

Each component is measured by indicators (Brant 2007), which may vary depending on the time scale and the systems involved (social, economic and biological) (Smit and Wandel 2006). Often, vulnerability is ascertained as the starting point for the study of resilience.

This approach also allows monitoring vulnerability over time, though it is limited by the subjectivity of the selection, the variables considered, and the availability of data (Luers *et al.* 2003). The final considerations will therefore refer to the potential impacts of the drought forecast on the short- and long-term in accordance with the pre-existing scenarios of CC (Knutson *et al.* 1998). In the instance of the Chaco, we used 18 indicators from the Participatory Community Diagnosis (COOPI, OXFAM / PCI, ACH 2014), considering the data available.

The component E (exposure) included the impact of CC on the system, as explained below. For the study area, the system of liquid precipitation and drought that makes the system vulnerable represent the CC affect. Therefore, in this study, the analysis of drought is considered, supplemented by a detailed analysis of all other indicators that comprise the vulnerability analysis.

4.2.1.1 Methodology for Analysing Drought

To study drought in the Chaco, the SPI (Standardized Precipitation Index) was used, as recommended by the World Meteorological Organization (WMO 2012). This index measures probability relating to rainfall. The SPI is based on the probability of recording a given quantity of rainfall. The probabilities are standardised so that an index of 0 indicates the quantity of average rainfall. The index is negative in situations of drought, and positive for rainfall. The more arid or wet the conditions, so the index becomes more negative or positive. The SPI is calculated by the NCDC (National Climatic Data Center) on different time scales – from one to 24 months – with the aim of bringing forth variations both for the short- and long- term[15].

We decided to calculate the SPI for various points. From the NOAA (National Oceanic and Atmospheric Administration) website, we selected the weather stations that have continually recorded daily rainfall data for at least 20 years between 1949 and 2013. To choose the stations, a circle with a 250 km radius was traced with the meteorological station of Pozo Colorado as its centre (59° 18' 22" W, 23° 32' 43" S), and all 15 stations that fell within it were considered. Six stations showed incomplete or anomaly data (rainfall absent for months at a time or anomalous rainfall) (Figure 4.2). All rainfall data (recorded every six hours) was gathered from January 1, 1975 to December 31,2013 and monthly rainfall was calculated.

Once this phase was completed, the monthly, three-monthly, six-monthly and annual SPI were calculated. At this point, graphics were drawn up with the annual and six-monthly SPI (wet and dry seasons) to have an initial idea of its main variations and trend.

Subsequently, SPI maps were drawn up to show the variation of the index over the past 40 years. Following this, the six-monthly SPI was calculated for the dry season (from May to October) and for the wet season (from November to April).

4.2.2 The Analysis of Resilience

Resilience is defined as the ability of a system, a community or a society potentially exposed to risks to resist, absorb, encounter and recover from the effects of a hazard in a timely and effective manner, also through the conservation of its basic structures and functions (UN-ISDR 2009). This ability is determined by the degree to which the social system is able to increase its ability to learn from past disasters, increasing protection and reducing risk (African Development Bank *et al.* 2004).

Resilience is the opposite of vulnerability; the higher the level of resilience of a community, the lower its vulnerability. In other words, the probable impact of drought would increase the danger (measured, for example, by the number of people exposed and/or by the frequency and seriousness of the drought) and therefore the vulnerability of a community would be higher. Nonetheless, the drought risk of a community decreases as its resilience increases.

15 http://www.ncdc.noaa.gov/.

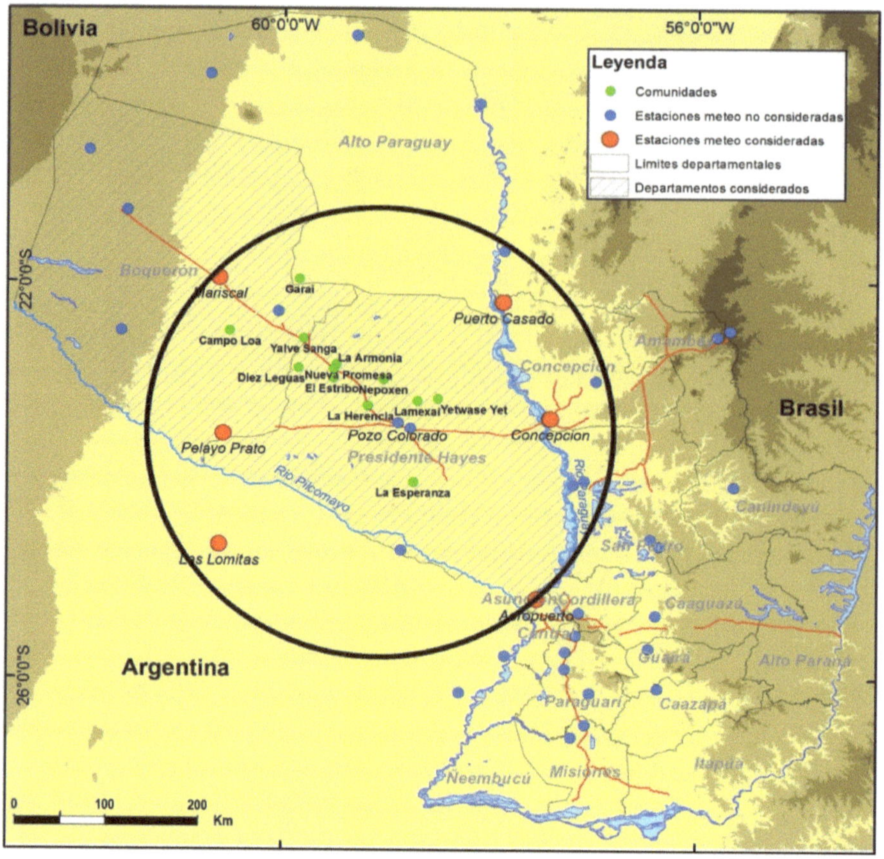

Figure 4.2: Paraguayan Chaco. Study area (circled) showing the comunidades (green dots), the weather stations considered (red dots) and those not considered (blue dots) (by E. Ponte).

Considering that communities are unable to reduce their exposure to drought, attention must therefore be focussed on looking for ways to reduce vulnerability and increase resilience. In fact, the resilience approach focuses attention on the components of a system and how they relate to each other (Berkes *et al.* 2003). But it is important to underline that resilience is not synonymous with adaptation (Walker *et al.* 2006). In fact, in some cases, high adaptability can unintentionally lead to a loss of resilience capacity.

The sources of information used are the Participatory Community Diagnosis and information gathered on site during a mission (April 2014). The main difficulty related to the tracing of representative indicators of resilience available for all the *comunidades* of the study area.

4.3 Characterisation of Atmospheric Drought

Although it may be considered a rare and causal event, drought is a component of climate. This is verified throughout almost all climatic zones, but its characteristics vary greatly from one region to the next (UN-ISDR 2009). Drought in itself is not an emergency situation, but it becomes such due to its impact on the population (UN-ISDR 2003).

Drought is "a dangerous physical phenomenon that may cause the loss of human life, injury or other effects on health, damage to property, loss of means of survival and services, social and economic disintegration and environmental damage" UN-ISDR (2009).

We can distinguish four types of drought: meteorological, agricultural, hydrological and socio-economical (UN-ISDR 2009). Meteorological drought is a natural event that has climatic causes that differ from region to region. It may also be defined as a lack of rainfall for a prolonged period of time, which is translated into a lack of water for a variety of activities. Agricultural, hydrological and socio-economic drought affects human and social aspects, especially the interaction between the climatic characteristics of meteorological drought and the human activities that depend on rainfall to provide enough water and products to satisfy social and environmental needs. This chapter reflects mainly the meteorological drought of the Paraguayan Chaco considering the strong impact on the indigenous population.

The United Nations Convention to Combat Desertification (UNCCD) and in part also the United Nations Framework Convention on Climate Change (UNFCCC) are particularly aimed at reducing the risk of drought.

The main objective of the analysis is to verify the trend of drought in recent years in the study area, and to evaluate the possibility of predicting it in advance.

The distribution of rainfall throughout the year (the Figure 4.3 shows the rainfall index derived from a dataset developed and maintained by NOAA) shows that the months with most rainfall are November, December, January and February (Ip11, Ip12, Ip01, Ip02), while the driest months are June, July, August and September (Ip06, Ip07,

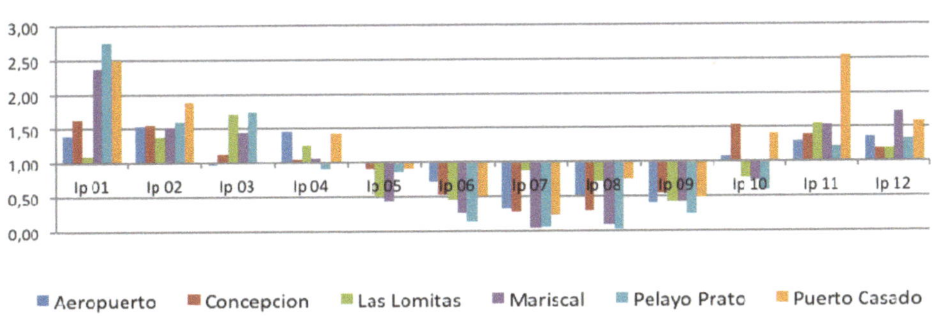

Figure 4.3: Paraguayan Chaco. Rainfall index for the six stations (source: NOAA).

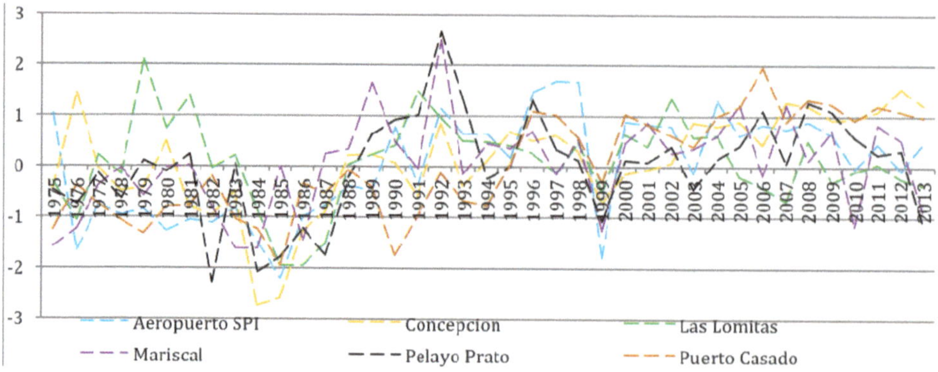

Figure 4.4: Paraguayan Chaco. Variation of the SPI for the six stations analysed.

Ip08, Ip09). The graph also reveals how drought is accentuated by the fact that for a good eight months of the year (from March to October), average monthly rainfall is less than 100 mm, and decreases to 50 mm from June to September.

The trend of the SPI for the six stations considered (Figure 4.4) may be summarised as follows:

- 1975–1984: slight increase in drought
- 1985–1994: slight decrease in drought
- 1995–2004: decrease in drought
- 2005–2013: decrease in drought

Observing the graph of the annual average SPI, we can note how anomaly values have been recorded in some years compared to the trend of the previous years: this is the case in 1992, a particularly wet year, or in 1999 which was very arid.

The next step is to understand whether the variations observed on the SPI maps have been influenced by global meteorological phenomena such as El Niño or La Niña.

These phenomena occur in the central Pacific Ocean. The former is characterised by warming and the latter by a cooling of currents. When the waters are warming (El Niño), the pressure of the western Pacific is high and when the waters are cooling (La Niña), the pressure of the western Pacific is low. These phenomena condition the rise of rainy phenomena in a large part of North and South America.

We used the annual climatological map[16] obtained using data averages from 1975 to 2013 (Figure 4.5). The map shows – according to the scale of colours from red to blue – the average variations of the sea temperatures.

16 http://www.cpc.ncep.noaa.gov/products/GODAS/

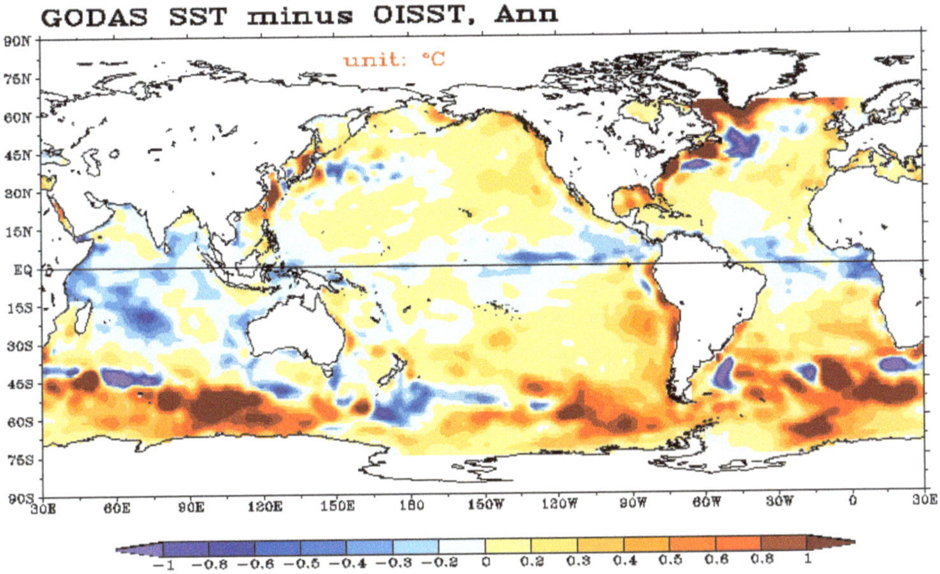

Figure 4.5: Average variation of the sea temperatures (1975/2013) (source: NOAA).

Subsequently, three different years were taken as examples, one per type of SPI map: 2003 (uniform map), 2013 (map with wet area to the south) and 1987 (map with wet area to the north).

4.3.1 Neutral Situation

To demonstrate this case, we have taken the year 2003 as an example. The SPI map (Figure 4.6.a) highlights a standard coloration if we exclude the area to the east where there is a very slight variation. This means that in 2003 there was minimal variation in humidity (violet colour, variation between 0 and 0.5), which was uniform throughout the territory under study.

If we observe at the same time the 2003 map of the Sea Surface Temperature (SST) (Figure 4.6.b), we can see how this differs little from the annual climatological map (Figure 4.5).

This similarity between the two maps can be explained by the fact that when there is no considerable variation between the annual climatological map (Figure 4.5) and the annual SST map, (Figure 4.6) we will have a uniform SPI map (Figure 4.6) and therefore there will be a minimal and uniform drought variation within the territory under consideration.

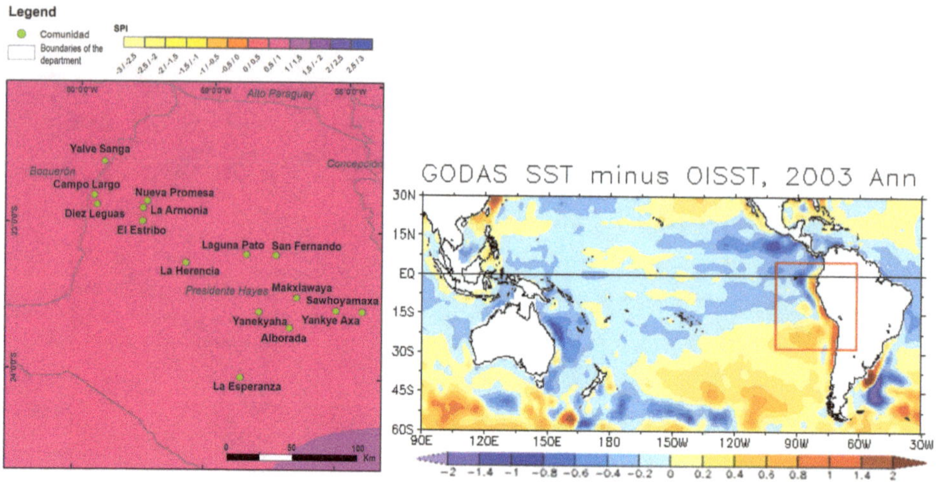

Figure 4.6: SPI in the Chaco in 2003 (left) and SST in 2003 (right).

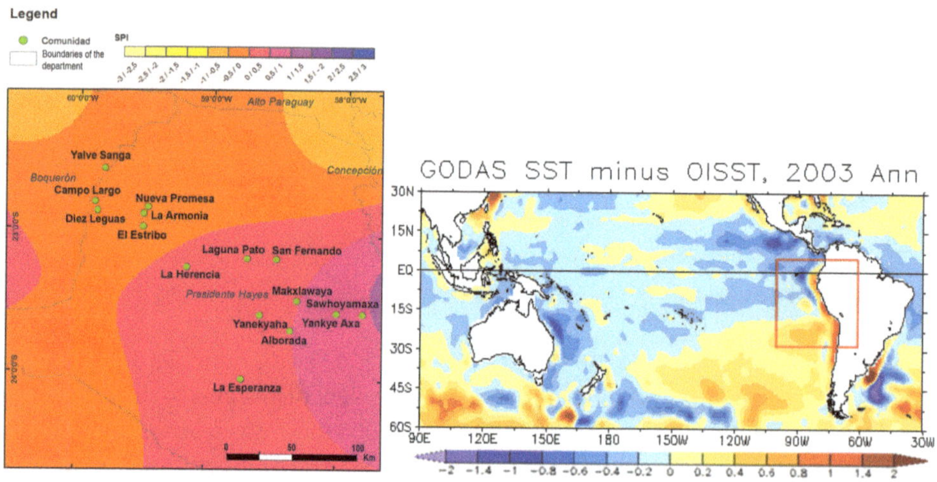

Figure 4.7: SPI in the Chaco in 2003 (left) and SST in 2003 (right).

4.3.2 The La Niña Situation

To demonstrate this case, we have used the year 2003 as an example. Observing the SPI map (Figure 4.7), we can see how there is a more arid area concentrated towards the North with points of variation that go as low as –1. This means that in 2013 there was a negative humidity variation to the north (orange colour, variation between –0.5 e –1) and a positive one to the south (violet colour, variation between 0 and 1).

If, at the same time, we look at the 2013 SST map (Figure 4.7), we can see how this differs from the annual climatological map (Figure 4.5) in two aspects: the presence of a wide cold area (in blue) at the centre of the Pacific Ocean and the almost complete absence of a warm front on the coasts of Ecuador and Peru.

This is what occurs to cause La Niña.

4.3.3 The El Niño Situation

To demonstrate this case we have used the year 1987 as an example. The SPI map (Figure 4.8) highlights a more arid area concentrated to the south (yellow) with points of variation that reach –1.5. The 1987 SST map (Figure 4.8) differs from the annual climatological map (Figure 4.5) in the absence of cold area (light blue) in the zone to the south of Panama, and in the consequent further extension of the zone to the north of Ecuador. This occurs with El Niño.

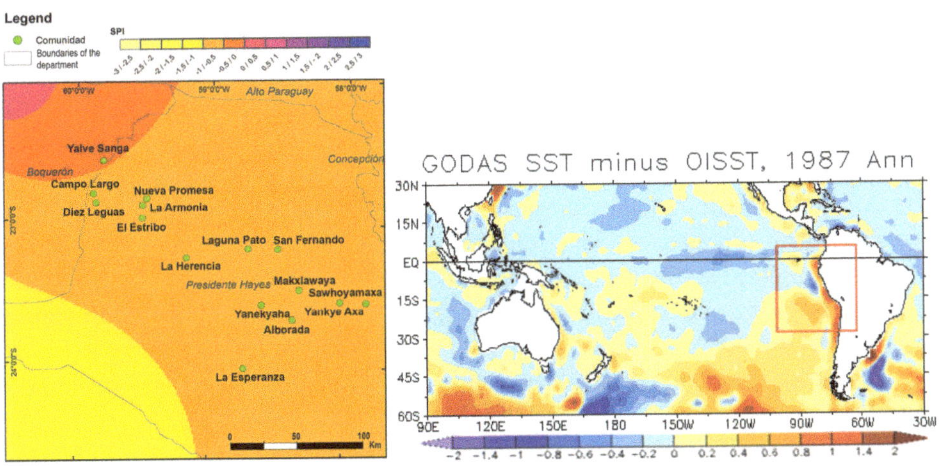

Figure 4.8: SPI of the Chaco in 1987 (left) and SST in 1987 (right).

4.3.4 Final Considerations

The three analyses reported above demonstrate the existence of a close relationship between the variation in temperature of the Pacific Ocean near the coasts of Ecuador and Peru, and the variation in SPI in the Chaco. These variations in temperature of the Pacific Ocean are influenced by the climatological phenomena of El Niño and La Niña. These phenomena can now be predicted and this may serve to warn the Chaco regarding those areas in which drought may occur in the subsequent months.

Based on these considerations, two thematic maps have been drawn up for each scenario (neutral, Niño, Niña). The first map represents a division of the drought by

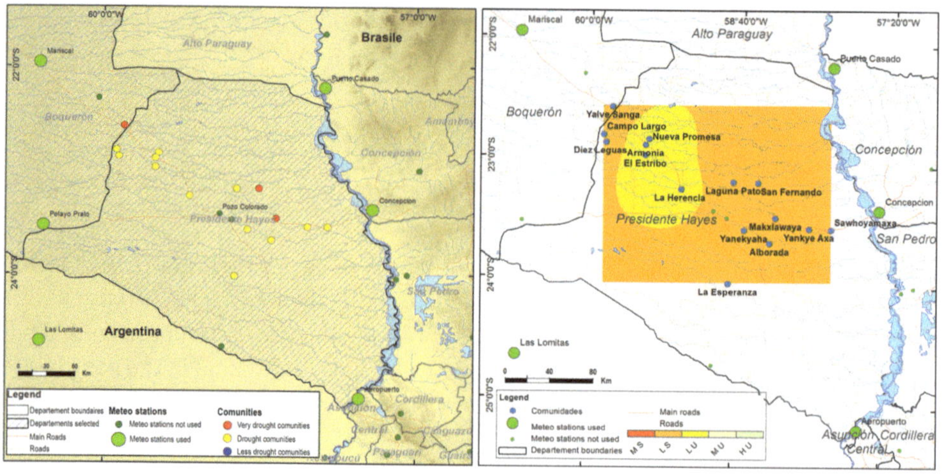

Figure 4.9: Paraguayan Chaco, neutral case: comunidad by level of drought (blue-humid, yellow-not very arid, red – arid) (left) and map of drought (right) (by E. Ponte).

comunidad while the second covers the entire territory. In Figure 4.9.a., which refers to a neutral scenario, the arid *comunidades* are represented with a red point, while those less arid and more humid are represented with yellow and blue dots, respectively. In Figure 4.9.b the arid areas are represented in red and humid ones in blue.

Observing the two figures given above, it can be noted as in the case of a neutral situation there is a general uniformity among the different Communities that appear to be mainly drought.

Regarding the scenario of La Niña, the situation changes. It seems obvious in both figures that a clear separation is found among the wetlands southeast, slightly damp areas in the centre, and driest areas in northwest.

Finally, regarding the scenario of El Niño, the situation is reversed, as in a mirror. It may be noted in both figures, as in this case, the wetlands are northwest, the little moist areas are in the middle and the driest areas are southeast.

This analysis, completed by the maps (Figures 4.10 and 4.11) outlined for the situation of Niña and Niño, covers a gap in literature, which does not present studies in drought in Paraguayan Chaco, and constitutes a starting point for future analysis. Alternatively, analyses of this type linked to national and regional evaluations of drought via the SPI are present in literature (Bonaccorso *et al.* 2003; Livada and Assimakopoulos 2007; Labedzki 2007; Michaelides and Pashiaridis 2008; Khan *et al.* 2008), demonstrating the current state of this type of research.

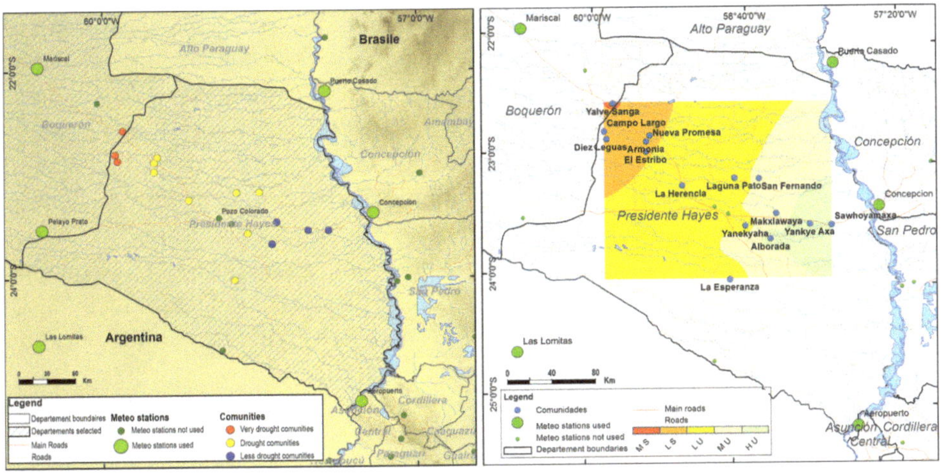

Figure 4.10: Paraguayan Chaco in the case of Niña: Comunidad by level of drought (left) and map of drought (right) (by E. Ponte).

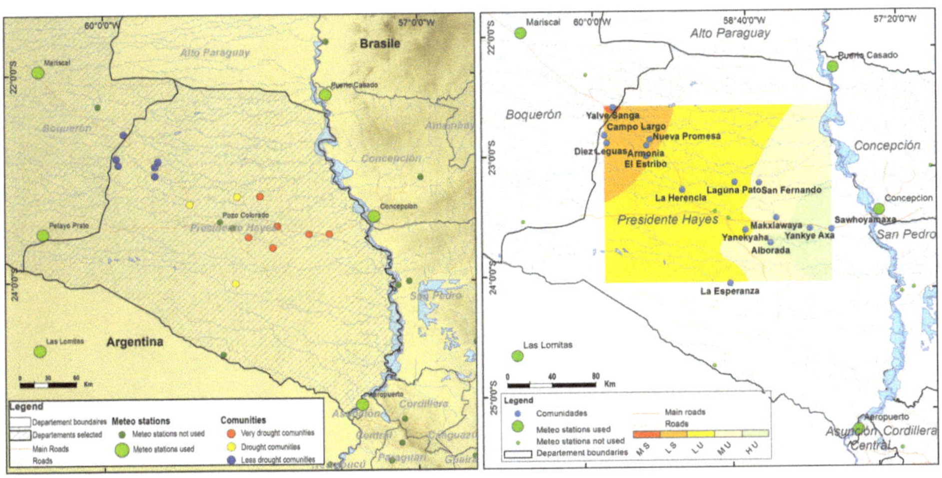

Figure 4.11: Paraguayan Chaco in the case of Niño: Comunidad by level of drought (left) and map of drought (right) (by E. Ponte).

4.4 Vulnerability to Drought

As mentioned in Section 2.1, vulnerability is measured through some indicators with the following equation: V = (E * S) / Ac; where V = vulnerability, E = exposure, S = sensitivity, Ac = adaptive capacity

The main source for developing this work phase is represented by the Participatory Community Diagnosis (PCD) (COOPI *et al.* 2014). This survey represents the systemisation of documented experience and gathers the methodological proposal of the Chaco consortium, made up of the following non-governmental organisations (NGOs): COOPI-Cooperazione Internazionale, Oxfam Intermon, Pro Comunidades Indígenas (PCI) and Acción Contra el Hambre (ACH). The availability of this resource has allowed us to identify and carry out activities for the preparation and mitigation of the vulnerability of 12 indigenous communities in order to deal with drought in Paraguayan Chaco.

Firstly, all the questions asked within the PCD were analysed, and subsequently questions that may be associated with one of the indicators of vulnerability (E, S, Ac and R) were identified (Table 4.1). Quality criteria were then applied for the selection (WRI & GIZ 2011). In this way, the following were identified:

- 2 indicators for exposure
- 7 indicators for sensitivity
- 5 indicators for the adaptive capacity
- 4 indicators for resilience (see chapter 5)

In this phase, we develop the analysis of exposure, sensitivity and adaptive capacity in order to determine a numeric value that may allow us to establish the value of vulnerability of each *comunidad*.

Table 4.1: Indicators chosen to determine vulnerability (source: COOPI *et al.* 2014).

#	Classification	Information
1	Exposure	Persons
2	Exposure	Drought analysis
1	Sensitivity	Distance from the paved road in km
2	Sensitivity	N. of people dedicated to self-consumption agriculture
3	Sensitivity	Time required to reach the nearest shops
4	Sensitivity	Duration of supplies of grown foods
5	Sensitivity	Transportation
6	Sensitivity	Types of soil: sandy, clay, other
7	Sensitivity	Availability of land to grow
1	Adaptation	Health centre
2	Adaptation	Agricultural tools
3	Adaptation	*Tajamar*
4	Adaptation	*Aljibes*
5	Adaptation	*Tanques*
1	Resilience	Community centre: yes or no
2	Resilience	Communications facilities: telephone, radio, other
3	Resilience	Existence of a community organization
4	Resilience	N. of people in the community with formal employment and medium-term

In the following sections, specific information for each indicator is given, including a rationale analysis of what it represents, and why it was chosen as an indicator.

4.4.1 Analysis of Exposure

Exposure to drought is expressed by both the physical phenomenon (drought) (DE) and the amount of population exposed (PE).

For the definition of a coefficient that may characterise the exposure to drought, based on the analysis presented in paragraph 2, indicators have been considered for the three scenarios presented: neutral situation, El Niño and La Niña (Table 4.2). In the first case, the SPI is between 1 and –1 throughout the area of interest. In the event in which we are in a Niño situation, there are humid areas to the north, with SPI values that may reach 2.5/3, and more arid zones to the south, with values that may reach –1/–1.5. Finally, in a Niña situation, we have humid areas to the south and more arid zones to the north, with values that may differ by as much as 2/3 points for the SPI value.

To define the correct coefficient to the *comunidades* in the three scenarios possible, the drought maps of the three scenarios have been used.

Table 4.2: Paraguayan Chaco. Coefficients of reduction to analyse drought.

Comunidades	Coefficients
Drought Comunidades	1
Less drought Comunidades	0.5
Humid Comunidades	0.1

In this case, the return time of the drought has been evaluated based on the connection between extreme events and the El Niño-Southern Oscillation (ENSO) cycle. In fact, there is a strong link between the extreme event, and consequently between the probability of occurrence and the ENSO cycle (see paragraph 3). As the ENSO cycle has a variable time deadline, it has been decided to simulate the probability of a disastrous event occurring based on the evaluation of the state of drought of the individual communities. This evaluation may be conducted while considering the ENSO cycle, and may be used both to develop the analysis as well as future forecasts. Consequently, the PE index has been assigned a numeric coefficient which comes within the calculation of the value of exposure (E). As can be seen in table 2, this value increases as the drought increases.

On the other hand, regarding the population affected, the information gathered from the PCD was used (COOPI 2014). Classes from 0.25 to 1 (Table 4.3) were defined.

Table 4.3: Paraguayan Chaco. Coefficients of reduction for the number of people in the comunidad.

Comunidad	Households	People per household	People	Coefficient
	n.	n.	n.	
Armonia	142	2.16	348[a]	0.25
Campo Loa	321	4.78	1,563[a]	1
Diez Leguas	306	3.95	1,210	0.75
El Estribo	468	3.71	1,736[a]	1
Garai	35	1.85	65	0.25
La Esperanza	311	3.96	1,230[a]	0.75
La Herencia	538	3.55	1,912	1
Lamenxai	149	4.8	716	0.5
Nepoxen	219	4.19	918	0.5
Nueva Promesa	240	3.33	800[a]	0.5
Yalve Sanga	671	2.72	1,830	1
Yetwase Yet	61	4.36	266	0.25

[a] N. households * Average n. of components per households.

Finally, to determine the value of exposure of the various *comunidades*, we used the algebraic sum of these two elements: E = DE + PE.

4.4.2 Analysis of Sensitivity

In order to determine sensitivity, seven different indicators were used, divided into two groups: an economic one (ES) and a physical one (PS). It thereby follows that:

S = ES + PS

Regarding economic sensitivity, the five questions associated with the indicators considered to analyse economic sensitivity, are:

- *Distance from the paved road (km) (ES1 indicator).* For each *comunidad,* a distance in kilometres has been identified to represent the distance of all the *aldea* (villages) which make up the Ruta Transchaco. Three groups have been identified: distance between 0 and 10 km (coefficient 0.1, minimum sensitivity), distance between 10 and 40 km (coefficient 0.5) and distance over 40 km (coefficient 1, maximum sensitivity).
- *Population dedicated to self-sufficiency (indicator ES2).* The higher the percentage of population affected, the higher the sensitivity, as in the event of rought, it will be harder to react (0 and 25, sensitivity coefficient 0.25; 25 and 50, sensitivity coefficient: 0.5; higher than 50, sensitivity coefficient: 0.75).
- *Time necessary to reach shops (ES3 indicator).* The times indicated start from 100 minutes for those *comunidades* that are nearest to the shops (sensitivity coef-

ficient 0.25), from 100 to 200 minutes in the intermediary range (sensitivity coefficient 0.5), and over 200 minutes for those *comunidades* that are furthest from shops (sensitivity coefficient 0.75). The distances are to be considered undertaken by foot.
- *Period during which the population uses cultivated foods (ES4 indicator).* This indicator, important in evaluating how sufficient the cultivated products are to cover annual food demand, is expressed in months. The values obtained range from 3.9 in the case of Campo Loa (sensitivity coefficient 0.25) to 1.3 in the case of La Esperanza (sensitivity coefficient 0.75) (Table 4.4).
- *Means of transport (ES5 indicator).* In the event of emergency, having motorcycles and cars available can drastically reduce the population's sensitivity. The more transport a *comunidad* has, the lower the vulnerability. The values are attributed according to the following classes: a single means of transport available coefficient 1, two 0.75, three 0.5 and four 0.25.

Table 4.4: Coefficients used for ES4 indicator.

Comunidad	Time (months)	Coefficient
Armonia	2.2	0.5
Campo Loa	3.9	0.25
Diez Leguas	2.6	0.5
El Estribo	2.6	0.5
Garai	1.7	0.75
La Esperanza	1.3	0.75
La Herencia	3.5	0.25
Lamenxai	2.8	0.5
Nepoxen	2.6	0.5
Nueva Promesa	2	0.5
Yalve Sanga	3.2	0.25
Yetwase Yet	3	0.25

Regarding physical sensitivity, on the other hand, the two indicators considered are:
- *Type of soil: sandy, clay or other (PS1 indicator).* This indicator expresses in general terms the agricultural type of soil: a clay land (sensitivity coefficient 0.1) is more suitable for agriculture than sand (sensitivity coefficient 1), which has a high capacity for water filtration and dries more easily. Clay land is given a lower value than sand as it is believed that in land with clay soils, agricultural activity is more developed. This results in a higher use of water for irrigation with a relative increase in the impact of the parameter on the vulnerability to drought of the *comunidades*.
- *Land available for cultivation (PS2 indicator).* Those *comunidades* that do not have land for agriculture, such as Armonia and Garai, have been assigned a coefficient

of 1; those that have land (Campo Loa, Lemenxai) the coefficient used is 0.1. In this case the possibility of cultivating land is considered positive for the population due to the possibility of having products for survival.

4.4.3 Analysis of the Adaptation

To prepare for drought, the following is necessary: monitoring of the indicators identified, an evaluation of the vulnerability and initial investments to plan adaptation (Engle 2009). Not only can this proactive approach contribute to reducing poverty, it also represents one of the most suitable tools for dealing with climate change. No direct parameters measure the process of adaptation in itself (UKCIP 2011), so *proxy* indicators are used.

For this study, those measures have been identified that are carried out by the population to adapt to drought. Also in this case, the main source for developing this analysis is the Participatory Community Diagnosis (COOPI *et al.* 2014).

The indicators identified are:

- *Health centres (Ac1):* in order to obtain a coefficient, the number of health centres for each community has been connected to the number of inhabitants thereby obtaining an amount of people per health centre. This number is then assigned a coefficient of adaptation: 0.1 for amounts above 1,000 inhabitants per health centre (minimum adaptation) and 0.8 for amounts lower than 500 people per health centre (maximum adaptation).
- *Agricultural tools (Ac2):* ranging from manual tools such as spades and hoes (minimum adaptation) to motorised ones such as tractors (maximum adaptation).
- *Tajamar (Ac3) (Figure 4.12.a):* this is a small dyke that collects rainwater in an artificial reservoir 2 to 3 metres deep. The bottom of the reservoir is covered in clay (Garcia *et al.* 2008). The water collected is then pumped out with a windmill into the canals that reach the surrounding villages. The untreated water is used to water animals and for washing. The tajamar have a water volume that varies from 5,000 to 100,000 litres (Table 4.5).
- *Aljibe (Ac4) (Figure 4.12.b):* holds rainwater collected from roofs. The water inside the tanks is protected. Nonetheless, the quality of the roofs of the houses must be studied accurately to guarantee the best possible collection (Junker 1999). This device is, in general, used both for community buildings, churches and schools (volume can be above 20,000 litres) as well as for homes (volume between 3,000 and 5,000 litres).
- *Tanque (Ac5) (Figure 4.12.c):* collect rainwater from roofs in plastic containers (volume up to 5,000 litres) and allow for a better control of water hygiene.

Table 4.5: Paraguayan Chaco. Coefficients used for the Ac3 indicator.

Comunidad	Litres	Litres/person	Coefficient
Armonia	51,600*	148	0.8
Campo Loax	43,100	28	0.1
Diez Leguas	74,400	61	0.4
El Estribo	112,120	65	0.4
Garai			
La Esperanza	64,000	52	0.4
La Herencia	52,500*	27	0.1
Lamenxai	20,000	28	0.1
Nepoxen	191,000	208	0.8
Nueva Promesa	109,400	137	0.8
Yalve Sanga	77,500	42	0.1
Yetwase Yet	7,200	27	0.1

Figure 4.12: Paraguayan Chaco, April 2014. Aljibe (left), tajamar (centre) and tanque (right) (photo by E. Ponte).

These measures increase the adaptation of the population in the event of drought in that they are able to provide a considerable amount of water. The adaptation coefficients are, therefore, expressed based on the calculated relationship of litres available per person (information presented in the PCD).

The coefficients of adaptation that have therefore been considered range from 0.1 (minimum adaptation) to 0.8 (maximum adaptation).

4.4.4 Results

In the case of the neutral scenario (Table 4.6 and Figure 4.13), a general uniformity of vulnerability emerged with eight out of 12 *comunidades* with a vulnerability between 2.5 and 7.5 in a scale from one to 10. The remaining four *comunidades* (Garai, Nueva

Table 4.6: Paraguayan Chaco. Vulnerability for the neutral scenario.

Comunidad	Ef	Ep	ES1	ES2	ES3	ES4	ES5	PS1	PS2	Ac1	Ac2	Ac3	Ac4	Ac5	V
Armonía	0.1	0.25	0.1	0.8	0.2	0.5	0.2	0.5	1	0.8	0.8	0.8	0.8	0.4	**0.6**
Campo Loa	0.1	1	1	0.5	0.5	0.2	0.8	0.1	0.1	0.4	0.1	0.1	0.4	0.3	**5.4**
Diez Leguas	0.1	0.75	1	0.5	0.5	0.5	0.8	1	0.1	0.1	0.4	0.1	0.1	0.4	**6.8**
El Estribo	0.1	1	0.5	0.2	0.8	0.5	0.8	0.5	0.1	0.4	0.4	0.1	0.4	1	**3.3**
Garai	0.1	0.25		0.2	0.8	0.8	1	0.3	1	0.1	0.4	0.8	0.8	0.4	**1.2**
La Esperanza	0.1	0.75	1	0.2	0.5	0.8	0.8	0.5	1	0.8	0.4	0.4	0.1	0.4	**3.9**
La Herencia	0.1	1	0.1	0.2	0.2	0.2	0.5	0.5	0.1	0.4	0.1	0.1	0.1	0.4	**3.6**
Lamenxai	1	0.5	1	0.2	0.5	0.5	0.2	0.5	0.1	0.4	0.1	0.1	0.4	0.4	**6.4**
Nepoxen	0.1	0.5	0.5	0.5	0.2	0.5	0.5	0.5	0.5	0.8	0.8	0.1	0.8	0.4	**1.3**
Nueva Promesa	0.1	0.5	0.1	0.5	0.5	0.5	0.8	0.3	0.5	0.4	0.8	0.1	0.4	0.4	**1.8**
Yalve Sanga	0.1	1	0.1	0.5	0.5	0.2	0.5	0.5	1	0.1	0.1	0.1	0.1	1	**5.2**
Yetwase Yet	1	0.25	1	0.8	0.8	0.2	0.8	0.5	0.1	0.1	0.1	0.4	0.4	0.4	**7.5**

Figure 4.13: Paraguayan Chaco, 2014. Charter of synthesis of the vulnerability of the comunidades for the neutral scenario (by E. Ponte).

Promesa, Nepoxen and La Armonia), had a vulnerability index of less than 2.5. There was, therefore an absence of *comunidades* with vulnerability values of over 7.5.

In the event of the scenario characterised by La Nina, most of the more vulnerable zones appeared to the west (Campo Loa, Yalve Sanga and Diez Leguas) with a value higher than 7.5. Only La Speranza to the south had a vulnerability index of over 5. All the remaining *comunidades* had a vulnerability index lower than 5.

In the case of El Niño, only La Esperenza had a vulnerability index higher than 7.5. With the exception of the *comunidades* of Garai, La Armonia, Nueva Promesa and Nepoxen which had a vulnerability index lower than 2.5, in all other cases all the *comunidades* had, in this scenario, vulnerability indexes between 2.5 and 7.5.

Observing all three scenarios together, we can see how the case of La Niña is the most critical in that drought affected the *comunidades* wherein there was already a more problematic situation. In fact, in this case, three *comunidades* have extreme vulnerability (coefficient higher than 7.5) while, for the El Niño scenario, just one *comunidad* had extreme vulnerability.

4.5 Resilience

The resilience approach considers that flexible institutions capable of reorganising themselves are needed to address future climate change (Folke *et al.* 2002). In particular, in Chaco Paraguayo, the resilience can be viewed as the extent to which a local disaster management group is connected to the disaster risk policy, depending on the availability of the data. Together with the vulnerability systems analysed, the resilience would help to shed more light on social networks and how individuals interact.

4.5.1 The Indicators

At present, there is no outline of the strategies and projects undertaken to increase resilience (Birkmann *et al.* 2008).

The indicators of resilience and adaptive capacity are difficult to discern, and it is not easy to use a list of 'off-the-shelf' indicators (Brooks & Adger 2005) as they vary from entity to entity, even in the same locality. In this case, the choice of indicators starts with the definitions given, and is determined by the data availability.

Consequently, the information coming from the PCD was used (COOPI *et al.* 2014) (Table 4.7):

- *Community meeting centre (Res1):* these centres often coincide with churches (Anglican or Evangelical) where general community meetings are organised. Wherever these centres are present, the *comunidad* is given a coefficient of 1, otherwise the coefficient assigned is 0.1.

- *Means of communication (Res2):* when the *comunidades* have a telephone or radio, the coefficient is 0.5; if they have both the coefficient is 1.
- *Existence of community organisation (Res3):* the coefficients used are 1 where organisations are present, and 0.1 where they are lacking.
- *Number of people that carry out formal activities beyond their own aldea (Res4):* this indicator allows us to evaluate the population's ability to carry out a work activity that may not be dependent on the site and therefore exposed to drought. To establish the coefficient, we have calculated the percentage of the population dedicated to this activity. In this way, three classes have been defined: less than 5% coefficient 0.1; 5–10% coefficient 0.3; higher than 10% coefficient 0.5.

Table 4.7: Paraguayan Chaco. Calculation of resilience.

Comunidad	R1	R2	R3	R4	R
Armonía	0.1	1	0.1	0.3	**1.5**
Campo Loa	1	0.5	0.1	0.1	**1.7**
Diez Leguas	1	0.5	1	0.5	**3**
El Estribo	1	1	1	0.3	**3.3**
Garai	0.1	0.5	0.1	0.5	**1.2**
La Esperanza	0.1	1	1	0.3	**2.4**
La Herencia	1	1	1	0.3	**3.3**
Lamenxai	1	1	1	0.1	**3.1**
Nepoxen	0.1	1	0.1	0.3	**1.5**
Nueva Promesa	0.1	0.5	0.1	0.3	**1**
Yalve Sanga	1	1	1	0.1	**3.1**
Yetwase Yet	0.1	0.3	0.1	0.3	**0.8**

There is an important difference between *comunidades* regarding resilience: from 0.8 for Yetwase Yet (slightly resilient) to 3.3 for La Herencia and El Estribo (highly resilient).

4.5.2 Vulnerability and Resilience Compared

We have previously explained the link that exists between vulnerability and resilience. Having reached this point, it is therefore necessary to try and compare the values obtained for the various *comunidades* to see how they relate.

For the neutral scenario (Table 4.8) all the *comunidades*, with the single exception of Yetwase Yet which presented higher vulnerability (higher than 6), had also developed a value of higher resilience (higher than 3). For the opposite case, there is also a relationship: all *comunidades* which have minimum vulnerability (less than 3), have also developed a value of lower resilience (less than 3).

Table 4.8: Paraguayan Chaco, 2014. Comparison between vulnerability and resilience for the neutral scenario.

Comunidad	V	R	V-R
Armonía	0.6	1.5	< 0
Campo Loa	5.4	1.7	3.7
Diez Leguas	6.8	3	3.5
El Estribo	3.3	3.3	< 0
Garai	1.2	1.2	< 0
La Esperanza	3.9	2.4	1.5
La Herencia	3.6	3.3	0.3
Lamenxai	6.4	3.1	3.3
Nepoxen	1.3	1.5	-< 0
Nueva Promesa	1.8	1	0.8
Yalve Sanga	5.2	3.0	2.2
Yetwase Yet	7.5	0.8	6.7

4.6 Conclusions

This chapter shows the multiple vulnerabilities to drought of the indigenous populations of Paraguayan Chaco.

The exposure of the communities of the Chaco to meteorological drought is linked to El Niño and La Niña, which can be forecast in advance in order to give a drought early warning in the Chaco.

It could prove interesting to develop an analysis of water balance of the study area. In this way, carefully studying the various components of soil, under-soil, water resources available, we could better understand which components are lacking the most.

The analysis of vulnerability highlights how in all three scenarios considered, La Armonia is by far the least vulnerable community. This is mainly due to its high level of adaptation.

Contrarily, no *comunidad* is the most vulnerable under all three scenarios. We can, however, confirm that La Esperanza and Diez Lugas are the two *comunidades* with the highest values, even if in some scenarios they are equal to others.

Observing the data in more detail, it emerges that the important investments for adaptation (especially in providing *aljibe*, *tajamar* and *tanque*) are not presently enough to lower the population's vulnerability, which is some cases reaches high levels.

The links between vulnerability and resilience indicate a good strategy for resilience that has effectively helped the most vulnerable *comunidades*.

The choice of indicators has also been made from the viewpoint of predicting future analysis monitoring as advised by the literature on the subject (Ponte 2014, GIZ 2012). Specifically, regarding the analysis of resilience, it emerges that for the indicators, it is not possible to assess any important information such as the opportunity for planning, organisation, mobilisation, or mutual aid to the individual communities. This information would certainly be useful: in fact, it is one thing to have the tools available, another is to use them to effectively become more resilient.

The integration of possible new information could allow us to increase the precision of the study. In particular, passage of scale-adjusted information from *comunidad* to *aldea* could prove very interesting. In fact, numerous *comunidades* show many differences between the various *aldea* that they consist of: distance of infrastructure, tools available, activities carried out. It could also prove useful to consider how the strategies and national policies for resilience and adaptation are implemented in Paraguayan Chaco (GIZ 2012).

In considering the results of climate research, the existence of a national Early Warning System (EWS) on the manifestation of El Niño and La Niña would certainly be of great importance for the Chaco.

In fact, an EWS could increase the adaptation capacity of the system and reduce its vulnerability. Moreover, if the system was nationwide, this would help the poorest communities increase their adaptation capacity without having to invest their capital in the construction of reservoirs for rainwater harvesting. A careful analysis of the water balance at atmospheric precipitation levels (carried out using appropriate climate indices), a forecasting system of atmospheric events and a detailed vulnerability assessment at the Community level would lead to an effective and efficient EWS for water balance prediction and water crises management.

Finally, the obtained results confirmed that the methodology applied in the presented research demonstrated how it is possible to obtain significant analysis at a local scale, even if the starting point is a study of the Climatological factors elaborated at global scale.

We can therefore conclude that the study can be replicated in similar situations where there is an impact of drought on communities living in remote areas.

References

African Development Bank, African Union, NEPAD, UNISDR. 2004. *Guidelines for mainstreaming disaster risk assessment in development*.

Behringer, D., and Y. Xue. 2004. Evaluation of the global ocean data assimilation system at NCEP: The Pacific Ocean. In 8th Symposium on integrated observing and assimilation system for atmosphere, oceans and land surface, AMS 84th meeting, Washington state convention and trade center, Seattle, 11–15 January.

Berkes, F., J. Colding, and C. Folke. 2003. *Navigating social-ecological systems: Building resilience for complexity and change*. Cambridge: Cambridge University Press.

Birkmann, J., P. Buckle, J. Jaeger, M. Pelling, N. Setiadi, M. Garschagen, N. Fernando, and J. Kropp. 2008. Extreme events and disasters: a window of opportunity for change? Analysis of changes, formal and informal responses after mega-disasters. *Natural Hazards* 55: 637–655.

Bonaccorso, B., I. Bordi, A. Cancelliere, G. Rossi, and A. Sutera. 2003. Spatial variability of drought: an analysis of SPI in Sicily. *Water Res. Management* 17: 273–296.

Brant, M. 2007. Assessing vulnerability to drought in Ceará, Northeast Brazil. University of Michigan.

COOPI, OXFAM/PCI, ACH. 2014. El diagnóstico comunitario participativo con enfoque en medios de vida y agua. Asuncion.

Engle, N. 2009. Preparation, not procrastination, for effective drought management. Development in a climage climate change: Making our future sustainable. http://blogs.worldbank.org/climatechange/preparation-not-procrastinationeffective- drought-management. Accessed 20 May 2010.

Folke, C., et al. 2002. Resilience and sustainable development: Building adaptive capacity in a World of transformations. Scientific background paper on resilience for the process of the World summit on sustainable development on behalf of the environmental advisory council to the Swedish government.

García Petillo, M., and P. Cánepa. 2008. Manual para el diseño y la construcción de tajamares de aguada. Proyecto producción responsable.

Grassi, B., A.M. Pastén and J. Armoa. 2005. Un análisis del comportamiento de la precipitación en el Paraguay: Informe final. Asunción: Facultad Politécnica, Universidad nacional de Asunción.

GIZ-Gesellschaft für Technische Zusammenarbeit. 2012. Adaptation made to measure - A guidebook to the design and results-based monitoring of climate change adaptation project. Eschborn.

GTZ-Gesellschaft für Technische Zusammenarbeit. 2005. *Atlas del Gran Chaco Americano*. Secretaría de ambiente y desarrollo sustentable de la Argentina y la Cooperación técnica alemana.

IDAEA-Instituto de Diagnóstico Ambiental y Estudios del Agua. 2012. Informe final de actuaciones y resultados. Estudio hidrogeológico de los acuíferos someros del Chaco Central paraguayo, Grupo de hidrología subterránea, unidad asociada de la Universidad Politécnica de Cataluña.

IPCC. 2001. *Climate change 2001: Impacts, adaptation, and vulnerability. Contribution of working group II to the third assessment report of the IPCC.* Cambridge: Cambridge University Press.

IPMPCC. 2011. Pueblos indígenas, poblaciones marginadas y cambio climático: vulnerabilidad, adaptación y conocimientos indígenas. Ciudad de Mexico.

Junker, M. 1999. Informe tecnico sobre las instalaciones de agua potable para asentiamentos de escasos recursos. Filadelfia.

Khan, S., H.F. Gabriel, and T. Rana. 2008. Standard precipitation index to track drought and assess impact of the rainfall on watertables in irrigation areas. *Irrig Drainage Syst* 22: 159–177.

Knutson, C., M. Hayes, and T. Phillips. 1998. How to reduce drought risk. Report of the preparedness and mitigation working group of western drought coordination council.

Labedzki, L. 2007. Estimation of local drought frequency in central Poland using the Standardized precipitation index SPI. *Irrigation and Drainage* 56/1: 67–77.

Livada, I., and V.D. Assimakopoulos. 2007. Spatial and temporal analysis of drought in Greece using the Standardized precipitation index. *Theor. Appl. Climatol.* 89: 143–153.

Luers, A.L., D.B. Lobell, L.S. Sklar, C.L. Addams, and P.A. Matson. 2003. A method for quantifying vulnerability, applied to the agricultural system of the Yaqui Valley, Mexico. *Global Environmental Change* 13: 255–267.

Michaelides, S., and S. Pashiardis. 2008. Monitoring drought in Cyprus during the 2007–2008 hydrometeorological year by using the Standardized precipitation index. *European Water* 23/24: 123–131.

Nunez, M.N., S.A. Solman, and M.F. Cabré. 2009. Regional climate change experiments over South America. II: Climate change scenarios in the late twenty-first century. *Clim. Dyn.* 32: 1081–1095.

Pacheco, B.D. 2012. Características sociales y productivas del Chaco boliviano, en cambio climático, sequía y seguridad alimentaria en el chaco boliviano. Ed. Universidad de la Cordillera - Fundación de la Cordillera.

Pasig, R. 2005. Origen y dinámica del agua subterránea en el noroeste del Chaco sudamericano. Tesis doctoral, http://www.hydrogeologie-wuerzburg.de/abstract/hu34_abs_esp.htm.

Pasten, A. M. 2007. Análisis de eventos meteorológicos extremos en el Paraguay. Facultad Politécnica, Universidad nacional de Asunción.

Ponte E. 2014. Risk assessment to floods and sea level rise in the municipality of Maputo, Mozambique. In *Climate change vulnerability in southern African cities. Building knowledge for adaptation*, ed. S. Macchi and M. Tiepolo, 187–204. Springer. doi: 10.1007/978-3-31900672-7.

PNUMA-Programa de la ONU para el Medio Ambiente. 2013. Estudio de vulnerabilidad e impacto del cambio climatico en el Gran Chaco Americano. Centro de conocimiento para el Gran Chaco Americano y cono sur.

PSAC-Proyecto Sistema Ambiental del Chaco. 1998. Inventario, evaluación y recomendaciones para la protección de los espacios naturales en la región occidental de Paraguay. Cooperación técnica paraguayo-alemana, Dirección de ordenamiento Ambiental-DOA y Bundesanstalt für geowissenschaften und rohstoffe-Instituto federal de geociencias y recursos naturales.

REDIEX-Red de Inversiones y Exportaciones. 2009. Atlas geográfico del Chaco paraguayo. Asunción: Unidad GIS-REDIEX.

Smit, B., and J. Wandel. 2006. Adaptation, adaptive capacity and vulnerability. *Global Environmental Change* 16: 282-292.

UKCIP-United Kingdom Climate Impacts Programme. 2011. AdaptME toolkit - Adaptation monitoring & evaluation. Oxford.

UNISDR-United Nations International Strategy for Disaster Reduction. 2003. Drought living with risk: An integrated approach to reducing societal vulnerability to drought. ISDR ad hoc discussion group on drought. Geneva: UNISDR.

UNISDR. 2009. Drought risk reduction, framework and practices: Contributing to the implementation of the Hyogo framework of action. Geneva: UNISDR.

USAID-United States Agency for International Development. 2007. *Atlas climatico del Chaco Paraguayo.*

Vila, C. 2010. La situación de los pueblos indígenas del Chaco paraguayo en cuanto al acceso a la justicia. Pueblos originarios y acceso a la justicia, CIPAE.

Walker, B., L.H. Gunderson, A. Kinzig, C. Folke, S. Carpenter, and L. Schultz. 2006. A handful of heuristics and some propositions for understanding resilience in social-ecological systems. *Ecology and Society* 11: 13.

WMO-World Meteorological Organization. 2012. *Standardized precipitation index – User guide.* Geneva: World Meteorological Organization.

WRI-World Resources Institute and GIZ. 2011. *Making adaptation count: concepts and options for monitoring and evaluation of climate change adaptation.* http://www2.gtz.de/dokumente/bib-2011/ giz2011-0219en-monitoring-evaluation-climate-change.pdfù.

Maurizio Bacci[17]

5 Characterization of Flood Hazard at a Municipal Level. A Case Study in Tillabéri Region, Niger

Abstract: The occurrence of flooding ranks top among natural disasters in terms of both the number of people affected globally, and the proportion of individual fatalities per case. The potential for flood causalities and damages is increasing in Africa due to social and demographic development raising pressure on land use. Furthermore, the evidence of increasingly heavy precipitation due to climate change supports the view that flood hazard is expected to increase in frequency and severity in many areas of Niger.

Niger is one of the countries most vulnerable to climatic risks, and the Tillabéri region is exposed to different typologies of floods: flash floods associated with short, high intensity rainfalls, mainly of convective origin and occurring locally, and floods on larger catchments, such as the Niger River, where dangerous episodes derive from a regional combination of precipitations and runoff effects. As such, flash flooding usually impact basins less than 1,000 km^2 with response times of a few hours or less, while large river basins have the response times of several days to months. Investigating these aspects in this area is difficult due to a lack of systematic observational data for flash floods, encompassing data on the flood-generating rainfall at the required space and time detail.

This case study aims to support local institutions and organizations in the design and implementation of disaster risk management policies in reducing Niger's vulnerability to recurrent natural disasters, as well as integrating the role of local authorities in building community disaster prevention and preparedness. The study proposes a methodology to classify each single municipality in the Tillabéri region based on its specific flood hazard characterization.

Keywords: Flood, natural disaster, climate change, weather extremes

5.1 Introduction

Niger is ranked as one of the countries most vulnerable to natural disaster (Beck *et al.* 2012), and in recent years, many floods have hit the country, affecting thousands of people (EM-DAT). At the same time, Niger has one of the poorest economies in the world; as such, it has difficulty in finding resources to support policies addressing

17 Institute of Biometeorology (IBIMET) National Research Council (CNR) of Italy, m.bacci@ibimet.cnr.it disaster prevention. Normally interventions are delegated to international organizations and NGOs which operate in a fragmentary way over the Niger territory, usually at a national or local scale. A major challenge for Niger is how to optimize and organize the limited resources available in the prioritization of flood risk interventions.

Climate change, human pressure and environmental degradation are the drivers of these floods, and it is expected that in the future floods will be more frequent, causing a rise in causalities and losses in this region (IPCC, 2012). An adaptation to this threat is urgent and expressed by the populace (Adger *et al.* 2003). This need is also strongly supported by politicians and decision makers as stressed in the *Programme d'Action National pour l'Adaptation aux changements climatiques* (PANA) of Niger (CNEDD, 2006). As such, the major concern of this paper it is to provide a methodology for flood hazard risk characterization and mapping finalized for operational use in the affected area. As defined by IPCC, hazard refers to "*the possible, future occurrence of natural or human-induced physical events that may have adverse effects on vulnerable and exposed elements*" (Cardona *et al.*, 2012). The choice of the municipality level is strategic due the need to provide appropriate information at the right scale for interventions. In the past, one of the major limits in the decision making process at a local scale was the absence of information at the municipal level. Typically Climate Change (CC) and Disaster Risk Reduction (DRR) studies refer to a global or national scale and do not take into account the specificities of municipal territories.

The Tillabéri region was chosen as a test case for the application of the proposed methodology to intercept and reduce the likelihood of a recurring meteorological event that would generate a flood. As determined by the mapped of frequency of floods recorded in the region in the last 5 years (Figure 5.1), it is evident that this area experiences a very high frequency of floods throughout the region.

The economy of the region is based primarily on agriculture and livestock (Table 5.1), with the population living mostly in rural areas (Table 5.2). So these territories, strongly dependent on the primary sector of the economy, are very vulnerable to natural hazards for the reason that floods impact directly on the main sources of subsistence of the population.

The hydrological network of the Tillabéri region is characterized by the presence of the Niger River, its tributaries, and in the east by the presence of the Dallols River. The most significant Niger River water supplies in the region come from the following tributaries: the Sirba, the Gouroubi, the Diamangou, the Tapoa, the Goroual and the Dargol. On both sides of the main stream, there are many smaller tributaries that drain significant runoff during the rainy season, as well as the fossil rivers: the Dallols Bosso and Maouri. The Dallols is a particularity of the region. The Dallols is a natural depression that collects rain but it doesn't exhibit surface water for most of the year. Surface groundwater and permanent ponds make the Dallols a suitable area for agriculture and livestock. This frequently results in flash floods when intense rain hits

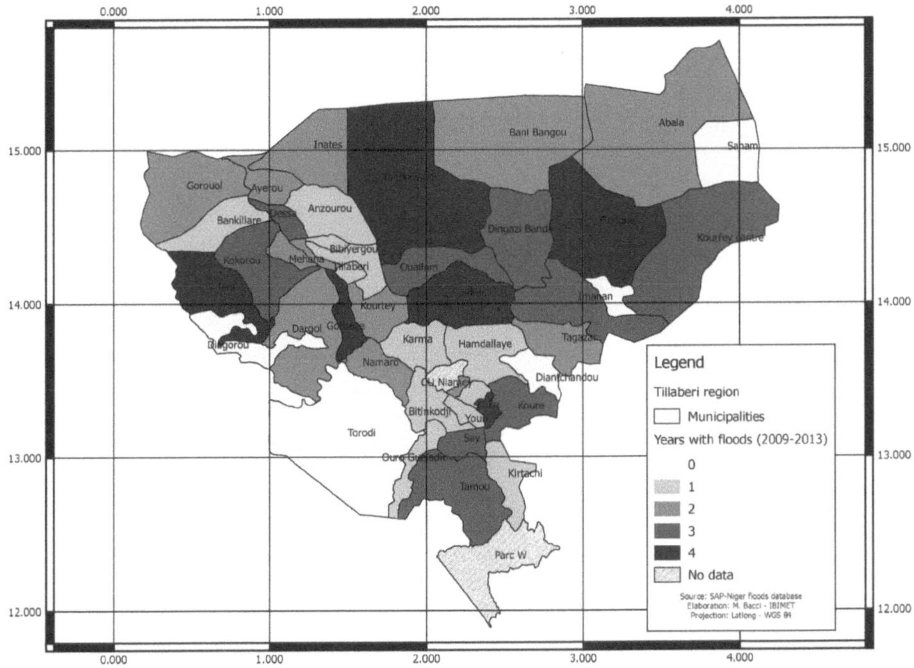

Figure 5.1: Observed flood frequency per municipality in Tillabéri Region – Period 2009–2013 (Data source: SAP database)

these basins. This is particularly dangerous because people live on or near the temporary watercourse.

In Niger, flood information is collected by the Prime Minister Cell of Coordination of Early Warning System and Disaster Prevention (*Cellule de Coordination du Système d'Alerte Précoce et de Prévention des Catastrophes CC / SAP / PC*) through the Regional Committees Prevention Disaster management and food crises (*Comités régionaux de Prévention Gestion des Catastrophes et Crises Alimentaires PGC CA*). The SAP maintains a database that records every flood that hits the Niger. Although this tool has registered floods since 1998, the start of systematic recordings coupled with the identification of the exact day of the flood has only been in place since 2006. For our case study, this yields 150 dates to be analyzed in the region.

The difficulties in completing a characterization of meteorological hazards in this region are: i) Lack of meteorological ground data for every municipality; ii) Absence of ground meteorological data at an hourly resolution; iii) Lack of a hydrometric measurements network; iv) No specific runoff model for the region; and v) No indication about flood typology recorded in the SAP list. The combination of these factors makes it difficult to describe the specific flood hazards and represents a current limit in performing an analysis that could bring useful information for decision makers. Currently there is no hazard assessment at a municipality level in the region.

Table 5.1: Agricultural and livestock production in Tillabéri region in 2010 (INS 2011)

Breeding	Heads (x1000)	Agriculture	Production (x1000 tons)
Cattle	2074	Millet	768.3
Sheep	1536	Sorghum	121.7
Goats	1767	Cowpea	287
Camels	91	Rice	21.9
Horses	19	Sesame	10.4
Donkeys	307	Peanuts	12.1

Table 5.2: Department population and urban and rural distribution at 2010 (INS 2011)

Department	Population
Filingue	537,715
Kollo	431,017
Ouallam	372,943
Say	307,622
Tera	563,506
Tillabéry	287,651
Total Region	*2,500,454*
Urban pop.	377,361
Rural pop.	2,123,093

The rain that falls upstream of Tillabéri, in the Malian and Burkina territories, contributes to the genesis of the Niger River floods. Daily ground data from these countries are not available for the Niger Meteorological Service, but, even if these data were available, there is no evidence as to how to introduce this information in flood characterization due the lack of hydrological models describing Niger River streamflow and the changes in the whole Niger river basin environmental conditions.

The literature well describes the changing environmental and climate conditions and their interaction with hydrology in a large region such as the Niger Basin (Descroix, 2009) responsible for constantly increasing vulnerability to flooding. Moreover, urban development affects flooding in many ways (Jha *et al.*, 2012), such as: i) removing vegetation and crop areas and constructing drainage networks increases runoff to streams from rainfall; ii) mismanagement of the urban stream channels which then limits capacity to convey floodwaters; and iii) roads and buildings constructed in flood-prone areas exposed to increased flood hazards. The combination of

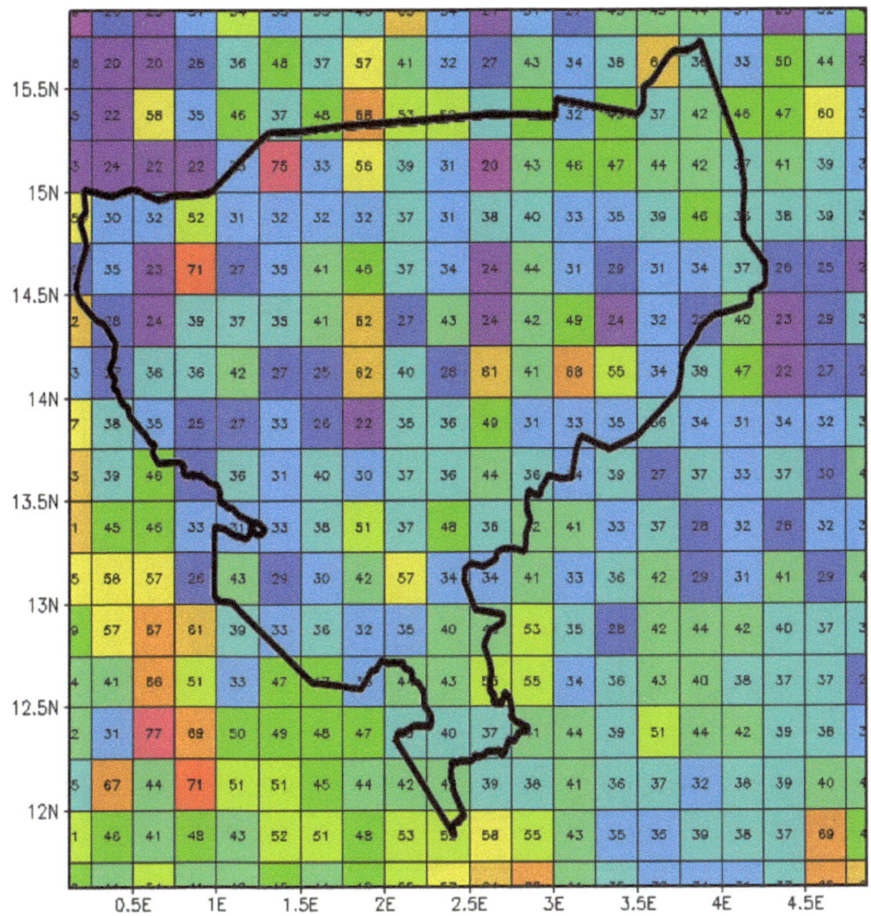

Figure 5.2: Maximum of hourly rain rate in the period 1998–2011 (Elaboration from NASA GES-DAAC TRMM_L3 TRMM_3B42 v6 three-hourly)

these three factors on flood risk in Niger urban settlements still remain uninvestigated by the scientific community.

To overcome these gaps, maximization of what little raw information was available for the characterization of meteorological hazards was chosen. The municipality's watershed typology is a good starting point to explain what type of flood can be expected. Even if flash floods are possible for all municipalities, in the administrative units affected by the Niger River and its major tributaries, the cause of a flood is driven by complex interactions between rainfall, soil water balance, water-holding capacity and river stream-flow over a vast territory. The local meteorological conditions partially explain the floods. The absence of hydrological models most of tributaries of the Niger River and the deficiency of information on the typology of floods affecting the surrounding villages shifts the focus towards a methodology that assumes some

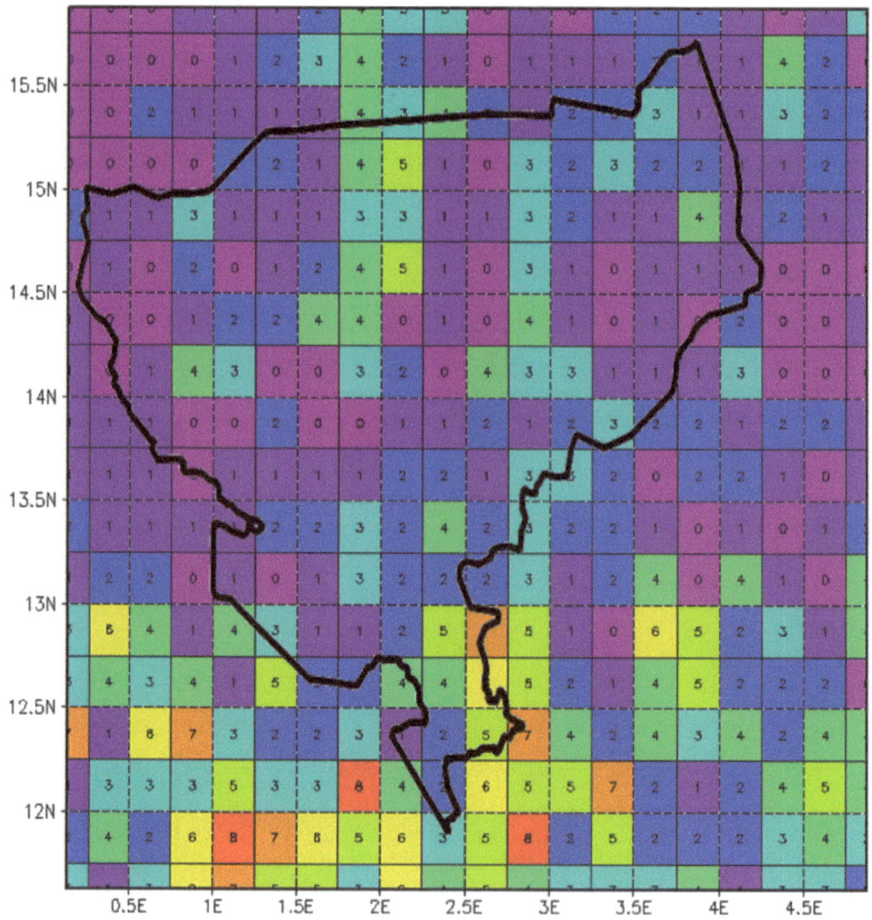

Figure 5.3: Number of episodes of RR > 90 mm/3-h (1998–2011) – Elaboration from NASA GES-DAAC TRMM_L3 TRMM_3B42 v6 three-hourly.

simplifications in order to reach a predictive value for the probability of flood hazard. Areas in the region where there are no main tributaries to the Niger River, such as in the eastern part of the region characterized by Dallols, the local meteorological conditions are better determinants for flash floods, and therefore it is possible to give a more accurate description of the phenomenon.

The absence of data at high temporal resolutions on the ground forces the use of a rainfall estimation dataset in this study to complete the process of analysis. With these data it is possible to increase temporal resolution in the phenomena details and cover the entire region. In addition, it is possible to map the rainfall extremes recorded during 1998–2011, as showed in Figure 5.2 and Figure 5.3, and to highlight rain distribution on the ground in order to demonstrate the necessity for carrying out an analysis at the municipality resolution.

The proposed methodology uses past flood episodes to perform an ex–post analysis to intercept the meteorological conditions that cause flooding. The characterization of hazards for each municipality is based on its historical rain distribution and hydrological configuration.

As far as performing an exact quantification for the probability of a flood event, the methodology aims to propose a hazard characterization highlighting where urgent interventions are needed among the municipalities in the region.

This methodology has been proposed and applied in the framework of the ANADIA Project to support disaster prevention from national to local institutions in the design and implementation of disaster risk management strategies.

5.2 Materials and Methods

The specificities of orography and hydrology generate different typologies of floods and rain and are not homogeneously distributed over space and time. Considering these two aspects on the municipality level, it is necessary to classify the region into homogeneous areas.

The characterization of municipalities by specific watershed typology is done by classifying the administrative units, based on the local experience of a group of experts from the National Hydrological and Meteorological Services (ANADIA 2014), with respect to the dominant watershed in the municipality. The classification distinguishes municipalities affected by 1) Niger River and primary tributaries, 2) Niger River, 3) Primary tributaries, 4) Secondary tributaries, and 5) Dallols (Figure 5.4). This classification is made by enhancing the differences among the critical rainfall duration for each of the representative drainage basins (Figure 5.5). The response of a watershed to a rain event is measured by the time of concentration. In hydrology it is defined as the time needed for water to flow from the most remote point in a watershed to the watershed outlet. It is assumed that in the little basins floods are better intercepted by rains of durations of a few hours, so they must be described by an hourly resolution dataset, contrariwise for large watersheds. With critical rain duration measured over the period of a day, the daily resolution is more useful to intercept critical rains.

In the definition of the meteorological critical conditions in municipalities bordering the Niger River and its primary tributaries, it was considered an acceptable simplification to analyze daily precipitation coming from rain gauge stations operated by the National Meteorological Service inside or surrounding the administrative units. For municipalities characterized by the presence of Dallols and/or small rivers basins, 3-hourly datasets *NASA GES-DAAC TRMM_L3 TRMM_3B42 v6 three-hourly* by Tropical Rainfall and Tropical Mission by NOAA (Huffman *et al.* 2007) were used to intercept the generation of flash floods in the absence of data collected on the ground at an hourly resolution.

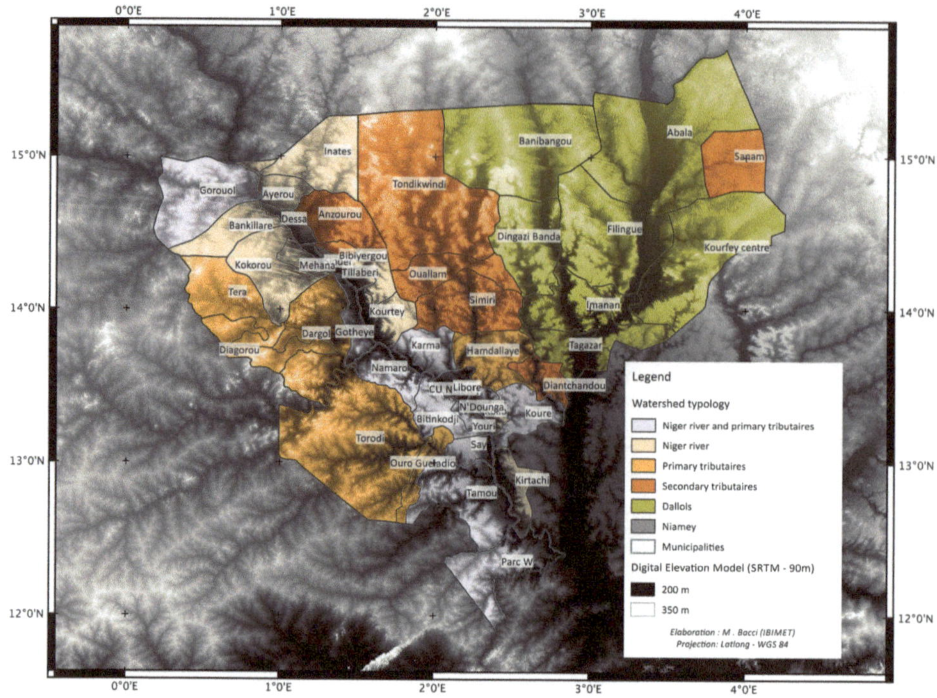

Figure 5.4: Characterization of Tillabéri's municipalities by specific watershed typology.

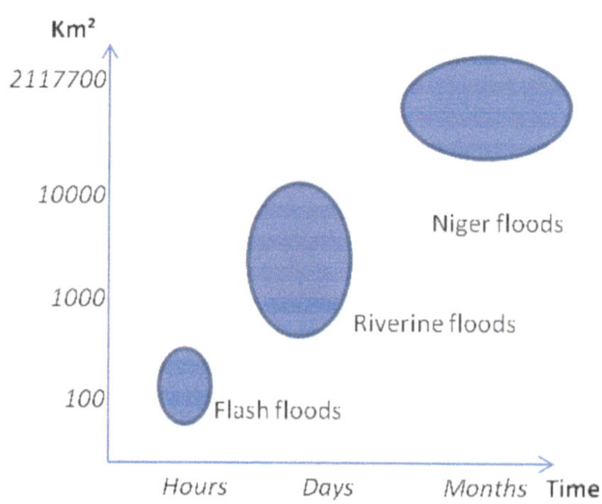

Figure 5.5: Relation between watershed surface and time of concentration.

The use of floods from the database produced by SAP allows for the identification of the date recorded for each episode. This list is not standardized, so it is possible to find very heterogeneous information recorded for each flood. The identification of the day of flooding began with regularity in 2005. Until this period, SAP usually record only the month when a flood occurred. In many cases, there is no information on the villages and the municipalities hit by the phenomenon. After a quality control of the database, only the cases where the day of the event and the affected municipality are clearly specified were selected. This process yields a selection of 110 acceptable test cases in the period from 2006 to 2012 (ANADIA 2014).

These selected dates are used to evaluate rainfall conditions in the 5 day period surrounding the flood in the two datasets available: the data from the rain gauge network, and by the TRMM. Before doing this, some preparation work was needed to utilize the two datasets.

The dataset TRMM is available from 1998 to June 2011; with Ncview (Pierce 2011) it is possible to extract the 3-hour value of the entire temporal series for a specific coordinate pair. It produced 45 extractions of the value of the grid point on the centroid of every municipality (44 + W park). So each municipality was characterized by a different meteorological series. If the resolution of the TRMM images (0.25' x 0.25') area is similar to the typical areal extension of the municipalities, then it was considered a good estimator for the study purposes. Moreover, the absence of a map with the watershed network of each single municipality in the region avoids areal extraction for each hydrological basin. In case of flood episodes beyond June 2011, only information from the rain gauge network was available.

The DMN database contains the daily data for all the meteorological stations in Niger. In Tillabéri and in its neighboring regions, localities with at least 30 years of records were selected. This choice is critical to producing a significant and homogeneous statistic on the return period for all stations. In the study area only 15 stations (13 in Tillabéri and 2 in the CU Niamey) were compliant with this restriction.

When the two rainfall datasets were ready, for each flood the values of rain that were potentially critical in the days around the flood were identified. As it was probable that a rainfall value higher than normal was responsible for the flood, the maximum value of rainfall in this period for the specific municipality was selected as a critical threshold. The process takes into consideration all the stations surrounding the municipality, normally prioritizing the upriver stations over the other stations closest to the flood. Carrying out this work, it is not always possible to extract, in the nearest station and/or in the rainfall estimation dataset by TRMM, significant values of precipitation that could explain the flood. It is not possible to estimate a priori the accuracy of the two datasets, so it was chosen to remove values less than 20 mm/day used to characterize critical rainfall.

At the end of this process there are two lists of critic rainfall thresholds for all the municipalities of the Tillabéri region, with the additional information of the reference rain gauge station and the precipitation date: one with the daily data, the other with the 3-hourly dataset (Table 5.3).

The next step was the selection of the significant rainfall thresholds for each municipality among the values extracted from the historical analysis. The choice

Table 5.3: List of floods episodes per municipality in Tillabéri region and rainfall. (Source ANADIA 2012).

Municipality	Flood date	Station	Rainfall date	Rainfall	TRMM/3h	TRMM Date
Abala	24/08/2006	Abala	24/08/2006	79.6	113.9	22/08/2006
Abala	01/07/2009	Abala	01/07/2009	35	13.8	30/06/2009
Ayerou	29/08/2008	Ayerou	29/08/2008	16	73.4	31/08/2008
Bannibangou	29/08/2006	Bannibangou	29/08/2006	2	58.6	28/08/2006
Bannibangou	08/09/2006	Bannibangou	08/09/2006	4	3.9	09/09/2006
Bannibangou	11/09/2006	Bannibangou	11/09/2006	0	5.8	10/09/2006
Bitinkodji	30/07/2012	Niamey Ville	30/07/2012	12	-	na
Dantchiandou	13/08/2007	Loga	13/08/2007	14.5	15.0	10/08/2007
Dantchiandou	03/07/2012	Loga	03/07/2012	14.5	-	na
Dantchiandou	05/07/2012	Loga	05/07/2012	0	-	na
Dargol	22/07/2007	Dargol	22/07/2007	61.2	6.6	22/07/2007
Dargol	04/08/2007	Dargol	04/08/2007	52	42.7	04/08/2007
Dargol	09/06/2010	Tera	09/06/2010	19.8	18.9	09/06/2010
Dargol	10/06/2010	Tera	10/06/2010	0	0.0	10/06/2010
Dessa	13/08/2011	Ayerou	13/08/2011	52	-	na
Filingué	15/07/2010	Filingué	15/07/2010	9	27.4	16/07/2010
Filingué	22/07/2010	Filingué	22/07/2010	28.5	7.6	21/07/2010
Filingué	06/08/2011	Chikal	06/08/2011	12	-	na
Filingué	06/08/2011	Toukounnous	06/08/2011	12	-	na
Filingué	12/08/2011	Filingué	12/08/2011	7.5	-	na
Filingué	19/08/2011	Filingué	19/08/2011	45	-	na
Filingué	03/08/2012	Filingué	03/08/2012	24	-	na
Filingué	04/08/2012	Filingué	04/08/2012	0	-	na
Filingué	13/08/2012	Filingué	13/08/2012	42.5	-	na
Filingué	15/08/2012	Filingué	15/08/2012	0	-	na
Filingué	19/08/2012	Filingué	19/08/2012	0	-	na
Filingué	20/08/2012	Filingué	20/08/2012	20	-	na
Filingué	21/08/2012	Filingué	21/08/2012	0	-	na
Gourouel	27/07/2007	Bankilaré	27/07/2007	30	66.2	27/07/2007
Gourouel	14/07/2008	Bankilaré	14/07/2008	0	12.3	11/07/2008
Gotheye	04/08/2007	Dargol	04/08/2007	52	28.0	04/08/2007
Gorouol	10/06/2010	Bankilaré	09/06/2010	5.4	7.5	09/06/2010
Hamdallaye	29/08/2006	Ouallam	28/08/2006	30	13.9	28/06/2006
Hamdallaye	03/09/2006	Ouallam	02/09/2006	0	4.3	03/09/2006
Imanan	24/08/2006	Filingué	24/08/2006	51	23.2	21/08/2006
Imanan	25/08/2006	Filingué	25/08/2006	0	19.2	22/08/2006
Imanan	09/08/2011	Filingué	09/08/2011	0	-	na
Imanan	19/08/2011	Filingué	19/08/2011	45	-	na
Imanan	02/07/2012	Filingué	02/07/2012	21	-	na
Imanan	04/07/2012	Filingué	04/07/2012	0	-	na
Imanan	20/08/2012	Filingué	20/08/2012	20	-	na
Imanan	21/08/2012	Filingué	21/08/2012	0	-	na
Imanan	03/09/2012	Filingué	03/09/2012	59	-	na
Imanan	04/09/2012	Filingué	04/09/2012	0	-	na

continued **Table 5.3:** List of floods episodes per municipality in Tillabéri region and rainfall. (Source ANADIA 2012)

Municipality	Flood date	Station	Rainfall date	Rainfall	TRMM/3h	TRMM Date
Imanan	19/08/2011	Filingué	19/08/2011	45	-	na
Imanan	09/08/2011	Loga	09/08/2011	13.5	-	na
Karma	02/07/2007	Dargol	02/07/2007	0	14.3	01/07/2007
Karma	04/08/2007	Dargol	04/08/2007	54	17.7	04/08/2007
Kourthey	09/08/2008	Tillabéri	08/08/2008	72.8	37.9	09/08/2008
Kourfeye	16/07/2010	Toukounous	15/07/2010	45	11.2	16/07/2010
Kourfeye	22/07/2010	Chikal	21/07/2010	38	26.5	20/07/2007
Kourfeye	08/08/2011	Chikal	08/08/2011	36	-	na
Kourfeye	03/09/2012	Chikal	03/09/2012	84.5	-	na
Kourfeye	04/09/2012	Chikal	04/09/2012	0	-	na
Kouré	08/07/2009	Niamey Aéro	07/07/2009	26.1	15.3	07/07/2009
Kouré	29/06/2011	Niamey Aéro	29/06/2011	0	-	na
Kouré	03/07/2012	Niamey Aéro	03/07/2012	17.8	-	na
Kouré	04/07/2012	Niamey Aéro	04/07/2012	0	-	na
Kouré	05/07/2012	Niamey Aéro	05/07/2012	0	-	na
Kokorou	22/08/2009	Téra	22/08/2009	52	61.8	22/08/2009
Kokorou	10/06/2010	Téra	09/06/2010	19.8	14.4	09/06/2010
Liboré	01/08/2012	Niamey Aéro	30/07/2012	15.3	-	na
Ouallam	08/09/2006	Ouallam	08/09/2006	15.5	11.5	08/09/2006
Ouallam	11/09/2006	Ouallam	11/09/2006	6.5	18.7	10/09/2006
Ouallam	30/07/2010	Ouallam	29/07/2010	36	19.8	29/07/2010
Namaro	14/06/2002	Niamey Ville	13/06/2002	68.5	2.6	14/06/2002
Namaro	02/08/2007	Niamey Ville	01/08/2007	58	18.3	02/08/2007
Sakoira	09/08/2008	Tillabéri	08/08/2008	72.8	86.3	09/08/2008
Sakoira	12/08/2008	Tillabéri	12/08/2008	0.4	5.3	11/08/2008
Sanam	16/06/2007	Abala	16/06/2007	0	5	16/06/2007
Simiri	16/08/2011	Ouallam	14/08/2011	30	-	na
Sinder	01/09/2010	Tillabéri	01/09/2010	0.2	3.2	30/08/2010
Tagazar	24/08/2006	Filingué	24/08/2006	51	46.8	22/08/2006
Tagazar	16/07/2010	Niamey ville	15/07/2010	28.2	30.9	16/07/2010
Tagazar	19/08/2011	Niamey ville	16/08/2011	3.5	-	na
Téra	22/08/2009	Tera	22/08/2009	52	8.7	09/06/2010
Téra	09/06/2010	Tera	09/06/2010	19.8	10.2	09/06/2010
Téra	10/06/2010	Tera	10/06/2010	0	0	10/06/2010
Téra	08/08/2010	Tera	07/08/2010	22.7	4.8	07/08/2010
Tillabéri	30/10/2003	Tillaberi	30/10/2003	0	7.1	27/10/2003
Tillabéri	09/08/2008	Tillaberi	09/08/2008	72.8	75.2	09/08/2008
Tondikiwindi	04/09/2006	Ouallam	04/09/2006	0	0	04/09/2006
Tondikiwindi	11/09/2006	Ouallam	11/09/2006	6.5	16.3	10/09/2006
Tondikiwindi	29/07/2010	Ouallam	29/09/2010	36	18.2	30/07/2010
Tondikiwindi	31/07/2010	Ouallam	31/07/2010	8	9	01/08/2010
Tondikiwindi	31/08/2010	Ouallam	31/08/2010	0	29.8	30/08/2010
Tondikiwindi	18/08/2011	Ouallam	18/08/2011	20	-	na
Tondikandia	24/08/2006	Filingué	23/08/2006	12	14.2	21/08/2006

continued **Table 5.3:** List of floods episodes per municipality in Tillabéri region and rainfall. (Source ANADIA 2012)

Municipality	Flood date	Station	Rainfall date	Rainfall	TRMM/3h	TRMM Date
Tondikandia	25/08/2006	Filingué	24/08/2006	51	21.4	22/08/2006
Tondikandia	09/08/2011	Filingué	09/08/2011	0	-	na
Tondikandia	03/07/2012	Filingué	02/07/2012	21	-	na
Tondikandia	28/07/2012	Filingué	28/07/2012	0	-	na
Tondikandia	03/09/2012	Filingué	01/09/2012	31.5	-	na
Tondikandia	04/09/2012	Filingué	03/09/2012	59	-	na
Youri	04/08/2007	Say	04/08/2007	75.5	61.7	04/08/2007

was made by an evaluation of all available values found for each municipality. This process extracts three different combinations: i) If there is a unique value it is considered as a critical rainfall threshold to characterize the administrative unit; ii) If there is more than one value, the lowest value is chosen; or iii) In case of multiple similar values, the average among the lowest values.

For municipalities where no significant rainfalls are found, 20 cases of 44, it was assumed that the hazard was similar to the nearest administrative units and the value from the upriver municipality with the same hydrological characteristics were used instead.

Given the critical threshold per municipality, the next phase was the identification of the specific return period of similar meteorological conditions. For each rain gauge station and TRMM municipality series, we evaluated the return period of the critical values by Gumbel distribution (Gumbel, 1954). The inverse return period is the probability in each year of having the same meteorological condition which, at least one time in the past, has generated a flood in the municipality. This process yields the same homogeneous reference period for describing the hazard. In this way, it is possible to make a relative comparison between the municipalities in the region in order to highlight which are most at risk. If we compare two stations with different temporal coverage, considering Climate Change, the return periods produce a different result influenced by the changing of the rainfall distribution over time. As in the case of the Ouallam station, where data have been collected since 1948, if we consider the whole period to calculate the return period, it produces skewed values, considering the higher precipitation in the 1950–1960 period compared to the driest period around the 1980s (Figure 5.6). So, for each station, the period considered was 1981–2010 to achieve a useful return period for critical rainfall.

Combining the information on the municipality watershed typology with the two different rainfall datasets, it is possible to choose the best indicator for each municipality's typology in order to best describe the probability of these critical meteorological conditions recurring in the future.

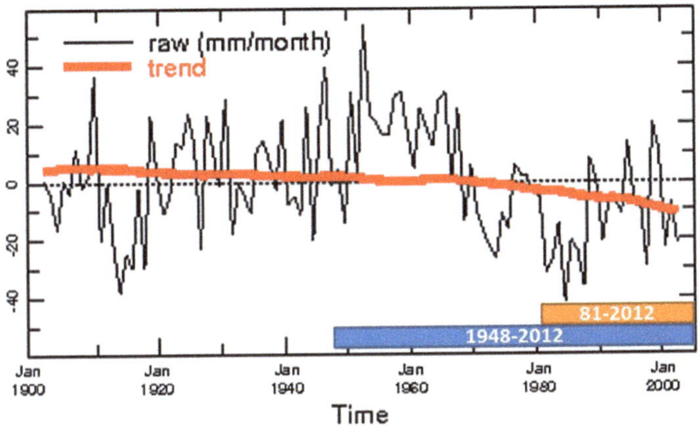

Figure 5.6: Precipitation anomaly Historical Series 1901-2012 – Maproom IRI, Data from monthly mean precipitation from CRU TS 2.1.

5.3 Results

The main result of this study is the identification of the different probabilities of critical meteorological conditions in the municipalities of Tillabéri region. The consideration that a single municipality is a part of a complex system characterized by specific hydrology and consequently by different critical rainfall distributions, leads to a differentiation of the territories, discriminating the zones most at risk and, at the same time, estimating the resulting value.

The process of reconstruction of weather conditions that characterized each event recorded by the SAP flooding database allowed the production of a synthetic picture of the critical rainfall for each municipality. As it has been described in the methodology, with the lack of high quality data available, it was necessary to interpret what happened in each event, which often resulted in the additional difficulty of explaining the discrepancies among the two meteorological datasets available for the same period.

Using the most objective and coherent analysis, a list of critical rainfall for each municipality with the two thresholds was produced, one from the rain gauge station network and one from TRMM (Table 5.4).

Using the Gumbel distribution for each rain gauge station and TRMM municipality extrapolation, the probability of the recurrences of flood hazard with the value of the return period of the critical rainfall was defined. This identification enabled us to define the distribution of meteorological hazard distributions in the region and define the probability of the intense events that can cause flooding (Table 5.4) to occur in the future. If we couple this information with the hydrological characterization of the municipality, it is possible to use one of the two probabilities found to assess the specific hazard.

Table 5.4: Critic threshold per municipality by rain gauge station network and TRMM (source ANADIA 2012).

ID	Municipality	Rain gauge	Reference Station	TRMM	Reference
1	Abala	79.6	Abala	113.9	
2	Anzourou	52	Ayerou	73.4	Ayerou
3	Ayerou	52	Ayerou	73.4	
4	Bani Bangou	79.6	Abala	58.6	
5	Bankillare	52	Ayerou	61.8	Kokourou
6	Bibiyergou	45	Tillaberi	75.2	Tillabéri
7	Bitinkodji	58	Niamey	61.7	Youri
8	Dargol	52	Dargol	75.2	Tillabéri
9	Dessa	52	Ayerou	73.4	Ayerou
10	Diagorou	52	Tera	75.2	Tillabéri
11	Diantchandou	51	Filingue	15	
12	Dingazi Banda	51	Filingue	10	
13	Filingue	42.5	Filingue	27.4	
14	Gorouol	30	Bankilaré	66	
15	Gotheye	52	Dargol	66	
16	Hamdallaye	30	Ouallam	13.9	
17	Imanan	45	Filingue	23.2	
18	Inates	52	Ayerou	73.4	Ayerou
19	Karma	54	Dargol	33.3	
20	Kirtachi	75.5	Say	61.7	Youri
21	Kokorou	52	Tera	61.8	
22	Kollo	75.5	Say	61.7	Youri
23	Koure	75.5	Say	61.7	Youri
24	Kourfey centre	40	Chikal	26.5	
25	Kourtey	72.8	Tillaberi	37.9	
26	Libore	58	Niamey	61.7	Youri
27	Mehana	72.8	Tillaberi	61.8	Kokourou
28	Namaro	58	Niamey	18.3	
29	N'Dounga	58	Niamey	61.7	Youri
30	Ouallam	36	Ouallam	19.8	
31	Ouro Gueladio	52	Torodi	75.2	Tillabéri
32	Parc W	75.5	Say	61.7	Youri
33	Sakoira	72.8	Tillaberi	86.3	
34	Sanam	79.6	Abala	113.9	Abala
35	Say	75.5	Say	61.7	Youri
36	Simiri	70	Ouallam	70	
37	Sinder	72.8	Tillaberi	3.2	
38	Tagazar	51	Filingue	46.8	
39	Tamou	75.5	Say	61.7	Youri
40	Tera	52	Tera	10	
41	Tillaberi	72.8	Tillaberi	75.2	
42	Tondikandia	59	Filingue	21.4	
43	Tondikwindi	36	Ouallam	29.8	
44	Torodi	52	Torodi	75.2	Tillabéri
45	Youri	75.5	Say	61.7	

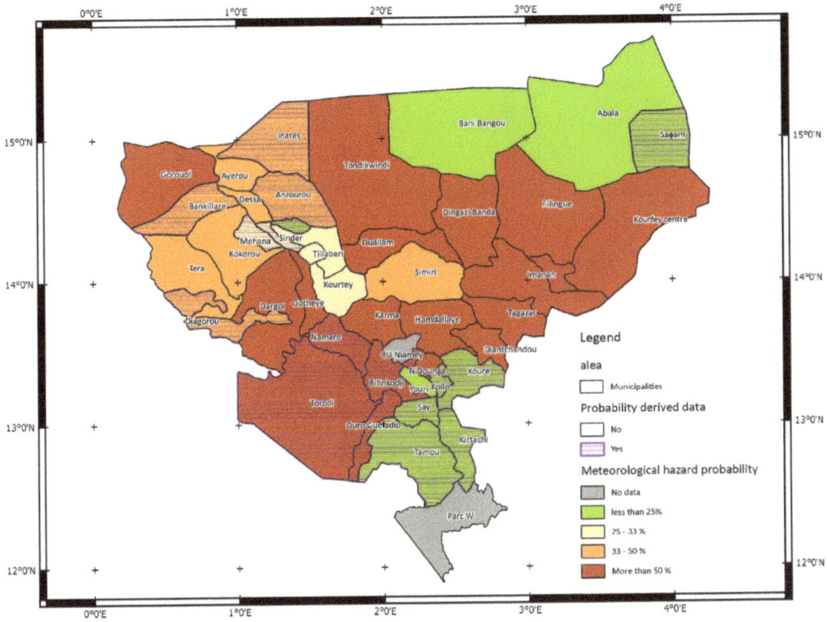

Figure 5.7: Meteorological hazard probability in Tillabéri Region.

In this case study it was chosen to associate i) to each municipality along the Niger River and bordered by the primary tributaries, the probability of the critical rainfall from the ground station network, and ii) the 3-hourly critical rainfall for the municipalities on secondary tributaries and the Dallols. Finally, a map was produced with the yearly probability distribution to record critical meteorological conditions in the entire region (Figure 5.7).

The results show hazard distributions in the Tillabéri region, mainly in four homogeneous areas. The southern part of the region has a lower probability of recurrence of floods. This is likely to be in part correlated with a higher rainfall distribution affecting the area, and therefore a greater adaptation of populations to floods. The other area with low probability of weather conditions favorable to floods is the north-eastern part of the region. In this case, however, the risk is lower due to lower rainfall affecting this arid area than to a less recurrent high intense rainfall phenomena.

The north-western part, also characterized by arid climate, has a mid-level chance of critical rainfall conditions. The hydrological system of the area is complex and the presence of the Niger River probably elevates the risk of floods due to the dynamics of river flow.

Finally, the central band of the region, from east to west, is characterized by a high risk of meteorological conditions favorable for floods almost every year. The causes are various and interconnected, but it is in this area that decision makers must address the need for flood risk prevention.

5.4 Conclusion

The analysis of weather hazards for disaster risk assessment in Tillabéri region aims to help decision makers in defining the probability of a critical rainfall that could cause flooding for each municipality. The impetus for this paper was to answer the question: What is the period for a dangerous weather phenomenon possibly recurring in the future?

It must be noted that logging critical rainfall does not immediately implie the occurrence of floods because exposure and vulnerability to floods are not static conditions. Also, timing of rain is a crucial aspect in flood risk. As evidenced by the TRMM 3-hourly data, a large part of dangerous floods match with heavy rains fallen during the night, when the population is not prepared to react to a flood.

As far as producing an exact prediction of the meteorological hazards in the Tillabéri region, the methodology moves its attention to identify areas most susceptible to dangerous events in the future. Normally rainfall distributions and/or the characteristics of the ground are not considered in the characterization of a municipality's flood hazard, and national authorities are not aided by information on the meteorological flood hazard at a municipal level.

The introduction of some elements in the methodology, which takes in to consideration the natural hydrological and rainfall distribution, is the contribution of this study, promoting a different approach to meteorological risk assessment tailored for Niger and replicable in all the West African countries for which disaster risk prevention is a priority in the immediate future. To do this, some assumptions are needed in order to produce the most accurate picture of the flood risk in the Tillabéri Region.

The lack of meteorological data and the limits of the non-standardized flood records present a great difficulty in flood hazard characterization. First of all, there is a need of appropriate spatio/temporal resolutions to interpret critical meteorological phenomena because floods could be generated by different types of convective systems and streamflows. This means that it is not possible to consider the territory as a homogeneous surface, but rather it is necessary to define, in the simplest way, the differences between different hydrological characteristics.

Second, each flood event (flash flood/riverine flood) is characterized by a different critical threshold that can be estimated by studying past disaster events. It is important to note that the brevity or lack of flooding events can be a reason why some administrative units are poorly represented in terms of data. In other words: they have not been affected by flooding or they do not have sufficient records to determine the critical rain-threshold. This aspect is mitigated by the high frequency of floods recorded in the region and the similar data retrieval methods associated with values of adjacent administrative units.

The use of two different meteorological datasets, one more appropriate to describe hazards in large river basins and one for small river basins and the Dallols, allows the differentiation of specific critical thresholds and the specifics return periods in each municipality, thus classifying the meteorological hazard in each municipality.

Adding the information about flood hazards with exposure and vulnerability assessment studies could help decision makers in the prioritization of prevention and preparation plans in the Tillabéri region using an objective approach.

5.5 Acknowledgements

This work was performed in the framework of ANADIA project. I would like to thank the project team and in particular Vieri Tarchiani, Aissa Sitta, coordinators of the project. My precious colleagues Katiellou Gatpia Lawan, Moussa Mouhaïmouni, Ousman Baoua and Abdourhamane Daouda who helped me in producing the analyses of the hazard. My thanks also go to Hassimou Issa SAP Niger, Prof. Maurizio Tiepolo and Sarah Braccio of Politecnico di Torino for the collection and organization of data on flooding. Finally, I would also thank the Director of the National Meteorological Niger Mr. Labo Moussa.

References

Adger, W.N., S. Huq, K. Brown, D. Conway, and M. Hulme. 2003. Adaptation to climate change in the developing world. *Progress in Development Studies* 3(3): 179–195.

ANADIA Niger-Projet Adaptation au changement climatique, prévention des catastrophes et Développement agrIcole pour la sécurité alimentaire. 2014. Evaluation du risque inondation à l'echelle communale en region Tillabéri, Niger, Report n. 3.

Beck, M.W., C.C. Shepard, J. Birkmann, T. Welle, M. Witting, J. Wofertz, and K. Radtke. 2012. World risk report 2012. http://www.ehs.unu.edu/file/get/10487.pdf. Accessed 4 May 2014.

Cardona, O.D., M.K. van Aalst, J. Birkmann, M. Fordham, G. McGregor, R. Perez, R.S. Pulwarty, E.L.F. Schipper, and B.T. Sinh, 2012. Determinants of risk: exposure and vulnerability. In *Managing the risks of extreme events and disasters to advance climate change adaptation. A Special report of working groups I and II of the IPCC*, ed. C.B. Field, V. Barros, T.F. Stocker, D. Qin, D.J. Dokken, K.L. Ebi, M.D. Mastrandrea, K.J. Mach, G.-K. Plattner, S.K. Allen, M. Tignor, and P.M. Midgley, 65–108. Cambridge: Cambridge University Press.

CNEDD-Conseil National de l'Environnement pour un Développement Durable. 2012. Programme d'action national pour l'adaptation aux changements climatiques. Niamey: Secrétariat exécutive du CNEDD. http://unfccc.int/resource/docs/napa/ner01f.pdf last access 04/05/2014.

Descroix, L. et al. 2009. Spatio-temporal variability of hydrological regimes around the boundaries between sahelian and sudanian areas of West Africa: a synthesis. *Journal of Hydrology*. doi:10.1016/j.jhydrol.2008.12.012.

EM-DAT-The OFDA/CRED international disaster database. www.emdat.be.

Gumbel, E.J. 1954. Statistical theory of extreme values and some practical applications. Applied mathematics series 33. U.S. Department of commerce, National bureau of standards.

Huffman, G.J., R.F. Adler, D.T. Bolvin, G. Gu, E.J. Nelkin, K.P. Bowman, Y. Hong, E.F. Stocker, and D.B. Wolff. 2007. The TRMM multi-satellite precipitation analysis: Quasi-global, multi-year, combined-sensor precipitation estimates at fine scale. *J. Hydrometeor.* 8(1): 38–55.

INS-Institut National de la Statistique du Niger. 2011. *Tillaberi en chiffres – Edition 2011*. http://www.stat-niger.org/statistique/file/Regions/TillaberienChiffres2011.PDF. Accessed 4 Mai 2014.

Jha, A.K., R. Bloch, and J. Lamond. 2012. *Cities and flooding: a guide to integrated urban flood risk management for the 21st century*. Washington: The World Bank. doi: 10.1596/978-0-8213-8866-2.

Pierce, D. 2011. Ncview 2.1.1 - a netCDF visual browser. http://meteora.ucsd.edu/~pierce/ncview_home_page.html. Accessed 4 May 2014.

Giuseppe Sappa[18], Giulia Luciani[19]

6 Sensitivity of Dar Es Salaam Coastal Aquifer to Climate Change with Regard to Seawater Intrusion and Groundwater Availability

Keywords: Groundwater recharge, Seawater intrusion, Climate change

6.1 Introduction

This paper presents the initial results of three years of investigation activities, carried on in the Dar es Salaam coastal plain (Tanzania) by the Adapting to Climate Change in Coastal Dar es Salaam (ACC-DAR) project, a cofounded research project, granted by the European Union, led by the Sapienza, University of Rome, in cooperation with Ardhi University of Dar es Salaam. The ACC-DAR project activities will enhance the capacities of Dar's municipalities by increasing their understanding of adaptation practices, and by developing methodologies for integrating adaptation activities into strategies and plans for Urban Development and Environment Management (UDEM) in unplanned and unserviced coastal settlements. In order to provide a series of enhanced methodologies for improving municipal activities related to climate change (CC) issues in the water management sector, the specific environmental phenomenon of seawater intrusion was investigated. This phenomenon is already contributing, and will increasingly contribute as CC progresses, to the degradation of those natural resources on which a large part of Dar es Salaam's peri-urban livelihoods depends. The target of this study was to investigate groundwater availability changes in Dar es Salaam's coastal aquifer as a consequence of seawater intrusion and urbanization processes in the framework of CC effects, with the aim to set up an integrated approach to evaluate CC effects on groundwater resources in coastal plains affected by seawater intrusion, and to better manage these important natural resources. As such, geological and hydrogeological characterization of the area is part of the study,

---

18 Giuseppe Sappa is an Associate professor in Engineering Geology and Applied Hydrogeology at the Department of Civil, Building and Environmental Engineering, Sapienza University of Rome. He worked as Hydrogeologist Senior in ACC-DAR project, as he is an expert in sustainable exploitation of groundwater resources with special regard to coastal aquifer management, giuseppe.sappa@uniroma1.it

19 Giulia Luciani is a PhD student at DICEA – Department of Civil, Building and Environmental Engineering, Sapienza University of Rome, where she graduated in Environmental Engineering. She participated at the activities of ACC-DAR project on a scholarship. Her research interests include assessment of impacts of climate change on water resources, sustainable groundwater management and artificial recharge of groundwater.

Authors thank Silvia Macchi for involving them in this really interesting and attractive experience.

as lithological properties of outcropping geological formations and their main hydrogeological settings, as well as chemical groundwater characterization also depend on them, but they were not the target of the study.

6.2 The Study Area

The United Republic of Tanzania is a country in Sub-Saharan Africa. It borders Kenya to the North, Uganda, Rwanda, Burundi and the Democratic Republic of the Congo to the West, and Zambia, Malawi and Mozambique to the South. The city of Dar es Salaam is the largest urban center of Tanzania, with a population of about 3 million people and a growth rate of 4.3%. The region of Dar es Salaam is divided into the three districts of Ilala, Kinonondoni, and Temeke. In the last two decades, the increased demand of water due to growing urbanization and uncontrolled groundwater abstraction due to scarcity of surface water resources has placed the principal coastal aquifer of Dar es Salaam in a growing danger of salinization. The possible presence of various sources of salinity in the area has been documented. These include: seawater intrusion and anthropic contamination due to overexploitation, and water ascending from deep marine sediments (Mjemah 2007). The study area covers a surface area of approximately 260 km². It extends along a 40 km stretch of coastline to the north of the city center, and includes a part of the city center, and some peri-urban areas.

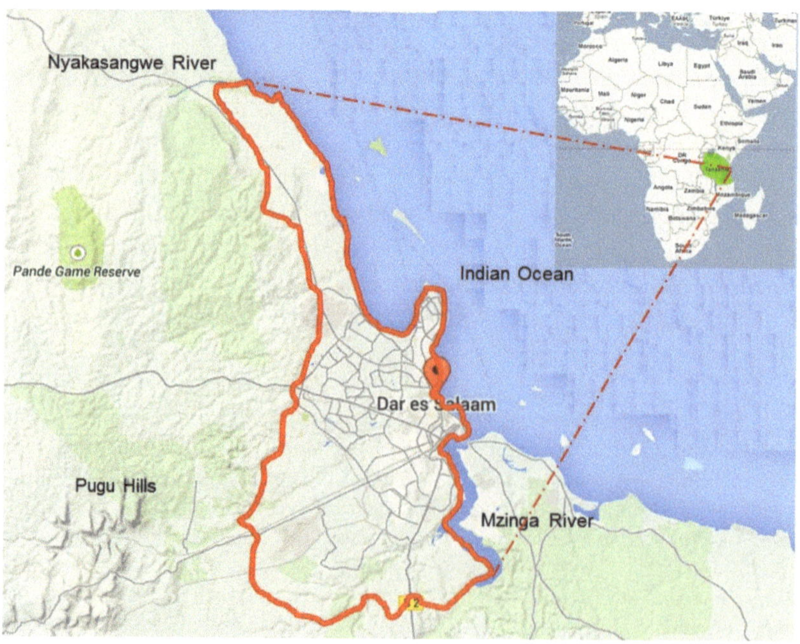

Figure 6.1: Study area.

The study area was selected on the basis of hydrogeological boundaries: the eastern boundary is the Indian Ocean, the western boundary is the Dar es Salaam Plateau, which rises west of the ocean along the entire study area up to the Pugu Hills, the southern and northern hydrogeological boundaries are, respectively, the Mzinga River and the Nyakasangwe River (Sappa *et al.* 2013a).

6.3 Geological and Hydrogeological Framework

The geological structure of Tanzania reflects the geological history of the whole African Continent. The oldest rock basement in coastal Tanzania includes the Precambrian sediments of the Archean age. The continental Karoo sequence forms the basal part of the sedimentary sequence in the coastal basin of Tanzania. Alternating periods of regression and transgression characterized the Jurassic, Cretaceous and Tertiary (Paleogene and Neogene) ages. In regression periods, clays, silty layers, and silty limestone dominated. In transgression periods, sediments mainly consisted of calcareous, sandy, and shelly limestones. The Neogene period (Miocene and Pliocene) in the study area was characterized by important tectonic activities, which defined the present topographic features.

Neogene sandstone formations, interbedded with siltstones and mudstones, occupy the upland zone south and west of the city center. Within the Neogene formations, several distinct varieties are recognizable. Sandstones occupy over three quarters of the region and comprise a variety of main types. The most important outcrop of the Miocene is the kaolinitic sandstone in the Pugu Hills. The massive terrace sandstone is the bedrock that limits the extent of terraces (Msindai 2002). The Pugu sandstones comprise massive, kaolinitic, and cross-bedded sandstones. Calcareous sandstones also occur on back reef areas of the uplands.

During the Quaternary period, deposition and erosion processes occurred in relation to tectonics, sea level, and climate change. In the coastal region of Dar es Salaam, the quaternary deposits can be divided into three geological layers: alluvial, coastal plain, and coral reef limestone deposits. Alluvial deposits fill the valleys of the Mzinga, Kizinga, and Msimbazi rivers. They consist of an alternation of fine and coarse-grained sands, clay, and sometimes gravel and pebbles. The coastal plain consists of unconsolidated sediments, predominantly sandy, with evidence of several marine intercalations. The presence of coralline limestone is found along the coastal strip. Carbonate rocks are present as fringing reefs and raised reefs. The northern part of the region has few fringing reefs, while raised reefs dominate the western margins of the upland.

The geological setting of the study area comprises unconsolidated sediments of Neogene and Quaternary ages (Figure 6.2). The alluvial deposits and coastal plain deposits are of Pleistocene to Recent age and are found mainly moving from the coast towards the mainland within the river valleys. The main part of the study area cor-

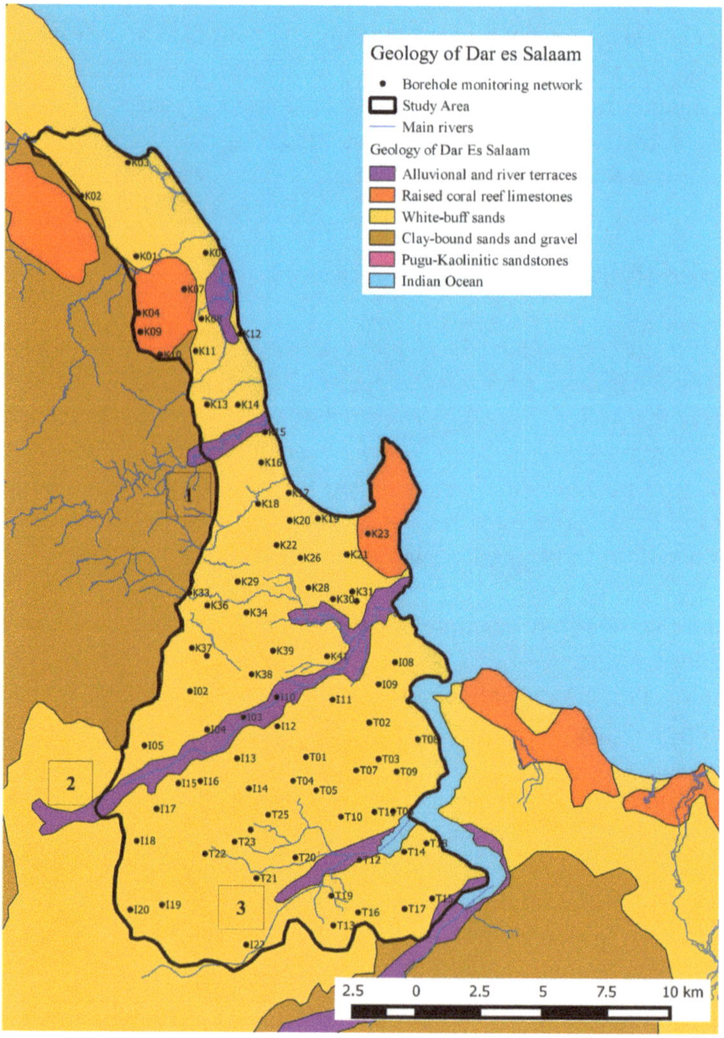

Figure 6.2: Geology of Dar Es Salaam. Indicated are the Mbezi River (1), Msimbazi River (2) and Kizinga River.

responds to such a valley. These deposits consist of sand, clay, gravels and pebbles. Fine to coarse-grained sands occur widely within valley creeks, deltas and mangrove sites of the Mzinga, Kizinga and Msimbazi Rivers.

The groundwater reservoir is located within the coastal plain in the quaternary sediments, having higher hydraulic conductivity than the underlying and surrounding Miocene sequence, and includes clay intercalations. The aquifer system of the study area is made of two main aquifers, both from Quaternary: an upper unconfined sand aquifer and a lower semi-confined sand aquifer, separated by a clay aquitard. The sediment type for both aquifers is almost the same, and consists of quaternary deposits from Pleistocene to the Recent age. The unconfined aquifer consists of fine to medium sand, with varying percentages of silt and clay. Due to the high percentage of sand, it is defined as a sand aquifer, which is considered the most important aquifer for water supply. The lower, semi-confined aquifer consists of medium to coarse sands. The lower aquifer overlies the substratum formed by Mio-Pliocene clay-bound sand and gravel and kaolinitic Pugu sandstones. The different hydrogeological formations characterizing the groundwater system in the area of concern are described in Table 6.1 below (Mjemah 2007).

Table 6.1: Hydrogeological scheme.

AQUIFER	PERIOD	EPOCH	LITHOLOGY	THICKNESS (m)
Unconfined	Quaternary	Pleistocene to recent	Fine to medium sand with silts and clay, coral reef limestone and calcareous, alluvial clay, silts and gravels	5–50
Aquitard	Quaternary	Pleistocene to recent	Clay, sandy clay	10–50
Semiconfined Aquifer	Quaternary	Pleistocene to recent	Medium to coarse sand and gravels with clay	100
Aquitard	Neogene	Mio-Pliocene	Clay-bound sands	≈ 1000

6.4 Methodology

The current state of groundwater quality in the coastal aquifer of Dar es Salaam was studied through the implementation of various methods, tailored to the available set of climatic and hydrogeological data (both historical and current data).

A review of available literature was conducted to assess the geological and hydrogeological sketch framework of the Dar es Salaam coastal plain. Climatic and anthropogenic influences on hydrogeological dynamics were investigated through the analysis of temporal evolutions of the groundwater recharge. Seawater intrusion was evaluated by hydrochemical methods, through physical and chemical testing of a monitored network of representative boreholes.

6.5 Data Collection and Analysis

Two main activities were carried out: collection of historical data from a variety of existing sources, and the execution of two groundwater monitoring campaigns in June and November 2012.

The output of these activities is a set of historical data that includes the climatic parameters precipitation and temperature data for 3 gauges with reference to the last 50 years, hydrogeological characteristics of the coastal aquifer, and the physical and chemical parameters of the groundwater for nearly 400 boreholes located in the Dar es Salaam region. These data have been collected and organized in a database, ACC-Dar Borehole Monitoring Database (http://www.planning4adaptation.eu/042_Maps.aspx).

A subset of boreholes located in the study area was chosen for the monitoring network from the database of 400 georeferenced boreholes, with consideration for uniformity of spatial distribution: the network consists of 79 boreholes, uniformly distributed with a frequency of about 1 borehole per 3 km^2. The study area and the borehole monitoring network are shown in Figure 6.3. The boreholes concerned with a monitoring campaign are commonly used to satisfy civil private demand. For many of them depth is unknown, and as they are all equipped with a pump, the collected samples can't be referenced to a specific depth, therefore they are representative of the average thickness of the aquifer crossed through.

6.6 Climate Change Effects on Active Groundwater Recharge

Hereinafter are presented elaborations and results regarding evaluation of contribution to Average Annual Groundwater Recharge (AAGR) coming from direct infiltration in the area under study. Although the values discussed may only be part of the total active groundwater recharge of the area, this is the majority of it, and so, this aspect should not affect the conclusions drawn.

AAGR was evaluated using the Hydrogeological Inversed Budget Method (Civita and De Maio 1999) as modified by Sappa, Trotta, and Vitale (Sappa *et al.* 2014). This method involves a spread parameters approach, based on the discretization of the study area in cells, and for each cell the estimation of climatic, topographical and hydrological input parameters usually available and involved in the evaluation of the hydrogeological budget.

The values of Average Annual Precipitation (AAP), to be assigned to each cell, were determined from the Annual Average Precipitation values, and recorded in the gauging stations through appropriate techniques of spatial interpolation. Due to the absence of temperature values necessary for the calculation of evapotranspiration, as well as the land cover properties, the application of the method was performed, as suggested by the authors, enclosing the effects of evapotranspiration and runoff, directly in the potential infiltration factor (PIF). This coefficient represents the amount of rainfall reaching the subsoil and contributing to groundwater recharge, ranging between 0 and 0.55, depending on land cover characterization. The values of Average Annual Infiltration (AAI), given to each cell and expressed in mm, are calculated as a product of the PIF and the AAP. The correspondent average annual volumes infil-

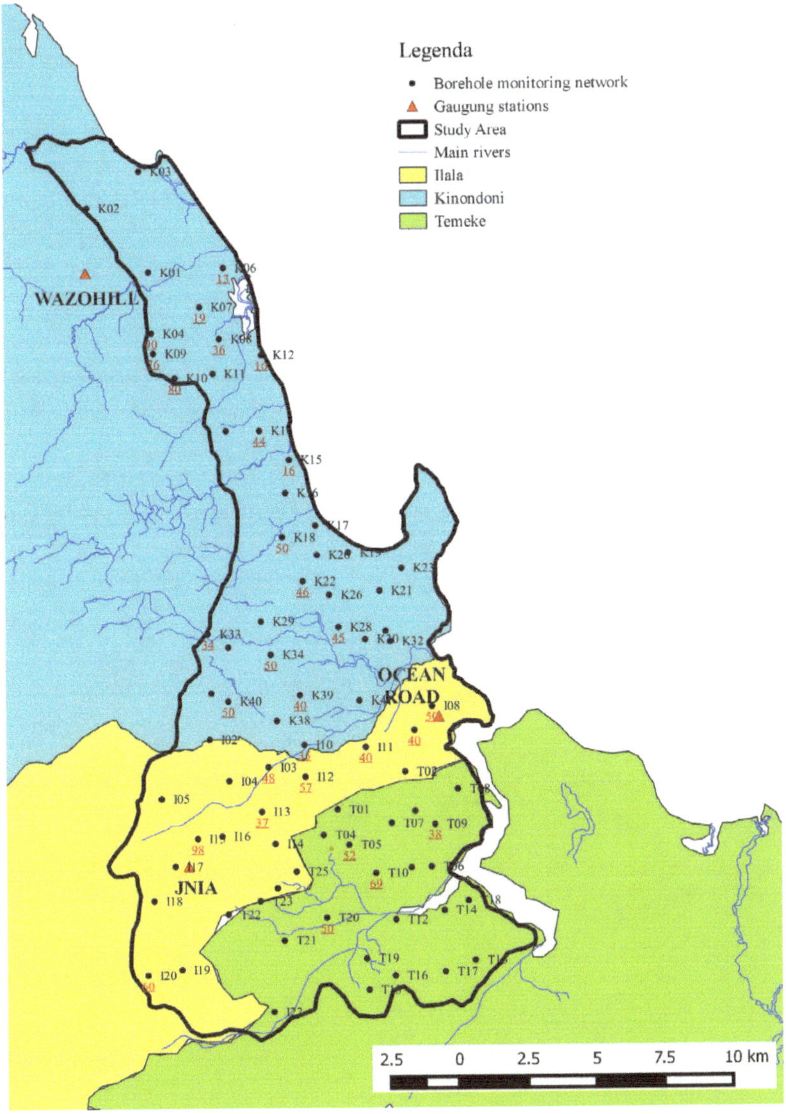

Figure 6.3: The study area and the monitoring network in 2012. Where known, the depth of the boreholes is indicated.

trated are obtained as a product of the AAI and each cell area. Finally, the amount of AAGR, referred to the whole study area is obtained by the addition of values assigned to each cell.

The study area was divided in 500m x 500m squared meshes. Precipitation records for the previous 50 years, from 1960 to 2010, were collected in 3 meteorological stations in Dar es Salaam: JNIA, Wazohill, and Ocean stations.

Since only the JNIA station, located close to the international airport of Dar, had data for every one of the 50 years, statistical elaboration processes were applied in order to create the missing data for the other stations. To determine these data, three different methods were used: (i) between-station, method based on the average of the data registered in adjacent different stations; (ii) within-station method, based on the average of the earlier and later data; and (iii) regression following the WMO approach (WMO 2011), which is based on an important result: as concurrent values of rainfall in relation to two stations are compared, their difference (temperature) or their ratio (precipitation, wind speed) tends to be constant.

Afterwards, as the missing data series were recreated, data's adaptation to theoretical distributions through the method of moments was verified. Two different statistical tests, Pearson and Kolgomorov Smirnof, were conducted. Only the Gaussian distribution passed both tests.

For each station, referring to the Gaussian distribution, the rain values, which define the range corresponding to a 68% of probability of occurrence, defined by the mean (μ) and the std.dev (σ) as (μ-σ) and (μ+σ), were chosen as representative of the minimum and maximum probable rain value for the station. These values were used for successive elaborations and are shown in Table 6.2.

Table 6.2: Average Annual Precipitation (AAP) values for the three meteorological stations (1960–2012).

Jnia	μ+σ	1410 mm
	μ-σ	854 mm
Ocean Road	μ+σ	1249 mm
	μ-σ	801 mm
Wazo Hill	μ+σ	1091 mm
	μ-σ	727 mm

In order to spread these values over the study area, the Inverse Distance Weighting method (IDW) was used, which is deterministic for multivariate interpolation. The IDW method was applied three times to determine the minimum, mean and maximum of rain values for each station, respectively.

In the present study, the values of PIF were assigned to each cell on the basis of land cover interpretation data, coming from elaboration of the analysis of satellite images (Congedo *et al.* 2013), carried on by another research team involved in the same ACC-DAR project. The interpretation of these data permitted assessment of the evolution of land cover in the study area for a period of ten years (2002–2012), and allowed us to associate PIF values to the different land cover properties outcropping in the study area, as it is reported in Table 6.3.

Table 6.3: Potential Infiltration Factor (PIF) values, given to the different land cover class.

Land cover	PIF
Out of city	0,15
Unclassified	0,15
Full vegetation	0,3
Mostly vegetation	0,4
Continuos urban	0,1
Discontinuos urban	0,2
Soil	0,3
Water	0,55

Data on land cover evolution in Dar Es Salaam (Congedo *et al.* 2013) were elaborated, and for each year a PIF value was given to each 500m x 500m square cell.

During the 2002–2012 period, an impactful demographic growth occurred, involving substantial changes in urbanization and land cover. The analysis of land cover evolution between 2002 and 2012 shows that urban land, as both continuous urban and discontinuous urban land, increased from 40% to 65%, following a linear trend. Meanwhile, over the last 10 years, soil, which is the land cover type with the maximum PIF value, decreased by 20%, from about 47% to about 27%. The increase in urban areas transformed the region, rendering it less permeable to rainfall. As a result, many areas are now contributing less to the recharging process than before (Sappa *et al.* 2013a).

A range of possible values of AAGR for the study area was estimated. The estimate was made considering the land cover of 2011 and the AAP values calculated on the basis of the 52-year data set. AAP values are expressed in terms of minimum, medium

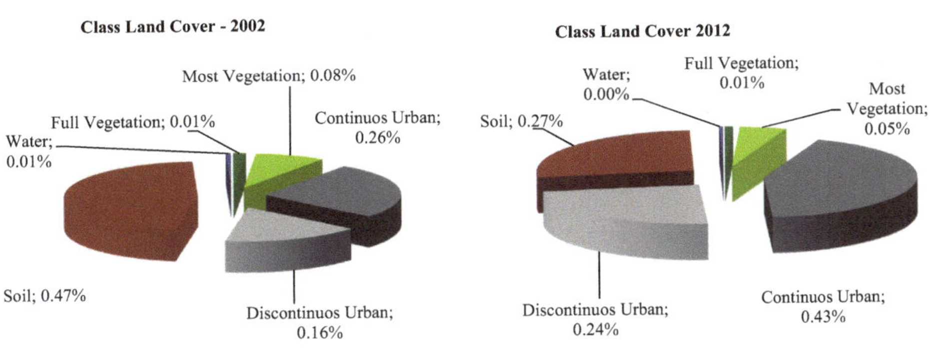

Figure 6.4: Percentage variations of land cover distribution (2002–2012)

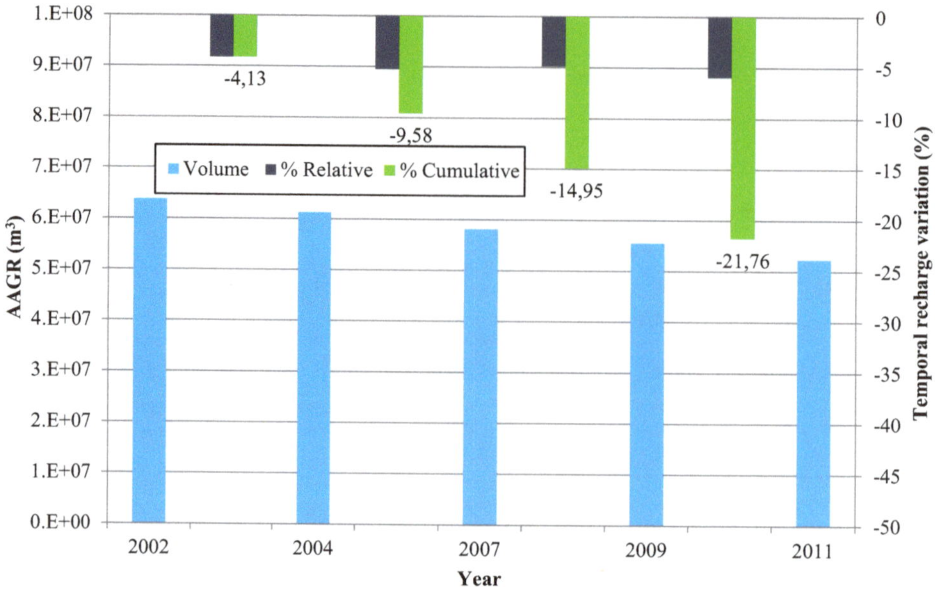

Figure 6.5: Average Annual Groundwater Recharge (AAGR) temporal variation in the period 2002–2011.

and maximum values, which correspond to the 0.7 decile result of the Gaussian distribution for the AAP data set. Land cover for the year 2011 was 29% soil, 65% urban, and 6% vegetation. The estimated corresponding volumes of infiltration (AAGR) are $4.0x10^7$, $5.2x10^7$ and $6.4x10^7$ m^3, respectively. Values obtained were compared with similar results coming from previous studies, and they seemed very similar (Van Camp *et al.* 2012).

The evolution of AAGR in the period from 2002 to 2011 was analyzed, considering the AAP and land cover of each year. The analysis showed a decrease of 20% in AAGR from 2002 to 2011, almost 5% for each year.

As a further investigation, analysis of the rainfall evolution (mm) in the last decade for the study area was carried out. It was determined that, although the AAP decreased over the last decade for all of the monitored stations (Figure 6.6a), the changes in precipitation amount were only slight in the last five years (Figure 6.6b), and AAP appears to be almost the same, with values between 900 and 1000 mm/y. Thus, in recent years most of the decreases in groundwater active recharge likely originated from land cover use modifications.

Based on the results of the previous recharge estimate, a prediction was made of estimated AAGR volumes from 2013 to 2020, assuming a constant rate of precipitation (the analysis was carried out using the mean estimated values for each station) and assuming a linear trend of urban land cover increasing.

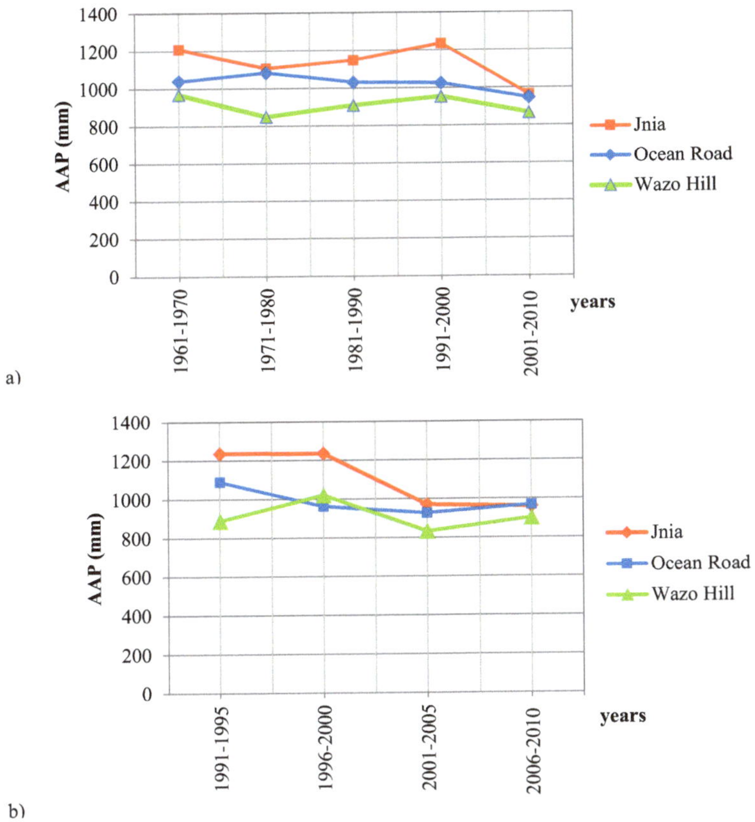

Figure 6.6: a) and b) Average Annual Precipitation (AAP) referred to any 10 and any 5 years.

In the diagram of Figure 6.7, the AAGR, calculated for the last decade, and the linear regression line relating to these values was drawn. From this interpolation, the corresponding evolution of AAGR from 2012 up to 2020 was extrapolated. For 2020, a recharge value of 5.0 x10^7 m^3 was estimated, representing a 14% decrease in AAGR across the timeframe considered, 2012–2020 (Sappa and Ioanni 2013b).

6.7 Seawater Intrusion Assessment

The main parameters that indicate the groundwater salinization rate are electrical conductivity (EC) and chloride (Cl^-) concentration. EC depends on total salt concentration, which is the major quality factor generally limiting the use of saline waters for crop production. FAO (Rohades 1992) and WHO (2003) guidelines indicate that waters with EC values higher than 2000 and 1500 μS/cm, respectively, are not suitable for drinking or for irrigation. Results of the chemical analysis of groundwater in June 2012 (79 samples), showed that 25.3% of the samples were not suitable for drinking

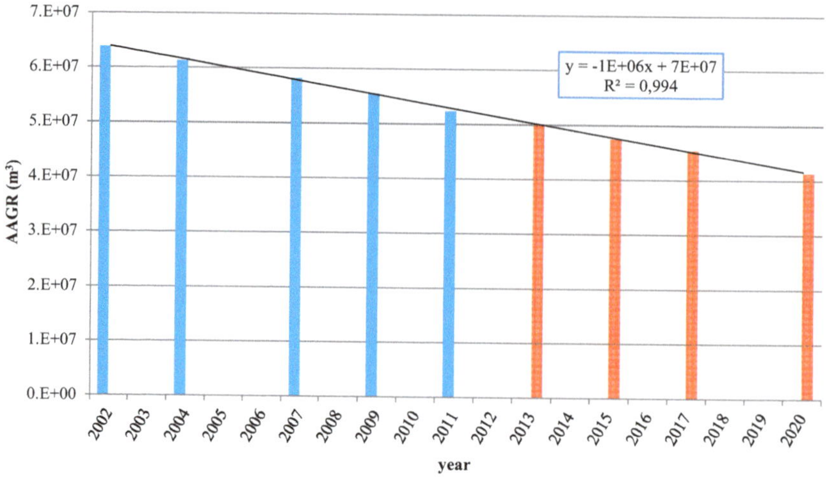

Figure 6.7: Estimate of temporal evolution of AAGR (m³) from 2012 to 2020.

or irrigation according to FAO guidelines for EC (36.7% according to the WHO classification).

The analysis of the temporal evolution of EC and chloride concentration in groundwater, in the period 2002–2012, is useful to assess the presence of seawater intrusion. In coastal aquifers an increase of EC can be attributed to different factors, such as seawater intrusion, the processes of dissolution of carbonate, the presence of anthropogenic pollutants, (i.e. nitrates, phosphates, sulphates) coming from civil activities or those related to agricultural and industrial sectors, and in some cases, by fossil saline water rising from the deeper layers of the aquifer. To assess whether increases in EC are related to seawater evolution, the parallel increase of chloride concentrations in groundwater were analyzed. Chloride is the most abundant ion in seawater (seawater contains approximately 35,000 mg/l of dissolved solids, which include about 19,000 mg/l of chloride, while fresh-water generally contains less than 30 mg/l). Moreover, chloride is chemically stable, as it is not involved in ionic exchange processes, and thus moves at about the same rate as intruding seawater. Therefore, an increased chloride concentration in a freshwater aquifer is the most commonly used chemical indicator of seawater mixing with freshwater. The analysis of the evolution of chemical composition of groundwater was carried out for boreholes where measures from different years were available in the period 2001–2012. The total number of wells available for this type of processing was 15, 10 of which were in the Kinondoni area, 2 in Ilala, 3 in Temeke.

Boreholes with the highest percentage increases in EC also had the highest percentage increases in chloride content. Boreholes K08 and K14, which are in the northern coastal area of Kinondoni, in the wards of Kunduchi and Kawe, respectively, showed the greatest percentage increase in EC. These areas also had high percent-

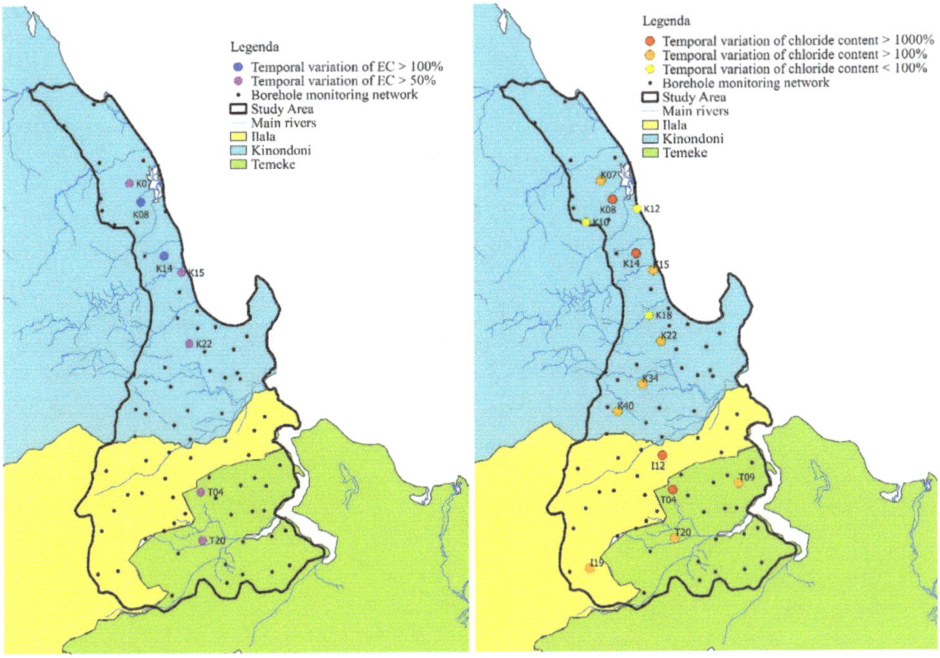

a) b)

Figure 6.8: Temporal evolution of EC (a) and chloride content (b) of groundwater in the period 2003–2012.

age variations in chloride content in groundwater, with slightly higher values in K22 (Mikocheni), K34 (Sinza) and in K40 (Mabibo). Other areas with high increases of chloride content were in the wards of Buguruni in Ilala (I12) and Chang'hombe in Temeke (T04). Moreover, areas affected by sensitive increases of chloride concentrations in groundwater are, from the north to the south, Miburani (T09), Yombo Vituka (T20) and Kitunda (I19).

In order to verify results from the previous analysis, the following diagrams were created. They show the comparison between ions mainly of marine origin ($Cl + SO_4$) and ions mainly of continental origin ($Ca + Mg$), expressed in milliequivalents, obtained from boreholes during in the period 2001–2012, as these were the available measurements relating to more than one year. In the following diagrams, the evolution of ($Ca+Mg$) and ($Cl+SO_4$) content in the boreholes is described. The larger symbols represent more recent data. All the graphs, particularly those referring to the area of Kinondoni, currently show higher concentrations of marine ions than those observed in the previous decade.

On the other hand, groundwater recharge mainly occurs during the long rainy season (March to May) and to a lesser extent during the short rainy season, i.e. between October and December, as shown by the analysis of average monthly values of precipitation registered at JNIA station during the last 50 years, which is represented in

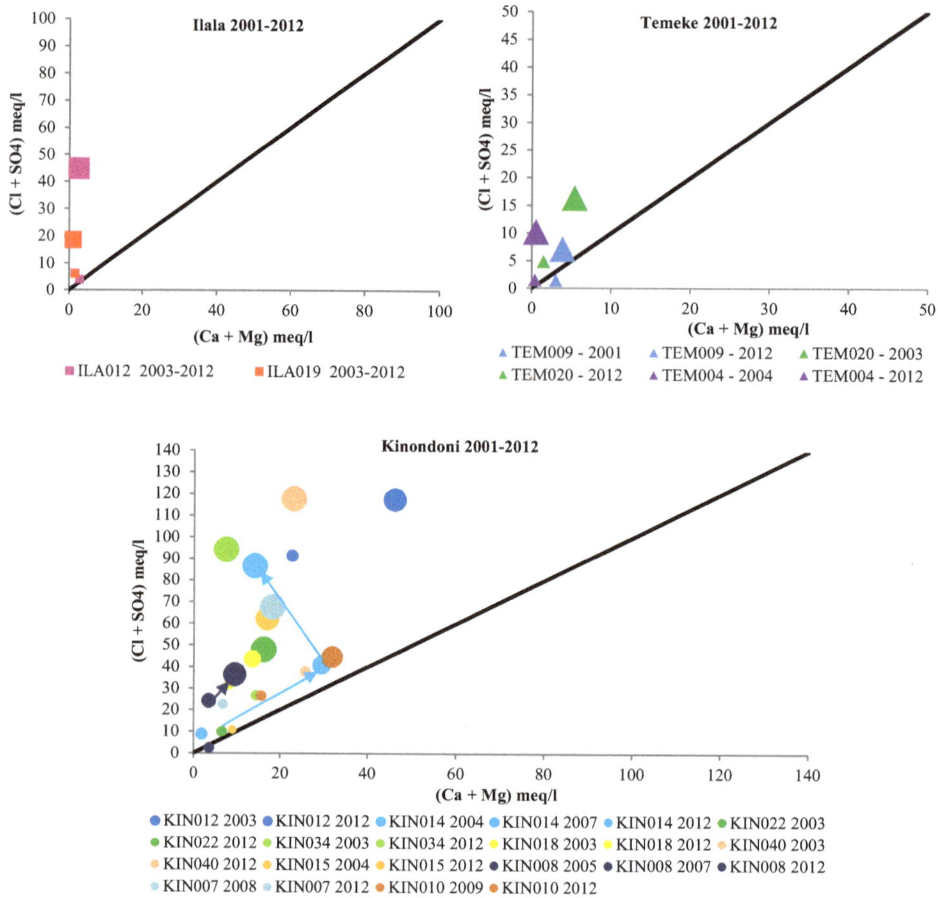

Figure 6.9: Diagram of (Ca+Mg) vs. (Cl+SO_4) for the three districts for the period 2003–2012.

Figure 6.11. Recharge during the long rainy season accounts for about 85% of the total annual recharge, with an important peak in April, whereas recharge occurring during the short rainy season contributes to only 15% of the total annual recharge, with a small peak in November (Mtoni *et al.* 2012).

Based on monitoring done in June and November 2012, a geochemical analysis enabled the identification of some features and some seasonal variations of groundwater chemistry, which are a possible indicator of the presence of various processes underway in the area. Chemical analysis were carried out on 79 samples collected in June 2012 (representing the end of the main wet season) and on 71 samples collected in November 2012 (representing the peak of the minor wet season). The interpretation was based on statistical parameters of principal ions present in groundwater, water type classification according to Stuyfzand classification system (1989) and scatter diagrams of principal ions.

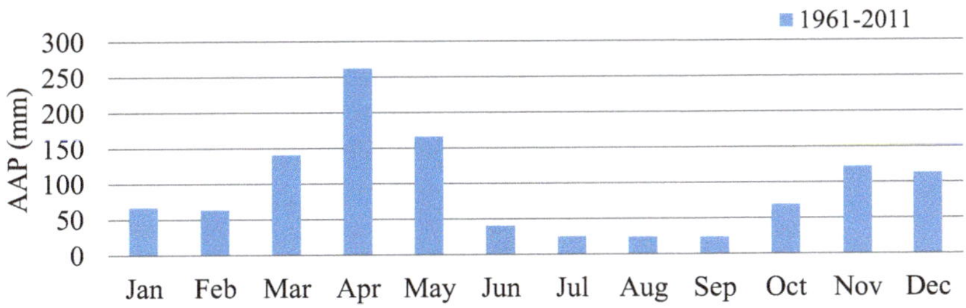

Figure 6.10: Average monthly rainfall for Dar es Salaam.

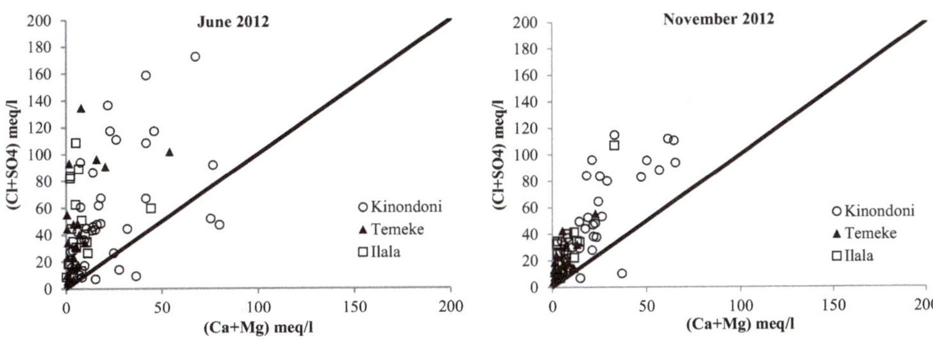

Figure 6.11: Diagram of (Ca+Mg) vs. ($Cl+SO_4$)

Stuyfzand classification subdivides water types in four steps, including chloride content, alkalinity, dominant ions and the parameter Base Exchange Index (BEX) (Stuyfzand 1989), which is:

BEX = (Na+K+Mg)corrected = (Na+K+Mg)measured – 1,061Cl (meq/l)

This parameter represents the difference between principal marine cations, which are found in the sample, and the expected values of these for seawater. Seawater intrusion leads to negative values of the BEX index (i.e. samples of class 1, according to the classification represented in Table 6.4). The definition of the BEX index is based on the description of the ionic exchange process that occurs when seawater intrudes in a fresh-water aquifer, which is described by the following reaction, where a,b,c,d represent molecular concentrations, with the balance in milliequivalents: 2a = b+c+2d:

aCa^{2+} + [bNa^{+},cK^{+},dMg^{2+}]-EXCHANGER ↔[aCa^{2+}]-EXCHANGER + bNa^{+} + cK^{+} + Mg^{2+}

In the early stages of seawater intrusion, an ion exchange process generally takes place. As such, when salt water intrudes an aquifer containing fresh water, sodium replaces calcium on the aquifer's clay particles through ion exchange before significant chloride increases are observed. This involves a decrease in Na^{+} concentration

Table 6.4: Stuyfzand classification.

Stuyfzand classification			
Main type	Cl^- (mg/l)	< 5	very oligohaline (G)
		5–30	oligohaline (g)
		30–150	fresh (F)
		150–300	fresh-brackish (f)
		300–1000	brackish (B)
		1000–10000	brackish-salt (b)
		10000–20000	salt (S)
		> 20000	hyperhaline (H)
Type	Alk (meq/l)	< 1/2	very low (*)
		1/2–1	low (0)
		1–2	moderately low (1)
		2–4	moderate (2)
		4–8	moderately high (3)
		8–16	high (4)
		16–32	very high (5)
		32–64	extremely high (6)
		64–128	extremely high (7)
		128–256	extremely high (8)
		> 256	extremely high (9)
$(Na+K+Mg)_{CORRECTED}$ (meq/l)			Class
$(Na+K+Mg)_{CORR} < -\sqrt{0,5Cl}$ and $(Na+K+Mg)_{CORR} < -1,5(\Sigma cat - \Sigma an)$			1
$\Sigma cat = \Sigma an$ and $\lvert(Na+K+Mg)_{CORR} + (\sqrt{0,5Cl})*(\Sigma cat - \Sigma an)/\lvert\Sigma cat - \Sigma an\rvert\rvert > \lvert 1,5(\Sigma cat - \Sigma an)\rvert$			2
$(Na+K+Mg)_{CORR} > \sqrt{0,5Cl}$ and $(Na+K+Mg)_{CORR} > 1,5(\Sigma cat - \Sigma an)$			3

in groundwater, and a parallel increase in Ca^{2+}. Conversely, when seawater is flushed by freshwater, most abundant ions in freshwater such as Ca^{2+} expel and replace the Na^+, absorbed in the solid matrix, and sodium is released into the water. Therefore a decrease in the molar ratio of sodium to chloride can be used as an indicator of seawater intrusion, while an increase in the ratio can be used as an indicator of flushing of the aquifer by freshwater (Stuyfzand 1992, Walraevens 2004).

Scatter diagrams (Ca+Mg) vs. ($Cl+SO_4$) allow the comparison between relative concentrations of typical marine origin ions and typical continental ions. In samples from both months of investigation, Cl and S04 ion content prevails over (Ca+Mg) content. This is a possible indicator of the presence of seawater. For boreholes of Ilala and Temeke, a predominance of marine ions is clearly indicated, while samples coming from Kinondoni boreholes appear dispersed and show high a content of Ca and Mg. The presence of water with these prevailing ions could be due to carbonate dissolution, related to calcareous formations outcropping, and ionic exchange processes. The boreholes of known depth, which show the higher concentrations of marine ions are K33, K39, K40, K34, T10, K12, K14, I13 and I15 in June, and K09, K06, K04, K33, K39, K40 and K14 in November.

Some of the ionic ratios are characteristic of the water types. As seawater is rich in Cl^- ions (chloride) and continental waters are rich in HCO_3^- (bicarbonate), a diagram of Cl vs. Cl/HCO_3 is used to evaluate the level of salinization. The values of the Cl/HCO_3 ratio given for fresh water are between 0.1 and 5, and for seawater between 200 and 500 (Custodio 1987). In the diagrams of Figure 6.12, the lower line represents a typical value of the Cl/HCO_3 ratio expected for fresh waters, which is 0.5 (Al Farrah and Walraevens 2011, Mtoni 2013). When fresh waters exhibit ratio values exceeding this, they have to be considered as being affected by salinization. The line at 6.6 represents a typical value indicating highly salinized waters (Al Farrah and Walraevens 2011, Mtoni 2013). Analysis of Cl vs. Cl/HCO_3 diagram (Figure 6.12) shows that samples from both months of investigation are all affected by salinization. Considering boreholes of a known depth, as in June and November 2012, the most salinized samples presenting with higher chloride concentrations are borehole KIN033, KIN034, KIN039, KIN040, all located in the center of the study area, and KIN006 and KIN014 in the northern coastal zone in Kinondoni. On the contrary, boreholes ILA013, ILA015 and KIN012 exhibit high salinization from the June investigation only, while boreholes KIN004 and KIN009 show high salinization only in November.

Other samples with high salinization, but with a lower content of chlorides are those from boreholes I03, I08, I09, I10, I11, I12, I20, in the SW of the study area

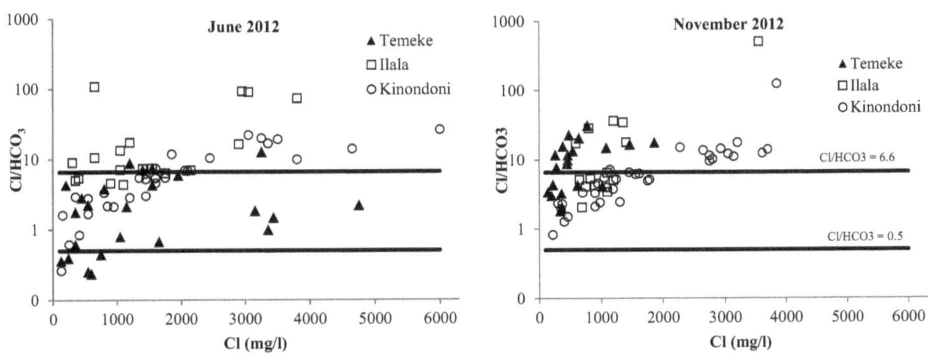

Figure 6.12: Diagram of Cl^- vs. (Cl^-/HCO_3^-)

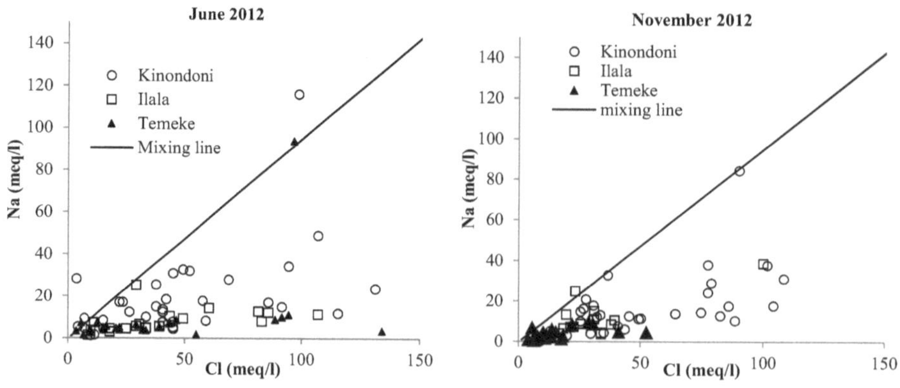

Figure 6.13: Diagram of Cl^- vs. Na^+

(respectively in wards of Tabata, Upanga West, Jangwani, Mabibo, Mchikichini, Buguruni and Yombo Vituka), and boreholes K07, K10 (wards of Kunduchi in the northern coastal area of Kinonondoni), K15 and K22.

In June, for boreholes in the northern coastal area of Kinondoni a possible trend in salinization can be identified, as salinization increases going toward the north, along the coastal strip: between the two lines in the diagram, from the bottom upward, with some exceptions, boreholes K23, K28, K17, K04, which are localized in an orderly way from the south to the north along the coastal strip, follow a growing trend of salinization. In June boreholes of Temeke present the minor salinization and chloride content. For these boreholes, low values of Cl/HCO_3 are due to high concentrations of HCO_3 ions. Borehole T10 exhibits high concentrations for all ions, which could be ascribed to associated anthropogenic contamination. In November all boreholes are affected by salinization.

As chloride can be considered like a conservative tracer, because it is not involved in the ionic exchange processes, it is representative of the proportion of seawater intruding into the aquifer. The analysis of graphic relationships between Cl^- and the other major ions, like Ca and Na, relative to the simple mixing lines, is useful in identifying additional processes, such as ionic exchange, annexed to the mixing phenomenon. A simple mixing line can be drawn in the diagrams, starting from the known composition of two extremes, freshwater and seawater. These dilution lines represent the theoretical gradual (linear) transition from one water-type to the other that occurs in the transition zones in coastal aquifers. If samples follow the dilution lines, the source of the plotted ions is fresh groundwater mixed with the seawater. The mixing lines that appear in the diagrams were constructed by the two extremes of freshwater and seawater collected in October 2013. The concentrations of major constituents present in the sampled waters were compared with corresponding values of dilution lines. In diagram of Cl vs. Na (Figure 6.13), almost all samples are under the dilution line, meaning that sodium concentrations are lower than the expected quantities in

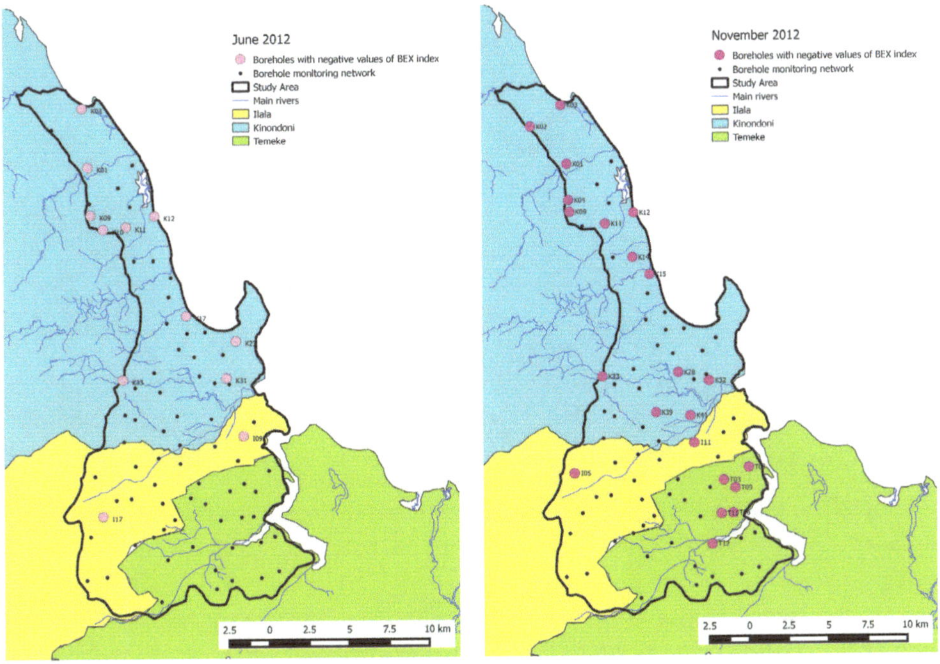

Figure 6.14: Boreholes with negative values of BEX index in June and November.

the presence of simple mixing with marine waters. This result is a possible indicator of the process of ionic exchange associated with seawater intrusion. For both months of investigation, samples from Kinondoni appear scattered, while the same samples of Ilala and Temeke seem more grouped along defined directions. It can be distinguished from the group corresponding to boreholes of the central area (K33, K34, K39, K40, with I13, I15 and T10), which presents sensitive deviations from dilution line. For the other samples, from boreholes for the most part from Ilala and Temeke and the rest of boreholes referred to Kinondoni, chloride concentrations are lower, but regardless they present deviations from the dilution line, indicating the presence of ionic exchange associated with seawater intrusion, and thus suggesting simple dilution of seawater into freshwater.

Concerning the Stuyfzand classification, analytical results show a slightly different distribution of samples in various classes for the two months of investigation. A decrease in samples classified as brackish-salt (72.7% in June and 60% in November) can be seen in November, but also a major number of samples with negative values of BEX index (21% in June, 33.3% in November), signifying an increase in the number of boreholes with signals of seawater intrusion (Sappa *et al.* 2013c). This occurs in many boreholes in Temeke's districts in particular, and in some boreholes of the southern districts of Kinondoni, in the centre of the study area.

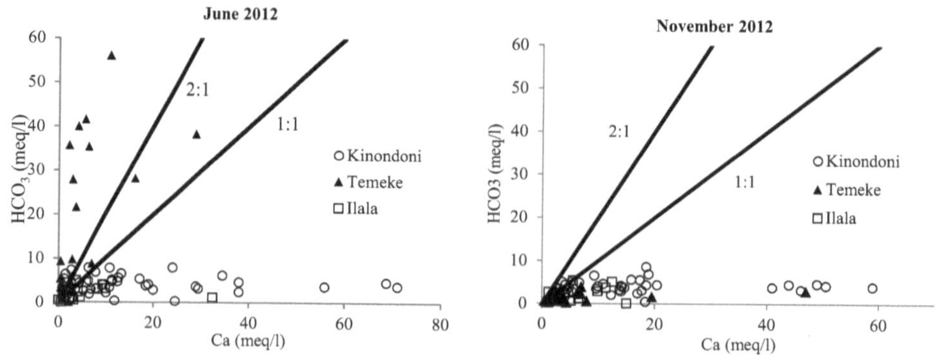

Figure 6.15: Diagram of Ca^{2+} vs. HCO_3^-

The quaternary deposits, underlying the coastal plain of Dar es Salaam also include coral reef limestone near the coast. Calcareous sandstones also occur on back reef areas of the uplands. Dissolution of carbonates releases calcium and bicarbonate into the groundwater; this process can be added to the other factors, determining a TDS increasing. Scatter diagrams of Ca vs. HCO_3 (Figure 6.15) have been used to show the influence of these processes on the chemistry of groundwater in the study area in both months. Lines in the diagram represent the relationship between $HCO3^-$ and Ca^{2+} corresponding to the following reaction:

$$CaCO_3 + CO_2 + H_2O \rightarrow Ca^{2+} + 2HCO_3^-$$
$$CaCO_3 + H^+ \rightarrow Ca^{2+} + HCO_3^-$$

The first one is the typical dissolution reaction of calcite by CO_2 in the unsaturated zone; the second one represents calcite dissolution by other acids apart from CO_2, such as humic acids and protons released from cation exchange (Mtoni *et al.* 2013). In the month of June, it can be seen that a first group of points is arranged according to a slope of less than 1:1, while a second group to a slope greater than 2:1. The first case represents samples from boreholes of Kinondoni, and indicates the presence of an additional process, which add up to dissolution of carbonates. This fact agrees with previous observation about depletion in Na, thus an excess of Ca can be explained as a consequence of both the dissolution of carbonates and the ionic exchange process associated with seawater intrusion. Boreholes of known depth, which present the highest values of Ca and Mg, are K04, K09, K10, K12, K33 and K39. The second group of samples aligns to a slope greater than 2:1, showing enrichment of groundwater in HCO_3. High values of HCO_3 from samples collected in the month of June can be explained as a consequence of the dissolution of carbonates associated with the meteoric recharge. The depth of most of these boreholes is unknown ($K_2$0, T07, T08, T09, T15, T23). Some boreholes of Ilala and Temeke with low concentrations of calcium, follow the dissolution lines.

6.8 Conclusions

The present study shows that AAGR has decreased over the past decade due to the combined effects of climate changes and anthropogenic causes. Although recent IPCC assessment reports have concluded that very little is known about the relationship between groundwater and CC (IPCC 2001, 2007, 2008), it is recognized however that CC usually acts as an effects multiplier in already altered hydrogeological systems. The impacts of CC generally lead to a decrease in groundwater quantity and quality. The changes in climatic conditions, which involve significant trends of rainfall decrease, from about 1200mm/year in the 1960s to about 1000mm/year in 2012, and mean temperature increases, have contributed to the decrease of local freshwater resources and an additional demand of groundwater (Sappa *et al.* 2014). Growing urbanization, socioeconomic development and growth of wellness and hygienic demands in cities of semi-arid areas can be considered as an indirect consequence of climate change. In fact, mean temperature rise and rainfall pattern changes due to CC, as agreed by most international climate studies' agencies, are major causes of degradation and unproductivity of cultivated land, and thus causes of abandonment of rural areas by populations, which relocate to big cities, increasing their social stresses and environmental impacts.

Regarding the anthropogenic causes of the AAGR decrease, they can be recognized as both fresh-water resources consumption growth and land cover changes. It has been possible to quantify the changes in land use, occurring in the period 2002–2012. These resulted in an increase of urban areas (from 40% to 65%) and a decrease of soil and vegetation areas (from 55% to 32%), which then considerably reduced the extent of potential infiltration areas.

The analysis of historical chemical data on groundwater composition provided a clear indication of the enlargement of the areas affected by salinization in the period 2001–2012. The origin of this salinization is difficult to determine because it depends on several factors, but the geochemical analysis performed allows us to highlight that a significant contribution to this phenomenon comes from seawater intrusion.

Geochemical analysis showed that chemical composition of groundwater of the most important aquifer of Dar es Salaam is influenced by the seasonal alternation of rainy periods, during which freshwater recharge occurs, and dry seasons, during which evaporation and excessive withdrawals for private water supply increased the degree of salinization of the aquifer. Anthropogenic pressure on groundwater resources follows the cyclical alternation of seasons and emphasizes seasonality effects. In fact, in the dry season, when most of the streams and canals that make up the surface water resources have low flow rates and soil has depleted its moisture content, the population has a greater need to use groundwater reserves to survive and sustain agriculture, and this results in a widespread and huge aquifer overexploitation.

Differences in groundwater chemistry in different areas could be due to both aquifer litology and land cover. Coral reef limestones are present in different locations along the coastal area. In Kinondoni, where calcareous rocks are found, most groundwater, in both seasons, contain calcium and magnesium.

Regarding land cover influence on different responses to seasonal weather patterns, it can be observed that in Kinondoni, which is the most populated amongst the districts and comprises residential areas provided with hygiene and social services, groundwater overexploitation and consequently salinization lasts all year. The same explanation probably can be applied in the area near to the city center, where the most salinized boreholes are found (K33, K39, K40, I13, I15). Moreover, these areas showed the biggest increase in EC and chloride content in the period 2001–2012. Conversely, in Temeke there are middle to low-income suburbs, characterized by poor settlement planning, low quality housing and social services, and areas used for urban agriculture. As such, urban land cover is less dense, and the effects of natural seasonal weather patterns on groundwater recharge are more evident, with the alternation of salinized and non-salinized composition, as highlighted by scatter diagrams of Cl/HCO_3 vs. chloride content, and by the different BEX index's spatial distribution in the two seasons. Regarding the fact that most salinized areas are relatively far from the coast, it can be said that higher salinity in boreholes of Temeke relatively far from the coast, is probably due to higher population densities and thus higher density of extraction wells, which is evidence of up-coning processes. Moreover, the highly salinized area of Temeke comprises the industrial district of the city, where the main manufacturing centers are located. This suggests that a further stress is applied on groundwater resources in this area. In addition, in Temeke most of the coastal area is covered by vegetation, and there are no boreholes analyzed in close proximity of the coastal area.

References

Al Farrah, N., K. Martens, and K. Walraevens. 2011. Hydrochemistry of the upper miocene-pliocene-quaternary aquifer complex of Jifarah plain, NW-Libya. *Geologica Belgica.*

Civita M. et al. 1999. Una metodologia GIS per la valutazione della ricarica attiva degli acquiferi. *Quaderni di Geologia Applicata* 1.

Congedo, L., M. Munafò, and S. Macchi. 2013. Investigating the relationship between land cover and vulnerability to climate change in Dar es Salaam. Working paper ACC DAR Sapienza University of Rome. http://www.planning4adaptation.eu/. Accessed 4 Dec 2015.

Custodio, E. 1987. Groundwater problems in coastal areas. A contribution to the International hydrological programme. Paris: UNESCO.

Rhoades, J.D, A. Kandiah, and A.M. Mashali. 1992. The use of saline waters for crop production. FAO irrigation and drainage paper 48. http://www.fao.org/docrep/T0667E/t0667e00.htm#Contents.

Mjemah, I.C. 2007. Hydrogeological and hydrogeochemical investigation of a coastal aquifer in Dar es Salaam, Tanzania. Ghent: Ghent University, Ph.D. thesis.

Mtoni, Y., I.C. Mjemah, C. Bakundukize, M. Van Camp, K. Martens, and K. Walraevwns. 2013. Saltwater intrusion and nitrate pollution in the coastal acquifer of Dar Es Salaam, Tanzania.

Mtoni, Y., I.C. Mjemah, K. Msindai, M. Van Camp, and K. Walraven. 2012. Saltwater intrusion in the Quaternary aquifer of the Dar es Salaam region, Tanzania.
Sappa G., A. Trotta, S. Vitale. in press. Climate change impact on groundwater active recharge in coastal plain of Dar es Salam (Tanzania). IAEG XII Congress, Turin, September 15–19, 2014.
Sappa, G., S. Vitale, M.T. Coviello, G. Faldi, and M. Rossi. 2013a. Analysis of the sensitivity to seawater intrusion of Dar es Salaam's coastal aquifer with regard to climate change. Working paper, Rome: Sapienza University: http://www.planning4adaptation.eu.
Sappa, G., F. Ferranti, S. Ergul, and G. Ioanni. 2013b, Evaluation of the groundwater active recharge trend in the coastal plain of Dar es Salaam (Tanzania). *Journal of Chemical and Pharmaceutical Research* 5(12): 548–552.
Sappa G., F. Ferranti, G. Luciani. 2013c. Effects of precipitations on groundwater salinization in Dar Es Salaam coastal plain (Tanzania), International conference on Frontiers of environment, energy and bioscience-ICFEEB 2013.
Stuyfzand, P.J. 1989. A new hydrochemical classification of water types. *J. Regional Characterization of Water Quality*, Proceedings of the Baltimore symposium, IAHS 182.
Stuyfzand, P.J. 1992. Behaviour of major and trace constituents in fresh and salt intrusion waters, in the western Netherlands. Proceedings 12th salt water intrusion meeting, Barcelona, nov.
Van Camp, M., I.C. Mjemah, N. Al Farrah, and K. Walraevens. 2012. Modeling approaches and strategies for data - scarce aquifers: example of the Dar es Salaam aquifer in Tanzania. *Hydrogeology Journal* 21(2) 341–356.
Van Camp, M., and K. Walraevens. 2004. Advances in understanding natural groundwater quality controls in coastal aquifers 18 SWIM. Cartagena, Spain: Ed. Araguàs, Custodio and Manzano.
World Health Organization. 2003. *WHO guidelines for drinking-water quality.*
World Meteorological Organization. 2011. *Guide to climatological practices.*

Jacques Piazzola, Gilles Tedeschi and Tathy Missamou[20]

7 Impact of the Construction of an Offshore Highway on the Sea-Spray Dynamics in La Reunion Island

Abstract: Sea-spray aerosols generated at the sea surface by breaking waves represent a major component of the atmospheric aerosol. In coastal areas, sea-spray particles are continuously produced by the breaking wave processes that occur near the shoreline. In La Reunion, a small island located in the Indian Ocean in the southern hemisphere, a road with an important traffic follows the northwest coast, which includes a cliff barrier. These cliffs, identified as a sensitive ecological space, undergo the influence of sea-sprays generated through breaking waves for prolonged periods of time. However, a new road project, which should be built offshore on the northwest coast of the island, raised some questions about possible impact of the increase of the distance between the hills and the sea-spray production sources on the introduction of exotic species. The present paper then proposes a study of the impact due to the new configuration of the shoreline on the sea-spray transport towards the hill barrier. To this end, simulations were made using a model for the marine aerosol transport in coastal areas, the model MACMod (Tedeschi and Piazzola 2011). This model was implemented on the study area on the basis of a survey of the aerosol size distributions measured on the La Reunion Island from July to September 2012. The influence of the distance to the shoreline on the local sea-spray dynamics was analyzed using a formulation for the sea-spray production in the surf zone through bathymetric processes. The transport calculations were then compared to aerosol data simultaneously measured at 10 meters height on the shoreline and at distance of 150 meters inland for cross-shore winds.

Keywords: Coastal aerosols, Sea-spray

7.1 Introduction

The role of atmospheric aerosols in climate change is well-recognized (IPCC 2013), but the estimation of their impact remains an important scientific challenge. In marine areas, the sea-sprays generated at the air-sea interface by wave breaking represent a major component of the natural aerosol mass (Jaenicke 1984; Andreae 1995; Yoon *et al.* 2007; Piazzola *et al.* 2009) and are therefore important in the Earth radiative budget (Laskin *et al.* 2003; Mallet *et al.* 2003; Mulcahy *et al.* 2008). A model of aerosol

20 Mediterranean Institute of Oceanography (MIO, UM n° 110)- University of Toulon, jacques.piazzola@mio.osupytheas.fr dispersion at local scale characteristics of the coastal area is important to a large number of applications linked to air quality and the survey of coastal ecosystems. Sea-spray results from primary production through breaking waves (e.g., Monahan *et al.* 1986) and also from secondary production issued from gas-particles conversion (e.g., Fitzgerald 1991). However, 92% of the aerosol mass deals with particles of radii larger than 0.5μm (Lewis and Schwartz 2004) which points out the importance of the primary production. Near the shoreline, the continuous production of sea-sprays, particularly through wave energy dissipation issued from bathymetric processes, can have a strong impact on a local climatology.

The island of La Reunion is a small island located in the Indian Ocean. On the northwest coast of the Island, a road connects the south of the island to the main urban zone of the Saint-Denis city, which is the economic center of the region. This road runs very closely along a hill barrier from which heavy rocks collapse several times a year, causing a large number of accidents and casualties. It was decided to move the automobile traffic further away from the hills. A new highway would then be built offshore, resulting in an increase of the distance between the cliffs lining the shoreline and the sea. These cliffs, identified as a sensitive ecological space, have been subjected the influence of sea-sprays generated through breaking waves for a long time, and questions about possible effects of the increase of the distance between the hills and the sea-spray production source, i.e. the sea surface on the development of intrusive exotic species, in particular at the foot of the cliff, have been raised. Some possible risks for the endemic flora are thought to result as the salt volume reaching the hill dramatically decreases. The present paper proposed a study of the impact of the new configuration of the coast on the sea-spray dynamics at a local scale. To this end, simulations were made using a model for the marine aerosol transport in coastal areas, the model MACMod (Tedeschi and Piazzola 2011). This model was implemented on the study area on the basis of a survey of the aerosol size distributions measured on the La Reunion Island from July to September 2012. The influence of the distance to the shoreline on the local sea-spray dynamics was analyzed using a formulation for the sea-spray production in the surf zone through bathymetric processes. This allowed for study on how marine aerosol transport towards the littoral zone is impacted by the distance between the shoreline and the coastal sites of a given height.

7.2 Field Site and Experiments

The experiments took place from the July 5th to September 2012 in the northwest coast of La Reunion island in the Indian Ocean (Figure 7.1a). The aim of the study was to predict the local transport of the sea-spray generated near the shoreline in the case of the construction of a new offshore road in the northwest coast (Figure 1b) which would result in an increase of the distance between the shoreline and cliff barriers lining the coast (Figure 7.2). To this end, measurements of aerosol size distributions

a)

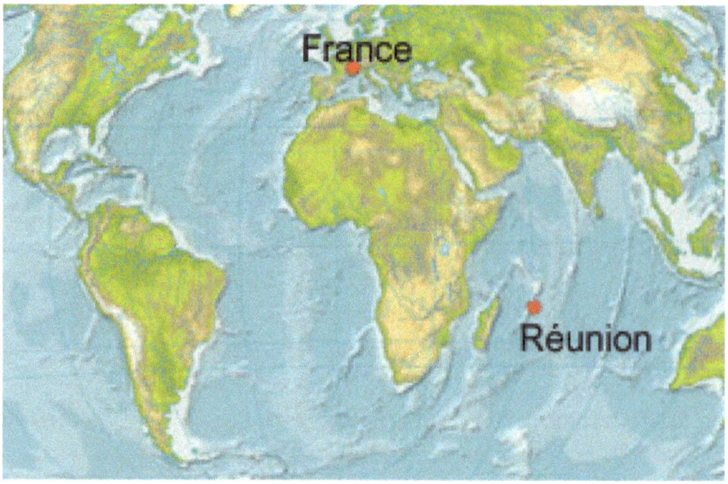

b)

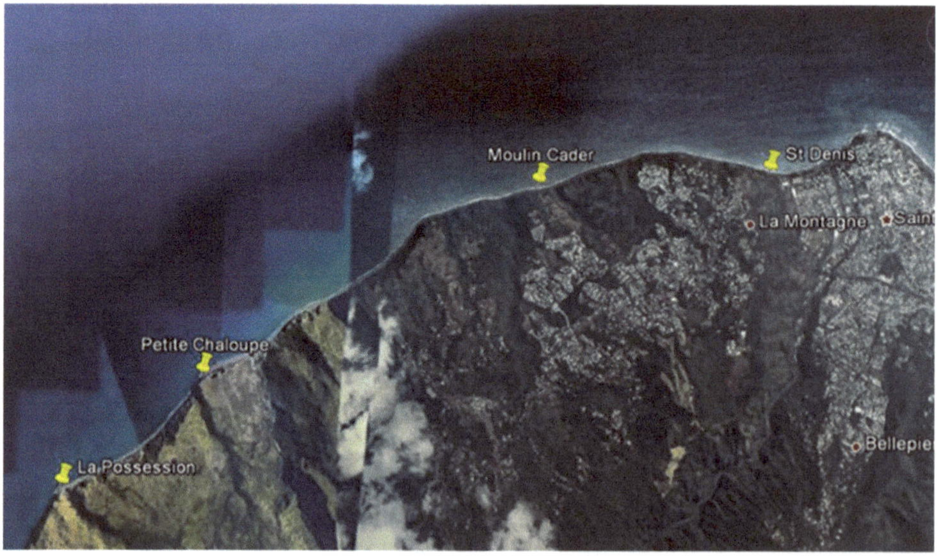

Figure 7.1: a) A map of the La Reunion Island located in the Indian Ocean; b) Zoom in on study area with the four measurement sites (yellow pins), i.e., "Saint-Denis," "Moulin Cader," "Petite Chaloupe" and "La Possession."

and meteorological parameters were taken at four different points along the northwest coast. The aerosol data were acquired on the shoreline lining the northwest coast of the island at each measurement station. On the shoreline, aerosol size distributions in the 0.1–95 µm size range were recorded using two Particle Measuring Systems

Figure 7.2: The project of the new highway which will be built offshore.

(PMS): the active scattering spectrometer probe (ASASP) and the classical scattering spectrometer probe (CSASP) located at an elevation of 10 metres above ground and located at 15 meters from the shoreline. The data accumulation period was 1 minute for the four probes, and the data were stored as the average over a 4 minute interval. Polynomial fits of 1st and 5th order in Log (dN/dr), the logarithm of the concentration versus the logarithm of the particle radius, Log (r) space were made for the distributions to facilitate the analysis. It should be noted that the PMS probes use a small inlet, which reduces losses during the sampling. The transport efficiency is considered optimal for aerosol sizes below 15 μm (Willeke and Baron 1993), which is the range of interest for the present study. Indeed, above this size, most of the sea-spray aerosols are mainly issued from the surface tearing processes which occurs for wind speeds larger than 10 ms^{-1} (Monahan *et al.* 1986), and we have not observed such wind conditions during Trade wind episodes in the study area. Prior to the experiments, the probes had been calibrated with latex particles of known sizes. For this type of optical counter, the calibration for sizing is relatively easy, thanks to calibrated latex spheres; however, the calibration for numbers remains a delicate operation and is more complicated to conduct. In addition, wave characteristics were recorded using a wave buoy that was moored one mile west off the northwest coast.

7.3 The Aerosol Transport Model, MACMod

The Marine Aerosol Concentration Model, MACMod, is a two-dimensional unsteady model developed to describe the evolution of aerosol concentrations in the marine area (Tedeschi and Piazzola 2011). The budget equation is integrated over a Cartesian grid (regular in the horizontal direction and stretched in the vertical direction), using the finite volume method (Patankar *et al.* 1972). Environmental data such as the wind velocity, friction velocity and air and sea temperatures, can be updated regularly with time (which is especially relevant when using a mesoscale meteorological model to drive MACMod). Subroutines were developed for each physical process, such as the aerosol source function, aerosol deposition on the marine surface, gravitational settling or turbulent dispersion. The user can select from various expressions to model the underlying physics. According to use, an empirical vertical profile of aerosol concentration can be applied as an input of the model following the work of Piazzola and Despiau (1997b).

7.4 Transport Calculations of Local Sea-sprays

Near the shoreline, the contribution of the bathymetric effects on aerosol production flux is generally substantial. In addition, when the waves reach the shoreline, they break and dissipate their energy. During Trade wind conditions, in the open ocean, the wind blows from the east to southeast direction in the southern hemisphere. The northwest coast of the La Reunion Island then constitutes the sheltered side of the island and hence, is exposed to rather low wave heights. This results in waves of low energy approaching the coast and implies a smaller local production of sea-spray aerosols due to the wind-wave interactions compared to open ocean. Close to the northwest coast of the island, the major contribution of the measured aerosol concentrations also deals with aerosols produced in the open ocean and then adverted into the study area. These particles tend to be dispersed by the wind during their atmospheric transport. This is confirmed by Figure 7.3 which shows the wind speed dependence of the aerosol concentrations of the 5 μm particles, as shown in Piazzola *et al.* (2013). The 5 μm particles are particularly representative for the marine contribution to the aerosols (Van Eijk and De Leeuw 1992). In accordance with the results reported in Piazzola *et al.* (2003), the negative slope of the line reported in Figure 7.3 clearly indicates that the local production of sea-sprays issued from breaking wave processes due to the wind wave interactions is negligible compared to the ones transported from remote ocean. This, by the way, confirms visual observations and hence we can consider a rather small local sea-spray production, except in the surf zone near the shoreline where the dissipation of the wave energy reaching the shore induces continuous foaming.

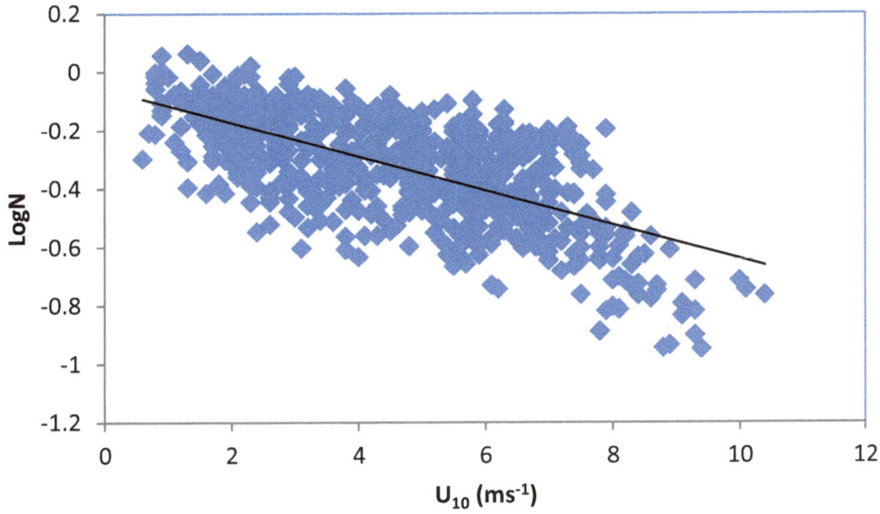

Figure 7.3: The wind speed dependence of the aerosol concentrations of the 5 μm particles.

7.4.1 The Local Sea-spray Flux

One of the study objectives is the determination of the local aerosol flux to be introduced on the transport calculations using MACMod. Our aim is to use a sea-spray production formulation which takes into account the wave propagation in the surf zone. For a soft bottom slope, the sea-spray production issued from the wave energy dissipation in the surf zone can be expressed by (Van Eijk *et al.* 2011):

$$\left(\frac{dF}{dD}\right)_{surf} = D^{c}.10^{\left[a(1-WED)^{b}\right]} \tag{1}$$

where $\left(\frac{dF}{dD}\right)_{surf}$ is the sea-spray flux, a = 10, b = −0.35 and c = −1.5, *D* is the particle diameter and *WED* is the wave energy dissipation.

As expected, the formulation reported in Eq. (1) depends on the wave energy dissipation, as noted *WED*, due to the breaking waves in the local zone of measurements. The model used for the wave energy dissipation (WED) near the shoreline is based on the work of Thornton and Guza (1983). The methodology is based on the energy conservation equation for random waves approaching a shore. The model equation expresses the budget between the energy flux of the incident waves and the dissipation terms, i.e., wave breaking and bottom friction (which will be neglected here). Among the different models based on this concept published in the literature, the work of Ruessink *et al.* (2003) provides a good description of the problem of plane slope beaches with an empirical improvement to the Battjes and Janssen

(1978) model using the a Rayleigh distribution of the wave spectrum as proposed by Baldock *et al.* (1998). They provided possible formulations for the energy balance for shore parallel contours. The local wave energy density and group velocity are usually given by linear water wave theory relationships, which have been proven to be remarkably accurate for narrow-banded waves with characteristic incident (in deep water) peak frequency $f_p = \omega p/2\pi$. We assumed that reflection of the wave energy can be neglected and that the cross-shore variations of the bottom profile occur on larger scales than that of the wave field. The bathymetry of the sea bottom near the shoreline of the four points investigated in the study area was determined using sonar techniques. The slope of the bottom was approached by a combination of linear equations. The wave height and period fixed to initialize the calculations were given by the wave buoy moored west off the northwest coast. Having the *WED*, the sea-spray flux was calculated for typical wave conditions on the northwest coast using Eq. (1).

7.4.2 Results

For the MACMod simulation, we needed to fix the length of the breaking zone. For the present calculations, except these particles produced along the shoreline, the atmosphere is considered initially as empty. For the calculations, we assumed that the aerosols produced on the shoreline were transported cross-shore on the flat land with small rugosity until the coastal sites, which constituted a hill barrier. Simulations were made for two different travelling distances of the particles from the shoreline to the coastal hills, i.e., 50 and 150 meters. These two values corresponded to the distance between the shoreline and the hill barrier of the northwest coast in the actual and future configuration of the coastline, respectively. Indeed, the new highway project will increase the distance between the shoreline and the cliffs up to 150 meters. The width of the breaking zone was chosen equal to 10 meters according to the observations made during the experimental period. Figure 7.4 shows vertical profiles of aerosol concentrations calculated using MACMod for a wind speed of 6 ms^{-1} and for particle radii of 1 and 5 µm. We can observe that when the distance between the hill barrier and the shoreline increases, the concentrations were smaller in the lower heights, but larger above the ten-fifteen meter height compared to simulations made for shorelines closer to the hill. This is probably due to the fact that, in the first case, the freshly produced particles have enough time to rise vertically until reaching greater heights during the transport at a local scale. This confirms that the role of particle loss through deposition processes was probably negligible in the decrease of the concentration in the first few meters for such short atmospheric transport, as predicted by the simulations reported in Figure 7.4. In contrast, the results reported in Figure 7.4 seem to show the importance of the vertical motion of the freshly produced aerosols.

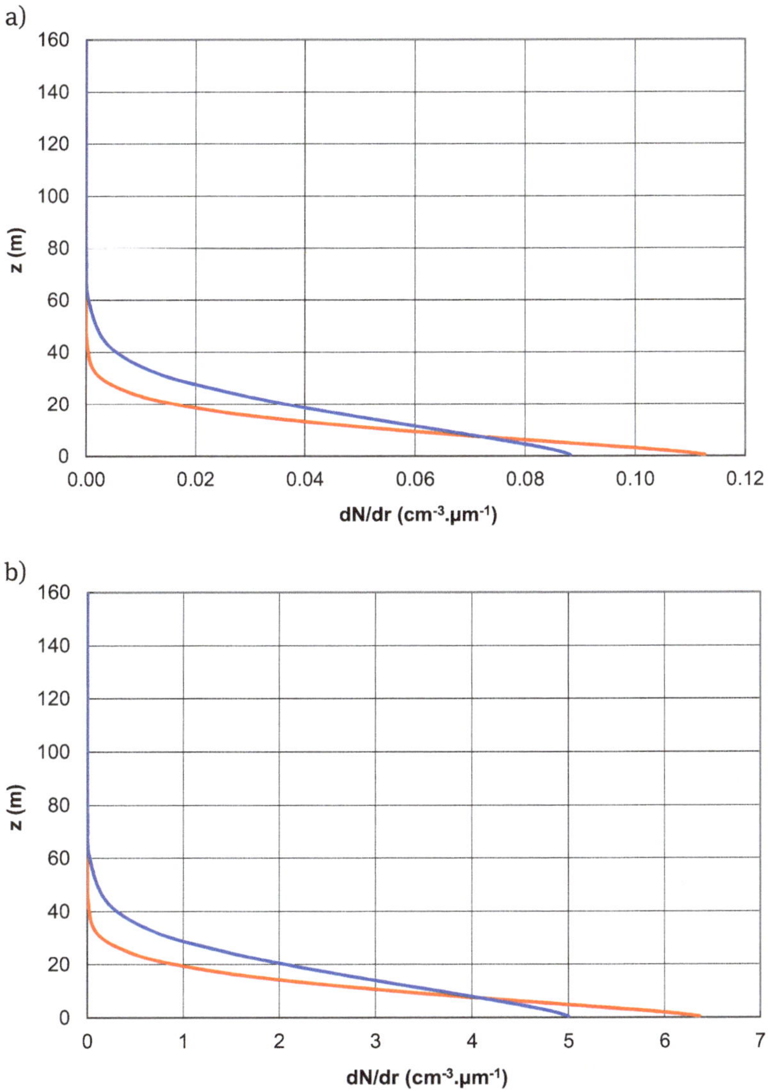

Figure 7.4: Vertical distribution of sea-spray concentrations along the vertical cliffs lining the northwest coast of the island for a local wind speed of 6 $ms^{-1}$and for different distances between the shoreline and the cliffs: 50 meters (the blue curve) and 150 meters (the red curve) for (a) 1 µm particles and (b) for 5 µm particles.

The averaged budget of the mass concentration of sea-spray aerosols (PM10 and PM20) in μgm^{-3} was calculated for the two configurations of shoreline and for Trade wind conditions. To this end, the aerosol concentrations were converted in mass concentration and integrated along the vertical axis using the following expression:

$$N_t = \frac{1}{L} \int_{z_{min}}^{z_{max}} \frac{dM}{dD} dz \qquad (2)$$

where N_t is the total aerosol concentration, dM/dD the mass concentration of diameter D, L the average height of the vertical axis, z_{min} and z_{max} the minimum altitude and the maximum altitude, respectively.

This was done for each experimental location and Table 7.1 shows an example corresponding to integration along the vertical axis of a hill of about 100 meters height. We can note in Table 7.1 that that the global mass concentrations of sea-spray aerosols that will reach the hills should increase after the construction of the new road.

Table 7.1: Averaged budget of the mass concentration of sea-spray aerosols (PM10 and PM20) in µgm^{-3} for the two configurations.

	Actual configuration	Project
PM10	456	475
PM20	789	833

7.5 Conclusion

The aim of this work was to study the impact of the increase of the distance between the cliff lining the northwest coast of La Reunion island and the shoreline in case of the construction of an offshore highway. To this end, a study of the local transport of the sea-spray particles produced near the shoreline was proposed on the basis of the implementation of the model MACMod on the study area. For source term characteristics of the local sea-surface production on the shoreline, we introduced a sea-spray source function characteristic of the surf zone proposed by Van Eijk *et al.* (2011) for low wave energy dissipation. The simulations clearly showed the importance of the distance between the source and the coastal site potentially impacted by the aerosols. In addition, the results allowed for a better understanding of the respective influences of the deposition processes and the vertical motion of aerosols as a function of their sizes. In particular, they show that the total concentration integrated along the vertical axis in the littoral zone is larger if the distance between the source and the site of impact increases.

7.6 Acknowledgements

The authors wish to express their gratitude to the staff of Nortekmed for their contribution to the experimental procedure and the DDE of St Denis for its administrative help.

References

Andreae, M.O. 1995. Climate effects of changing atmospheric aerosol levels. In *World survey of climatology. Vol. 16-Future climate of the World*, ed. A. Henderson-Sellers, 341–392.

Baldock, T.E, P. Holmes, S. Bunker, and P. Van Weert. 1998. Cross-shore hydrodynamics within an unsatured surf zone. *Coastal Engineering* 34: 173–196.

Battjes, J.A, and J.P.F.M Janssen. 1978. Energy loss and set-up due to breaking of random waves. In Proc. 16th Int. conf on Coastal engineering, ASCE, New York: 570–580.

Fitzgerald, J.W. 1991. Marine aerosols : a review. *Atmos. Environ*, 25°: 523–545.

Intergovernmental Panel on Climate Change. 2013. Working group I Contribution to the IPCC Fifth assessment report-climate change 2013: The physical science basis- Final draft underlying scientific-technical assessment, accepted but not approved in detail by the 12th Session of working group I and the 36th Session of the IPCC on 26 September 2013 in Stockholm, Sweden.

Jaenicke, R. 1984. Physical aspects of atmospheric aerosol. In *Aerosols and their climatic effects*, ed. H.E. Gerbard, and A. Deepak, 7–34.

Laskin, A., D.J. Gaspar, W. Wang, S.W. Hunt, J.P. Cowin, S.D. Colson, and B.J. Finlayson-Pitts. 2003. Science reactions at interfaces as a source of sulfate formation in sea-salt particles. *Science* 301 (5631): 340–344.

Lewis, R., and E. Schwartz. 2004. Sea salt aerosol production: mechanisms, methods, measurements and models-a critical review. American geophysical union.

Mallet, M., J.C. Roger, S. Despiau, O. Dubovik, and J.P. Putaud. 2003. Microphysical and optical properties of aerosol particles in urban zone during ESCOMPTE. *Atmospheric Research* 69 (1–2): 73–97.

Monahan, EC, D.E. Spiel, and K.L. Davidson. 1986. A model of marine aerosol generation via whitecaps and wave disruption. In *Oceanic whitecaps and their role in air-sea exchange processes*, ed. E.C. Monahan, and G. MacNiocaill, 167–174. Reidel Hingham MA.

Mulcahy, J.P., C.D. O'Dowd, S.G. Jennings, and D. Ceburnis. 2008. Significant enhancement of aerosol optical depth in marine air under high wind conditions. *Geophysical Research Letters* 35, L16810. doi: 1029/2008GL034303.

Patankar, S V, and D.B. Spalding. 1972. A calculation procedure for heat, mass and momentum transfer in three-dimensional parabolic flows. *Journal of Heat Mass Transfer* 15: 1787–1806.

Piazzola, J., G. Tedeschi, A. Dolle, E. Richard, and L. Vincent. 2013. Rapport final sur l'étude de la dynamique des particules d'aérosols sur la côte nord-ouest de l'ile de la Réunion, Direction de l'équipement Région Réunion.

Piazzola, J., P. Forget, C. Lafon, and S. Despiau. 2009. Spatial variation of sea-spray fluxes over a Mediterranean coastal zone using a sea state model. *Boundary Layer Meteorology* 132, 1: 167–183. doi: 10.1007/s10546-009-9386-2.

Piazzola, J., and S. Despiau. 1997a. Contribution of marine aerosols in the particle size distributions observed in Mediterranean coastal zone. *Atmospheric Environment* 31 (18): 2991–3009.

Piazzola, J., and S. Despiau. 1997b. Vertical distribution of aerosol particles near the air-sea interface in coastal zone. *Journal Aerosol Science* 28 (8), 1579–1599.

Ruessink, B.G, D.J.R. Walstra, and H.N Southgate. 2003. Calibration and verification of a parametric wave model on barred beaches. *Coastal Engineering* 48: 139–149.

Thornton, E.B, and R.T. Guza. 1983. Transformation of wave height distribution. *Journal of Geophysical Research* 88, C7: 9779–9789.

Tedeschi, G., and J. Piazzola. 2011. Development of a 2D marine aerosol transport model, application to the influence of thermal stability in the marine atmospheric boundary layer. *Atm. Res.* 101: 469–479.

van Eijk, A.M., J. Kusmierczyk-Michulec, J. T. Francius, M.J. Tedeschi, G. Piazzola, J. Merritt, D.L. Fontana, and J.D. Fontana. 2011. Sea-spray aerosol particles generated in the surf zone. *J. Geophys. Res.* 116, D19210. doi:10.1029/2011JD015602.

Van Eijk, A.M.J., and G. De Leeuw. 1992. Modeling aerosol particle size distributions over the North sea. *J. Geophys. Res.* 97, C9: 14417–14429.

Willeke, K., and P.A. Baron. 1993. *Aerosol measurement, principles techniques and applications.* New York: Van Nostrand Reinhold.

Yoon, Y.J., D. Ceburnis, F. Cavalli, O. Jourdan, J.P. Putaud, M.C. Facchini, S. Decesari, S. Fuzzi, K. Sellegri, S.G. Jennings, and C.D. O'Dowd. 2007. Seasonal characteristics of the physico-chemical properties of North Atlantic marine atmospheric aerosols. *Journal of Geophysical Research* 112, D04206. doi: 10.1029/2005JD007044.

Elena Cristofori[21,22], Adriana Albanese[21], Piero Boccardo[21,22]

8 Early Warning Systems & Geomatics: Value-added Information in the Absence of High Resolution Data

Abstract: Natural disasters such as flood or drought have a strong impact, especially in developing countries where most of world's population lives. Moreover, increased vulnerability is due to population growth in risk prone areas. Early warning systems able to address the risk of flooding with a global coverage may be particularly effective tools for humanitarian practitioners involved in emergency management. However these systems often use open-source precipitation data for which resolutions are not able to produce accurate alerts for taking action at a regional or at a local scale. Moreover, extreme precipitation information derived from numerical weather prediction models or satellites are often not integrated with spatial reference data. This makes it particularly difficult to geographically contextualize the warnings and therefore to provide an efficient support for risk decision-making.
This paper presents an automated system for the monitoring and forecasting of extreme rainfall events at global scales developed by ITHACA. The system is based on an open-source WebGIS platform which elaborates both satellite and numerical weather prediction precipitation data for the calculation of accumulated precipitation for different time frames. Accumulated precipitation values are compared with extreme precipitation thresholds in order to derive extreme precipitation alerts with a global coverage. The integration of extreme precipitation alerts with cartographic reference geographic data is used in order to generate added-value information before disseminating the final warnings to the end-users. This methodology has been applied during a service developed for the UN-WFP, and results are presented for the flooding event that affected Mozambique in January 2013.

Keywords: Extreme precipitation, Warning system, Flooding, GIS

8.1 Introduction

Scientists agree that during recent decades the trend in flood damage has been growing exponentially. This is a consequence of the increasing frequency of heavy rain, changes in upstream land-use and a continuously increasing concentration of population and assets in flood prone areas (WMO 2013).

21 ITHACA (Information Technology for Humanitarian Assistance, Cooperation and Action), Via Pier Carlo Boggio 61, 10138 – Torino, Italy, elena.cristofori@polito.it.
22 Politecnico di Torino, Corso Duca degli Abruzzi 24, 10129 – Torino, Italy.

Moreover the trend in urbanization is progressively accelerating, leading to increased exposure and vulnerability to extreme events. There has been almost a quintupling of the global urban population between 1950 and 2011 with the majority of that increase being in less developed regions (UN-Habitat 2011, IPCC 2012).

In general, less developed countries are the most vulnerable to floods, causing damage that significantly affects the national Gross Domestic Product (WMO 2013).

For overcoming these issues, the international communities agree on the importance of integrating Disaster Risk Reduction strategies into disaster management policies.

A number of plans describing in detail the work required from all different sectors and actors to reduce disaster losses have been proposed. In particular, the Hyogo Framework for Action (HFA) was developed and agreed upon with the many partners needed to reduce disaster risk (governments, international agencies, disaster experts and many others), bringing them into a common system of coordination by building the disaster resilience of nations and communities, with the main aim of reducing disaster losses by 2015. This means reducing loss of lives as well as social, economic, and environmental assets when hazards strike. The HFA outlines five priorities for action, and offers guiding principles and practical means for achieving disaster resilience.

In particular, among the five priorities for action, the second one is: "Identify, assess and monitor disaster risks and enhance early warning". This action highlights that disaster reduction strategies should take into account on the one hand, a deep knowledge of hazards and on the other hand, the physical, social, economic and environmental vulnerabilities to disasters that most societies face (ISDR 2007).

Early warning systems for floods plays a crucial role in Disaster Risk Reduction. Early Warning Systems providing alerts at global scales are especially helpful in producing an immediate overview of potential flooding hazards worldwide. This is particularly effective for developing countries where accurate systems enabling the timely anticipation of severe events at local or regional scales are often missing.

Data acquired from rain gauge stations are essential in order to perform the climatological analysis of extreme rainfall, and to provide accurate real-time hydro-meterological information during a severe weather event. While some regions of the world provide good coverage of ground rainfall data, many regions have inadequate or unavailable gauge infrastructures. Therefore precipitation data should be taken from available open-source data bases. Even though numerical weather prediction models and remote-sensing imagery have been demonstrated to be an invaluable source for precipitation information at global scale, these data generally have a coarse resolution ranging from 0.25° to 0.5°. As a result, the spatial resolution of extreme precipitation alerts is still too far from the requirements for flood risk management at regional or at a local scale. Geomatics and in particular Geographical Information Systems (GIS) provide essential spatial analysis capabilities for the identification of areas and main assets at risk. While Geomatics plays a crucial role in flood risk analysis and in post-disaster assessment (Boccardo 2013), the application of GIS technology to Numerical Weather Prediction models is still in an early stage.

The aim of this study is to propose a methodology for the integration of precipitation open-source data with reference geographical data-sets designed to produce meaningful value-added information addressing the risk of heavy rainfall and potential flooding to humanitarian practitioners who need to prioritize actions in areas where high resolution forecasting systems are not available. Some results are illustrated for the heavy rainfall event that affected the Limpopo river basin in Mozambique in January 2013.

8.2 Methods

The proposed semi-automated methodology relies on the use of the Extreme Rainfall Detection System (ERDS), a WebGIS application developed to provide exceptional near-real time rainfall alerts to WFP and to other humanitarian assistance organizations who need accessible information about flood hazards worldwide (Terzo *et al.* 2012). As well, the ERDS Near-Real System time has been upgraded by adding forecasting capabilities for extreme precipitation events with up to seven days lead-time.

The first step of the methodology is the calculation of the accumulated precipitation in order to produce both near-real time and forecasted quasi-global rainfall raster maps for different time frames (i.e. 24h, 48h, etc...).

The near-real time rainfall data processing uses the Tropical Rainfall Measuring Mission (TRMM) Multisatellite Precipitation Analysis (TMPA) 3B42RT dataset (Huffman *et al.* 2010). TRMM precipitation data are available in 0.25° spatial resolution, 3h temporal resolution, and covering the latitude band 50°N-50°S.

The forecast rainfall data processing uses the NOAA-GFS (Global Forecast System) model having a 0.5° resolution and a 6h temporal resolution at global scales. Firstly, TRMM 3h and GFS 6h rainfall datasets are processed in order to calculate daily accumulated precipitation matrices. Secondly, daily precipitation matrices are summed for the calculation of near-real time and forecasted accumulated precipitation for different periods of time, and up to 7 days lead-time. Finally, accumulated precipitation matrices are transformed into raster maps and displayed through the ERDS WebGIS application (Figure 8.1).

This allows the generation of particularly flexible information suitable for users interested in assessing both short-term heavy rainfall events and medium term persistent rainfall events (up to 14 days considering the aggregation of near-real time and forecasted precipitation data).

The second step of the methodology is devoted to the calculation of extreme rainfall alerts. This is done by applying rainfall thresholds to the accumulated precipitation values calculated before. Thresholds are defined as the amount of precipitation for a given duration over a specific climatological area. When thresholds

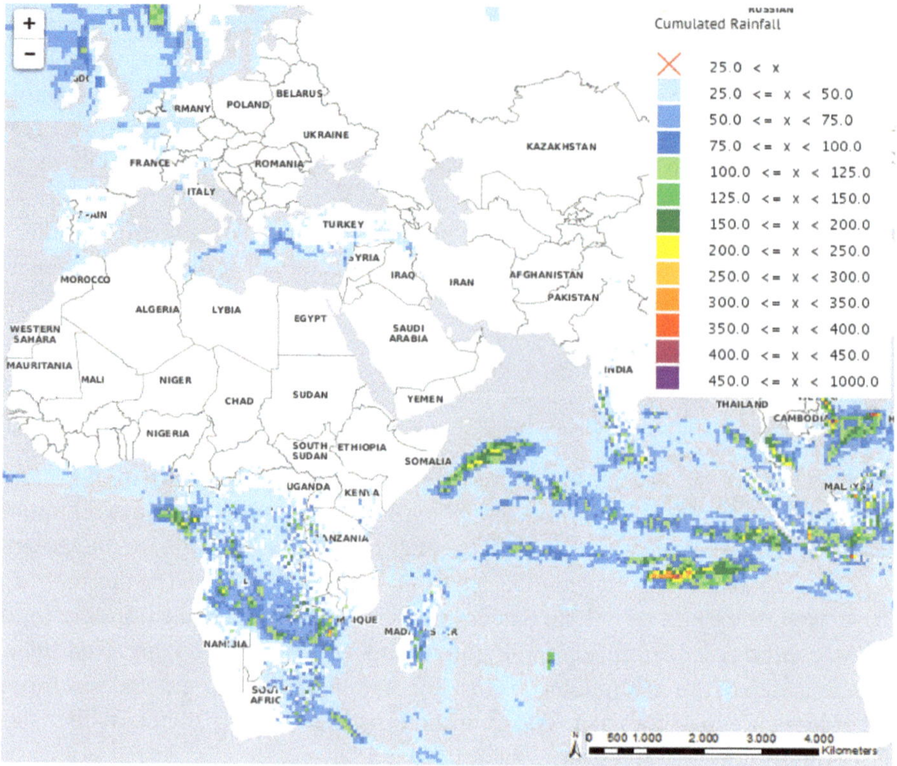

Figure 8.1: ERDS Web-GIS application, overview of 144h accumulated precipitation from GFS.

are exceeded, three alert levels (low, medium and severe) are produced. An accurate calculation of temporal and spatial distribution of climatological precipitation extremes is a key component in the generation of reliable heavy rainfall alerts (Fowler and Kilsby 2003; Ramesh and Teegavarapu 2012). On the other hand, meteorological information and warnings have traditionally been based on fixed thresholds for one parameter, often with a fixed time scale, e.g. precipitation rate over 24 hours. This traditional approach, currently adopted by a number of national and regional civil protection meteorological services, makes evaluation of the quality of forecasts easy and homogenous for a given area with climatically homogenous conditions (http://www.euroforecaster.org/newsletter11/Emma_2006.pdf).

Therefore two methods for the calculation of extreme rainfall alerts are applied in ERDS. The first one considers daily precipitation exceeding 80mm, 120mm and 150 mm as fixed thresholds for the three levels of alert (low medium and severe) and multiplies them by three coefficients (0.65, 0.8, 1) depending on the climatological classification derived by the new digital Köppen-Geiger world map on climate classification for the second half of the 20th century (Kottek *et al.* 2006). The map of the climatological coefficients that are used at global scales is presented in Figure 8.2.

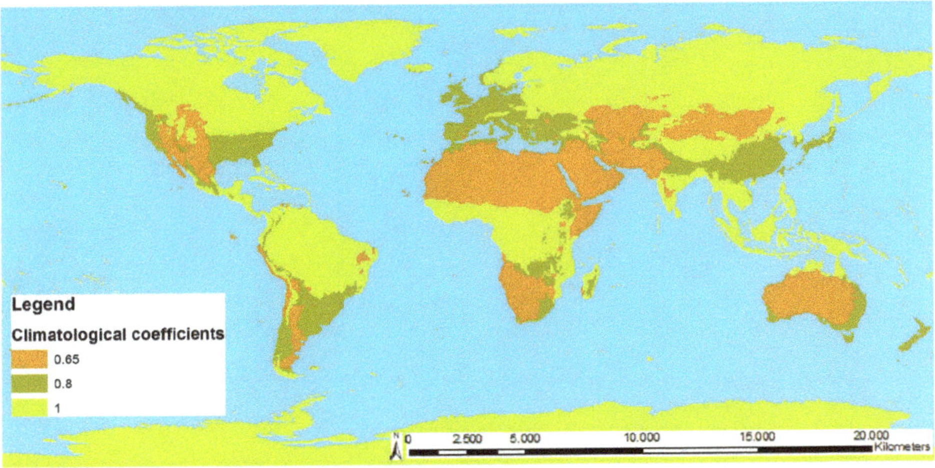

Figure 8.2: Climatological coefficients for the calculation of fixed rainfall thresholds.

The second method uses the climatological analysis of the TRMM-TMPA historical data set, which covers the period from 1998 to the delayed present (Huffman *et al.* 2007). For instance, the three levels of alerts are issued when the daily precipitation rate exceeds the 95th, 98th and 99th percentiles.

Warnings are disseminated through the ERDS WebGIS platform as raster maps displaying alerts at a pixel resolution (Figure 8.3).

As it was previously stated, the resolution of TRMM and GFSS data is not suitable for generating alerts with the level of accuracy required for a local scale risk assessment. Moreover, alerts displayed at pixel resolutions might not be meaningful for users who are not familiar with scientific information.

Therefore, the third step of the automated methodology illustrated in this paper consists of the integration of rainfall alerts with vulnerability data sets such as base map data, demographic data, etc. This allows for geographically contextualizing the event and producing value added information, such as an assessment of the damage grade of affected critical infrastructures, transport systems, affected population, aid and reconstruction logistics, etc.

Globally consistent datasets are generally available, even in the public domain: those data have the advantage of allowing the production of consistent output results, and therefore can be considered as a primary resource in the case of services intended to provide a wider and seamless overview of specific phenomena. Locally, more accurate and updated datasets may be available, but access to them is not granted, both due to technical and licensing issues. Other than the advantage of producing a more accurate value added analysis, the adoption of local and authoritative datasets has the positive consequence that post-event analysis and maps would more easily be integrated into the end users operational framework.

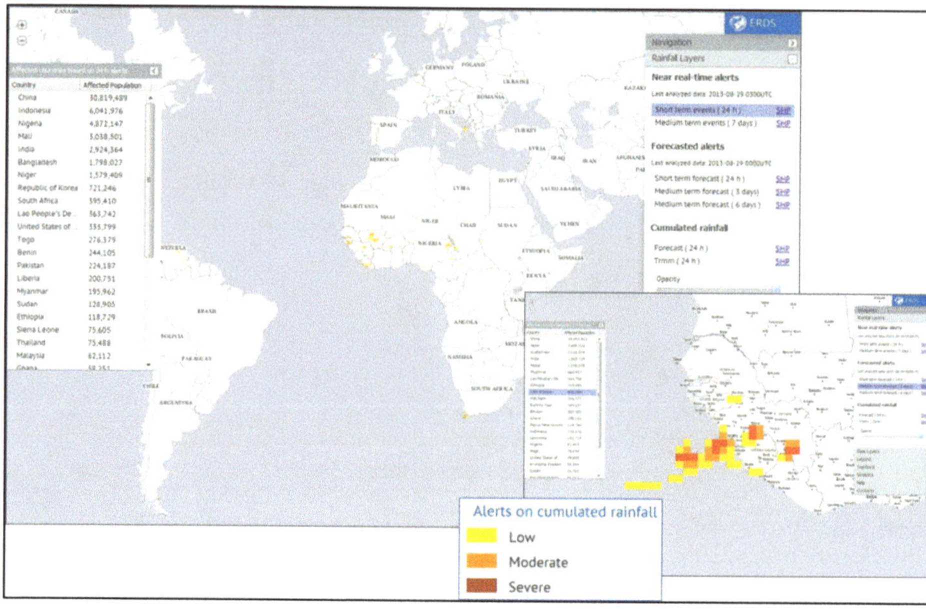

Figure 8.3: Example of ERDS warning map.

For the scope of the analysis described in this paper, the adoption of globally consistent reference datasets in order to estimate vulnerability was therefore considered the best option. In detail, the following data sources were used to derive vulnerability layers:

- Vector Map (VMAP) Level 0 is a vector-based collection of GIS data with global coverage (nominal scale 1/1.000.000) and entirely in the public domain. VMAP0 was used to derive major watercourses and administrative boundaries in Mozambique;
- LandScan ™ is a global population distribution dataset at approximately 1 km resolution, representing ambient populations (average over 24 hours);
- GeoNames is a geographical database that covers all countries and contains over eight million 'placenames' that are available for download free of charge.

8.3 Results

The main results presented here are obtained from the analysis of the heavy rainfall event that affected southern Mozambique during the second week of January 2013. This event was officially classified as a disaster meeting the EM-DAT criteria (http://www.cred.be), on January 21, 2013. For instance a GLobal IDEntifier number (GLIDE), which is a globally common Unique ID code for disasters proposed by the Asian Disaster Reduction Centre (UNISDR 2004), was issued for the Mozambique

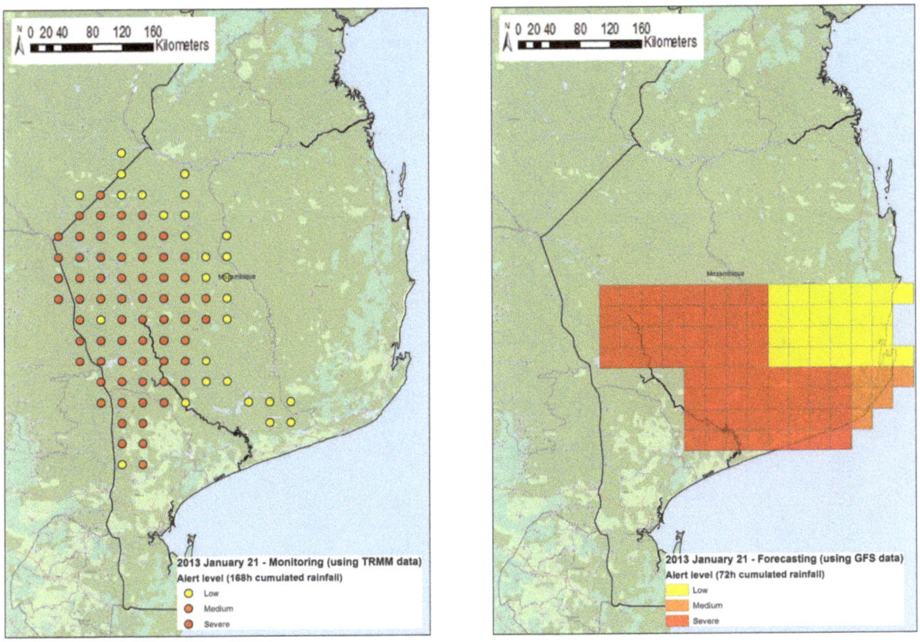

Figure 8.4: Rainfall situations as of January, 21th: a) On the left: alerts about accumulated rainfall over the past 7 days; b) on the right: alerts about accumulated forecasted rainfall for the next 3 days.

January 2013 event. The assigned code was FL-2013-000008-MOZ, composed of two letters to identify the disaster type (e.g. FL – Flood), the year of the disaster, a six-digit sequential disaster number and the three-letter ISO code for country of occurrence. Glide Number news reported that on January 12th the Mozambique authorities already declared an orange alert due to heavy rains that affected central and southern Mozambique since early January 2013. On 21 January, the population had already suffered losses and damage and small-scale flooding, resulting in nine deaths and affecting a total of 18,699 people throughout the country. 1,889 houses were destroyed, 985 damaged and 679 inundated.

The situation as of January 21st was assessed using the 1 week accumulated rainfall from TRMM. Figure 8.4 shows the extreme rainfall alerts derived from the comparison between accumulated rainfall and the fixed thresholds. Medium to severe extreme rainfall warnings were issued for the upstream areas of the Limpopo river basin (red and orange dotted alerts) based on the past 7 days of accumulated rainfall over these areas.

Figure 8.4 presents as well, on the right hand side, derived alerts comparing the GFS 72h accumulated rainfall with the fixed thresholds. Examination of this map shows that ERDS forecasted, for the following three days, heavy rainfall moving from the upstream area of Limpopo River towards the downstream areas. This resulted in an extended area of severe level alerts over south-western Mozambique.

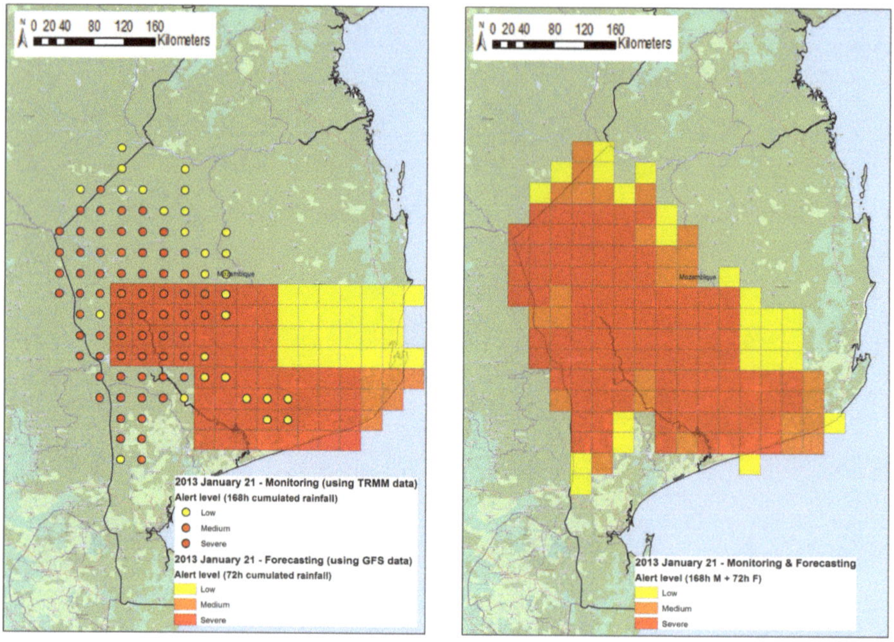

Figure 8.5: a) On the left: integration of the two different rainfall alerts, for monitoring and forecasting data; b) on the right: results of the integration.

In order to assess the effect of persistent rainfall over the same areas, alerts obtained from TRMM rainfall and alerts obtained from GFS rainfall were superimposed. New alerts were calculated based on the 1 week accumulated rainfall estimates, until the 21st of January, and on the forecasted rainfall until the 23rd of January. Results from this calculation are presented in Figure 8.5.

From the analysis of these maps it is evident that ERDS highlighted an extended area of severe alerts over southern Mozambique for the period 15th to 23rd of January. Since the severe alerts were spread on a very wide portion of territory, the results obtained were not easily usable by humanitarian practitioners who need to prioritize actions and areas of intervention. Therefore, rainfall alert maps were integrated with geographical reference data in order to identify the areas at highest risk.

The first analysis was made based on the idea that areas located in the proximity of major rivers have a higher risk of flooding. Therefore, an intersection of the rainfall alerts produced by the assimilation of monitoring and forecasting data with the main watercourses layer was performed. From the larger area of alerts identified before, only the areas at major risk of flooding were selected as criteria, considering the proximity to the Limpopo River (Figure 8.6).

Another important element to be considered is the population distribution, in order to make an assessment of the vulnerability in flood-prone areas. Using the LandScan layer, previously calculated alerts were derived, with coloured pixels as

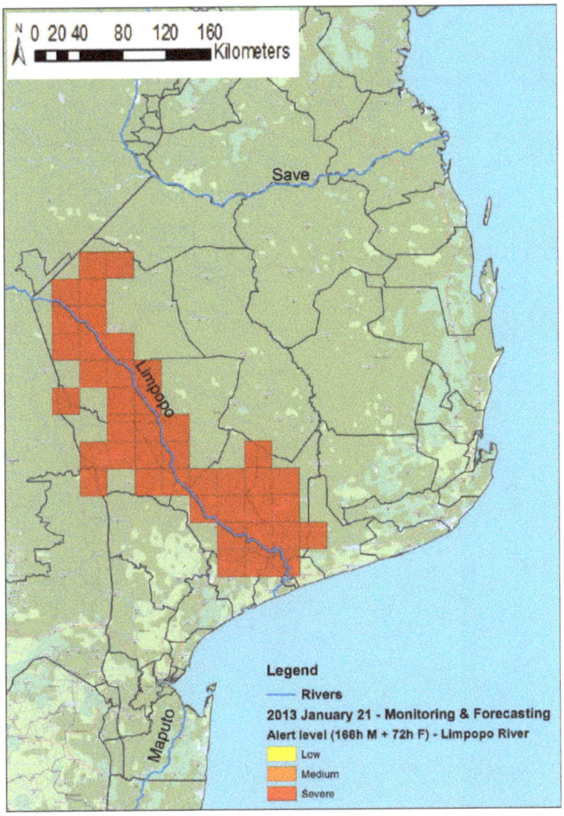

Figure 8.6: Selection of areas at major risk of flooding, obtained through the intersection of rainfall alerts and the main watercourses layer.

the criteria to represent the number of inhabitants living in each pixel. Moreover, thanks to the intersection with the GeoNames layers, the main cities located along the Limpopo River were superimposed onto the map. This allowed an immediate cross-reference between heavy rainfall information and land-marks.

The final map is presented in Figure 8.7 and highlights the higher risk areas as the ones where the most important cities are located and hence were a larger number of inhabitants are living. These areas are located along the Limpopo river, and in particular, downstream its course (pink-violet pixels).

As a final step, after having identified the main risk area, a flood-extent scenario was associated with the rainfall alert map. This is particularly useful for risk managers, in order to have information about the distribution of the past flood event effects and consequently to provide an assessment of the more probable floodable areas in case of an extreme rainfall event over the area identified in the map. In this example, the flood scenario was produced by processing the historical datasets of MODIS data (Figure 8.8) (Ajmar et al. 2010, 2012).

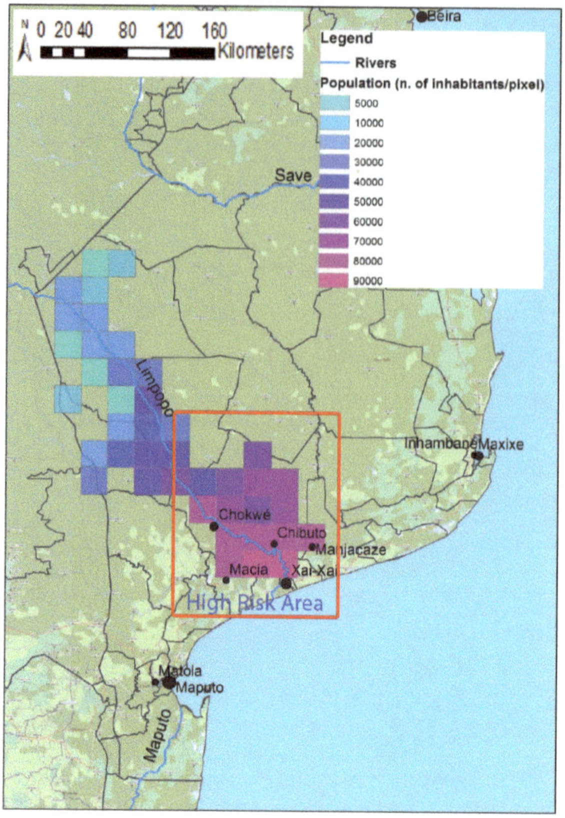

Figure 8.7: Integration of extreme precipitation alerts with LandScan and GeoNames layers.

The map produced by the United Nations Office for the Coordination of Humanitarian Affairs (OCHA) related to the Mozambique Provinces affected from 12 to 31 January 2013 (Figure 8.8), shows that, considering the southern Mozambique, the most affected area was the Gaza province. This is in fairly close agreement with the alert map obtained by the combination of TRMM and GFS rainfall alerts (Figure 8.4).

Moreover, in the report issued on 29th January 2013 by Reliefweb, it was stated that *'Multiple villages including the city of Chokwe are completely inundated by flood waters along the river. However flood waters appear to be receding in this area as they move south of Chokwe. Flood waters have reached as far as Xai Xai south of Chokwe city. East of the Limpopo River additional villages are inundated as flood waters progress south east from the river. Large sections of road and railroad are also affected.'*

This analysis strongly confirms that the most affected areas, within the Gaza province, were the ones around Chokwe and the east side of the Limpopo River as highlighted in Figure 8.6.

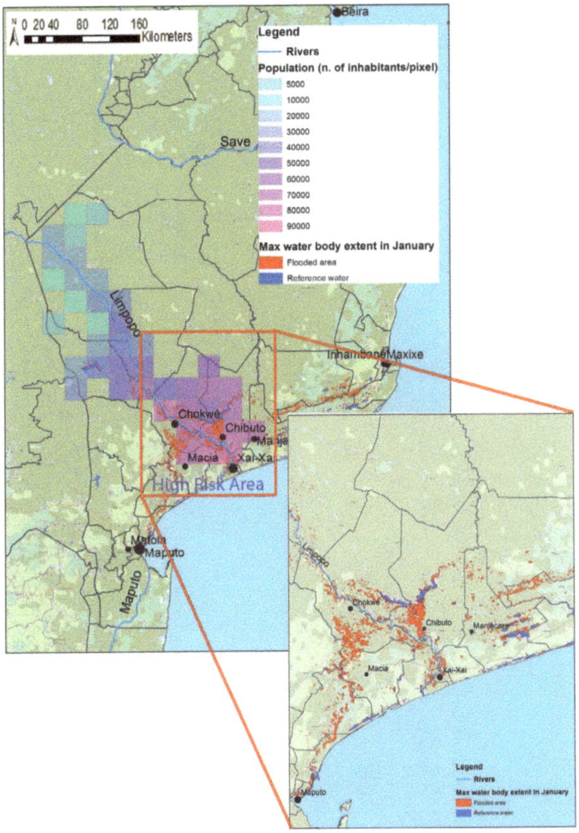

Figure 8.8: High risk flood scenario, produced processing the historical datasets of MODIS data.

8.4 Conclusion

With the aim of providing extreme rainfall and potential flood warnings for developing countries, a methodology that uses open source data at global scales has been illustrated. Firstly, near-real time and forecasted accumulated precipitation were calculated, starting from the most common open-source remote sensing and numerical weather prediction models rainfall data-set (TRMM, GFS). Then, extreme rainfall warnings were calculated by comparing the accumulated rainfall values with climatological rainfall thresholds. Moreover, the combination of past rainfall values with forecasted rainfall data was used to give information related to the persistence of heavy rainfall over the same area.

It should be noted that the use of extreme rainfall data for generating warnings about potential flood events may produce risk information that is not meaningful for non-scientific users. For instance, a high-level warning over an area with low population may become a low-level alert, depending on the specific end-user priorities.

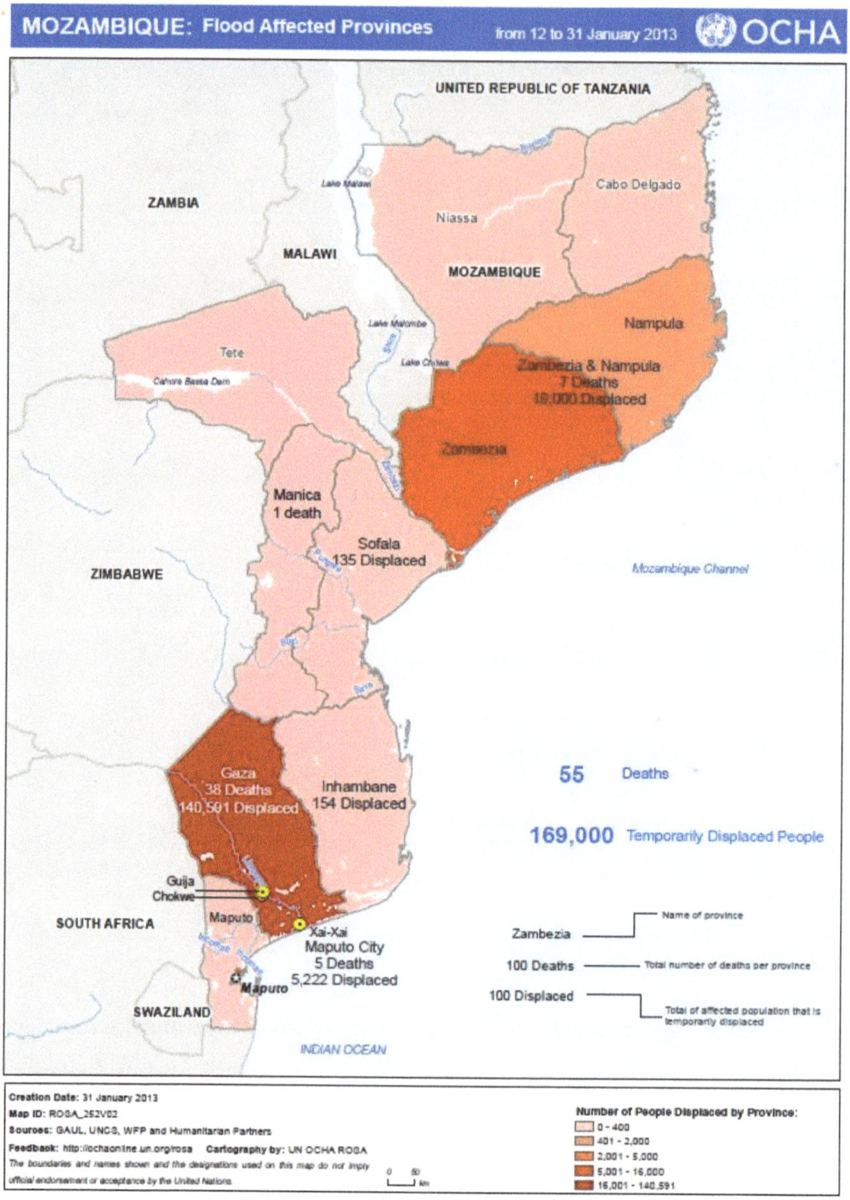

Figure 8.9: Map produced by OCHA, highlighting the Mozambique Flood Affected Provinces from 12 to 31 January 2013.

Hence, the integration of the above mentioned precipitation data with vulnerability data (e.g. reference data, global data sets of past flood events, etc.) is proposed.

A validation of this methodology for the case study of the Mozambique flooding event (January 2013) clearly demonstrated that this approach decreases the number of false alarms, generating value-added and event-specific information for areas at higher risk of flooding and main cities potentially affected, with a spatial scale that is relevant both at national and global levels.

References

Ajmar, A., P. Boccardo, F. Disabato, F. Giulio Tonolo. 2010. Near real time flood monitoring tool. Proceedings of the Gi4DM 2010 Conference. Turin, February 2–4, 2010. ISBN: 9788890313233.

Ajmar, A., A. Albanese, F. Disabato, F. Miotto. 2012. Integration of satellite and rainfall data for the identification of flood events in developing countries. *Italian Journal of Remote Sensing* 44(1): 167–180. ISSN: 2039-7879. doi: 10.5721/ItJRS201244113.

Boccardo, P. 2013. New perspectives in emergency mapping. *European Journal of Remote Sensing* 46: 571–582. doi: 10.5721/EuJRS20134633.

Fowler, H.J., and C.G. Kilsby. 2003. A regional frequency analysis of United Kingdom extreme rainfall from 1961 to 2000. *Int. J. Climatol.* 23: 1313–1334. doi: 10.1002/joc.943.

Huffman, G.J., R.F. Adler, D.T. Bolvin, G. Gu, E.J. Nelkin, K.P. Bowman, E.F. Stocker, D.B. Wolff. 2007. The TRMM Multi-satellite Precipitation Analysis: Quasi-Global, Multi-Year, Combined-Sensor Precipitation Estimates at Fine Scale. *J. Hydrometeor.* 8, 33–55.

Huffman, G.J., R.F. Adler, D.T. Bolvin, E.J. Nelkin. 2010. The TRMM multi-satellite precipitation analysis (TMPA). In *Satellite rainfall applications for surface hydrology*, ed. F. Hossain, and M. Gebremichael, 3–22. Springer Verlag. ISBN: 978-90-481-2914-0.

IPCC-Intergovernmental Panel on Climate Change. 2012. Managing the risks of extreme events and disasters to advance climate change adaptation. Special report of the Intergovernmental Panel on Climate Change. doi: 10.1017/CBO9781139177245.

ISDR-International Strategy for Disaster Reduction. 2007. Hyogo framework for action 2005–2015: Building the resilience of nations and communities to disasters. Extract from the final report of the World conference on disaster reduction (A/CONF.206/6), 1–28. Geneva: UNISDR.

Kottek, M., J. Grieser, C. Beck, B. Rudolf, and F. Rubel. 2006. World Map of the Köppen-Geiger climate classification updated. *Meteorol. Z.* 15, 259–263. doi: 10.1127/0941-2948/2006/0130.

Ramesh, S.V., and Teegavarapu. 2012. Floods in a changing climate: Extreme precipitation. Cambridge: Cambridge Univ Press. ISBN: 9781107018785.

Terzo, O., L. Mossucca, A. Albanese, P. Boccardo, N.P. Premachandra, and R. Vigna. 2012. Engineering of an extreme rainfall detection system using grid computing. *Systemics, Cybernetics and Informatics* 10, 5: 45–49. ISSN: 1690-4524.

UN-Habitat. 2006. State of the World's cities 2006/2007. Nairobi: United Nations Human Settlements Programme.

UNISDR. 2004. *Living with risk: a global review of disaster reduction initiatives*. 1. ISSN: 9211010640.

WMO-World Meteorological Organization. 2013. WMO disaster risk reduction work plan (2012–2015) updated as 26 September 2013, approved by 64th session of WMO Executive council (25 June – 3 July 2012). Geneva.

Enrico Ponte[23]

9 Planning the Adaptation of Coastal Cities to Climate Change: a Review of 14 Pilot Projects

Abstract: The natural disasters deriving from climate change have a growing impact in urban contexts and mainly affect the poorest people. Adaptation to climate change can significantly reduce the impacts if planned. The aim of this chapter is to present the state of adaptation planning, illustrating three international projects that promote it (Climate change initiative, Cancun adaptation framework, Hyogo framework of action) and 14 adaptation pilot projects for coastal cities. The latter are plans, strategies and guidelines produced over the past 5 years for coastal cities from different continents. These case studies are analysed and compared, paying attention to six key aspects: type of tool, aims, identification of the impacts deriving from climate change, time schedule and budget, role of the stakeholders in the planning process, identification of specific measures. It emerges that important steps have been taken in the adaptation to climate change, but there is still a long way to go before having valid adaptation planning tools.

Keywords: Urban adaptation, Adaptation pilot projects, Climate change, Natural disasters, Coastal cities

9.1 Introduction

The lives and means contributing to the sustenance of hundreds of millions of people over the next 5–10 years will be affected by what is (or is not) done in cities with regard to climate change (Satterthwaite *et al.* 2007). Cities are the most responsible for greenhouse gases (GHG) and for strategies to reduce their production, especially to reduce the dependency on carbon-based fuels (Romero Lankao 2007).

In urban contexts, natural disasters have devastating effects, especially on the marginal populations of Developing Countries (Wamsler 2004). Almost a billion people live in poor quality dwellings with little access to water, storm drainage and sanitation (UN Habitat 2003) in cites of Africa, Asia, South America and the Caribbean. Until now, the commitment to reduce emissions of GHG has not prevented global warming and the consequent climate change (CC) (World Bank 2010).

23 Enrico Ponte, planner PhD, holds a scholarship at the DIST Politecnico and Università di Torino enrico.ponte@polito.it
The author would like to thank Maurizio Tiepolo (DIST-Politecnico di Torino) for his suggestions while writing the chapter.

It becomes, therefore, a priority to understand which strategies to use in order to reduce the impacts of CC (Tiepolo 2014a). Among the latter, we ought to mention the increase in natural disasters. Without an effective adaptation on a local scale in the years to come, CC will represent one of the most important challenges of the century.

Developing Countries and communities may be at risk of sustaining severe damages due to their geographic position, low income and insufficient operational capacity by local administrations, as well as the major dependence of farming on climate conditions.

This chapter presents the approaches and problems of planning the main strategies for the prevention and reduction of CC impacts promoted by the major United Nations organisations over the last 10 years. It then closely analyses 14 pilot projects (plans, strategies and guidelines) produced from 2009 to the present day, in order to adapt coastal cities to CC, with the aim of revealing the main positions on the matter of prevention and limitations. In selected cases, they differ in numerous aspects, particularly the social, economic and political context in which they were produced.

The concept is now more up to date than ever before but, paradoxically, to date, few comparative studies have been dedicated to planning instruments for adaptation to CC. In one of these (Birkmann 2010), a second generation of urban adaptation plans was recommended to overcome the critical points of plans produced in the mid-2000s.

This chapter intercepts the second generation plans to identify the main changes which have been made.

9.2 Action of the Multilateral Bodies to Favour the Prevention of Natural Disasters

The 2007 can be considered as the year in which the leading multilateral organizations became aware of the urgency to make decisions to reduce the urban impacts of extreme natural phenomena.

UN-Habitat dedicates the third volume of the "Global report on human settlements" to different ways to reduce the impacts of disasters in cities. The most important measures and proposals put forward include disaster risk assessment, land use planning, early warning and reconstruction (UN-Habitat 2007). CC is not presented as such, but is presumed in the various cases (tropic cyclones, heat waves, floods, rising sea levels). The consequences triggered by CC include migrations from affected areas into the cities, and the resulting increase of informal settlements from flood prone areas.

The fourth report of the IPCC states that there are now "impacts for which adaptation is the only available and appropriate response" (IPCC 2007). The urgency of tackling the problems emerges, therefore, without there being "a clear vision of the limits of adaptation, or of the costs, partly because the effective measures of adaptation depend largely on specific, geographic and climatic risk factors, as well as on institutional, political and financial impediments".

Over the years that followed, numerous initiatives for adaptation to CC were launched, three of which deserve particular attention.

9.2.1 Hyogo Framework for Action (HFA)

In 2005, the United Nations convened the second world conference on the reduction of disasters in Kobe, Japan. On that occasion, the delegations taking part drew up an international agreement to reduce the disaster risk. This agreement was called the Hyogo framework for action (HFA). The general aim was to build up the resilience of countries and communities to disasters by substantially reducing the losses (human, social, economic or environmental) caused by disasters by the year 2015. The HFA highlights five priorities of action, the guidelines and the tools to achieve disaster resilience for vulnerable communities within the scope of sustainable development (UNISDR 2005).

After the adoption of the HFA, numerous efforts at global, regional, national and local levels were undertaken for a better systemisation of the reduction of the risk of disasters.

The HFA has three strategic aims:

1. Integration of disaster prevention in sustainable development policies;
2. Development and strengthening of the institutions, mechanisms and capacities at all levels, particularly at community level, to build up the response to risks;
3. Inclusion of the risk reduction in preparing for emergencies and in programmes for the reconstruction of the communities affected.

There are five priorities of actions:

- to ensure that the reduction of risks and disasters is a national and local priority with a strong institutional base for implementation;
- to identify, assess and monitor the risks of disaster ad strengthen the pre-alarm devices;
- to use awareness, innovation and traininging to build a culture of safety and resilience at all levels;
- to reduce the risk factors;
- to strengthen the preparation for catastrophes.

HFA has the guidelines, priorities for action and means to achieve resilience to disasters in vulnerable communities. HFA clearly highlights that collaborative international cooperation is required to provide knowledge, skills and incentives for the DRR (UNISDR 2007).

At one year after its expiry, the balance is definitely positive on paper, in that 168 countries have signed this agreement, but it is still early to pass judgement on what will effectively emerge from HFA.

9.2.2 City and Climate Change Initiative (UN-HABITAT)

In August 2008, UN-HABITAT launched "City and climate change initiative" (CCCI), with the support of the Norwegian government. The aim was to tackle the effects of CC on cities in Developing Countries. The project began in four cities, including Kampala and Maputo in Africa. Subsequently it was extended to Beira, Bobo Djoulasso, Kigali, Mombasa, Saint Louis, Vilankulo and Walvis Bay (UN-Habitat 2012).

CCCI aims to strengthen the activities of preparation and disaster mitigation of cities in Least Developed Countries, focusing on practices of good governance, responsibility and tangible initiatives for local governments and communities, developing a series of tools in support of decision-makers and professionals to tackle the impact of CC (adaptation) and to contribute to reducing emissions of GHG (mitigation).

Thanks to this initiative, the role currently played by cities to launch initiatives of adaptation and mitigation is sufficiently acknowledged in debates on CC. Moreover, the CCCI helps small and medium towns in Developing Countries, which are largely ignored by the international financial institutions and other agencies.

The fact that the project focuses mainly on the poor living in cities and other vulnerable groups is an element that distinguishes CCCI from other CC programmes.

At local level, the initiative has been very effective in helping cities to undertake actions with regard to CC, helping the formulation of strategies and plans of action.

The programme has also succeeded in increasing the awareness by local authorities of the effects of CC, generating an understanding of the importance of launching immediate and concrete actions within normal planning practices. In cities, the participative approach of CCCI, which involves residents and professionals, led to greater awareness and a strong sense of belonging to the activities of the Initiative among the various parties concerned, including the basic level. However, the importance of paying closer attention to understand the social-economic impacts of interventions on local communities also emerged.

A key component of CCCI is undoubtedly to encourage national – local communications, involving the main stakeholders in CC, including financiers. The assessments at the national level clearly indicate that this communication is vital in order to create and institutionalise a broad and deep support for CC interventions at the city level.

Six years after its launch, numerous initiatives have been accomplished, particularly in Asia, where about 50 cities have been involved. At the end of this phase, however, knowledge on CC affecting the urban environment continues to be vague and approximate: studies on Kigali, Maputo and Mombasa are far from adequately characterising CC (Tiepolo 2014a).

9.2.3 Cancun Adaptation Framework

During the Conference on Climate Change held in Cancun in 2010, numerous nations signed the Cancun Adaptation Framework (CAF). The aim of the CAF is to improve adaptation, also through international cooperation and the consistent consideration of matters relating to adaptation. In practice, the strengthening of adaptation actions aims to reduce vulnerability and to build up resilience in Developing Countries, taking into account their urgent needs.

The CAF is split into five groups (UNFCCC 2010):

1. Implementation:
   - Plans, priorities and interventions for adaptation through existing channels to provide information on the support provided;
   - Process for putting the administrations of Developing Countries in position to formulate and implement national adaptation plans, and to invite other countries to use the methods formulated to sustain such plans;
   - Programme of work which considers approaches to tackle losses and damages associated with the impacts of climate change.
2. Sustenance: Developed Countries sustain Developing Countries, taking into account the needs of those who are particularly vulnerable.
3. Involvement of institutions:
   - At the global level: institution of a regulation committee to promote adaptation consistent with the Convention;
   - At the regional level: the strengthening and, where necessary, the creation of regional centres and networks, particularly in Developing Countries;
   - At the national level: the strengthening and, where necessary, signing of institutional agreements at the national level.
4. Definition of principles:
   - follow a participative and completely transparent approach, taking into consideration the vulnerable groups, all the communities and the ecosystems;
   - base the definition on scientific foundations as well as the knowledge of the native peoples;
5. Involvement of stakeholders: multilateral, regional and national organisations, public and private sector, civil society and other parties concerned are invited to undertake and sustain adaptation at all levels.

9.3 Comparison of Local Adaptation Pilot Projects

In recent years, attention toward CC has increased. The IPCC has acknowledged that society is adapting to CC through spontaneous reactions and planned decision-making processes, although these are still not widespread (Adger *et al.* 2007). This attention can be attributed to three main factors:

- more awareness of the vulnerability of social and environmental systems to CC (Adger *et al.* 2007);
- growing awareness that extreme climate events are caused by anthropogenic activity (Hegerl *et al.* 2007; Trenberth *et al.* 2007);
- awareness of the inevitable consequences of CC, regardless of future emissions (Meehl *et al.* 2007).

At the national level, many Developing Countries have drawn up their own National adaptation program of action (NAPA), which defines the adaptation priorities. Developed Countries too have launched national adaptation plans.

The European Union (EU), for example, has developed the White Paper, indicating the possibilities for adaptation (CEC 2007), and some EU countries have developed national adaptation strategies (Swart *et al.* 2009).

In the United States, the National Research Council carried out an adaptation study through America's Climate Choices Initiative (NRC 2010) and the Interagency Climate Change Adaptation Task Force, set up in 2009 to develop useful recommendations for the adaptation policy, both at national and international levels.

Australia developed the National Climate Change Adaptation Framework (COAG 2007) and made significant investments in scientific research through the National Climate Change Adaptation Research Facility and the Commonwealth Scientific and Industrial Research Organisations (CSIRO).

To try to understand which measures the national programmes generate at the local level, we selected 14 planning pilot projects for adaptation to CC for coastal cities. Even if the national and regional levels provide the boundary conditions for planning and interventions at that level for CC, it is only on this scale that interventions can be planned, indicating specific adaptation measures, appropriate to the specific nature of the hazard and the area exposed to it. Coastal cities were chosen because they have a higher density in terms of population and key infrastructures exposed: harbours, warehouses, railway terminals, industry for the transformation of imported raw materials or docks for warehousing agricultural products before exportation.

The 14 cases (Figure 9.1) include plans, strategies and guidelines in Africa, Asia, Australia, Europe, North, South America and the Caribbean.

The plans considered include the cities of Cartagena (Colombia), 2012; Copenhagen (Denmark), 2011; Durban (South Africa), 2011; Esmeraldas (Ecuador), 2011; Ho Chi Minh City (Vietnam), 2013; Lagos (Nigeria), 2012; Melbourne (Australia), 2009; Montevideo (Uruguay), 2012; Port- au-Prince (Haiti), 2013; Santiago (Chile), 2012; Semarang (Indonesia), 2010; Sorsogon (Philippines), 2011; Surat (India), 2011; and Vancouver (Canada) 2012.

Starting from the criteria selected by Birkmann (2010), the analysis was developed to account for the six key factors:

- type of tool (plan, strategy and guidelines)
- general view and objectives of the tool

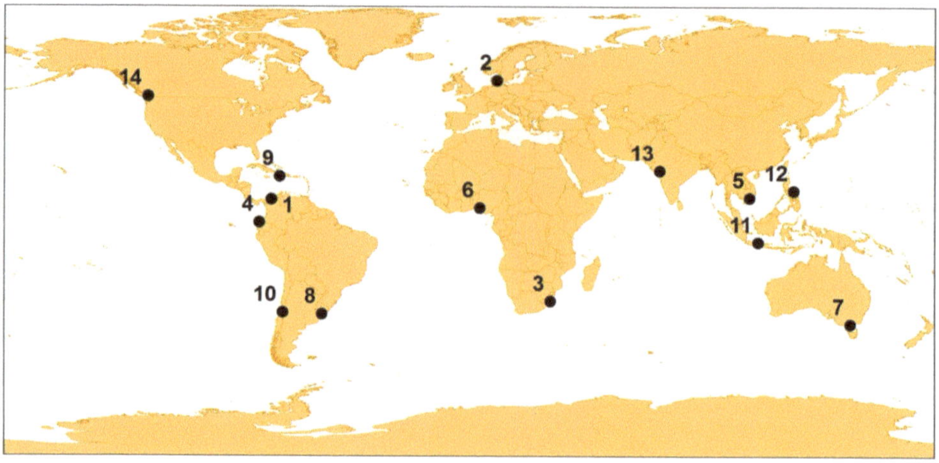

Figure 9.1: Localization of the 14 coastal cities considered: 1. Cartagena, Colombia, 2; Copenhagen, Denmark, 3. Durban, South Africa, 4. Esmeraldas, Ecuador, 5. Ho Chi Minh City, Viet-Nam, 6. Lagos, Nigeria, 7. Melbourne, Australia, 8. Montevideo, Uruguay, 9. Port-au-Prince, Haiti, 10. Santiago, Chile, 11. Semarang, Indonesia, 12. Sorsogon, Philippines, 13. Surat, India, 14. Vancouver, Canada (by E. Ponte)

- identification of the local impacts deriving from climate change
- time schedule and budget
- direct and indirect involvement of stakeholders in planning/implementation
- identification of structural and other measures

The cases[24] were identified to try to cover all continents in order to have a global vision. Evidently the contexts were different in terms of size of the urban area and population density. Santiago and Esmeraldas have an administrative territory in excess of 15,000 km^2, while Vancouver and Copenhagen cover just 115 and 77 km^2, respectively. The population density varies from 12 res./km^2 to almost 20,000 res./km^2 (Surat).

9.3.1 Type of Pilot Projects

The pilot projects analysed were 7 strategies, 5 plans and 2 guidelines (Figure 9.2).

24 Cartagena: CDKN (2012); Copenhagen: Copenhagen Carbon Neutral (2011); Durban: ERM (2011); Esmeraldas: UN-Habitat (2011); Ho Chi Minh City: VCAPS (2013); Lagos: BNRCC (2012); Melbourne: Australian Government (2009); Montevideo: PNUD (2012); Port-au-Prince: MICT (2013); Santiago: MMA (2012); Semarang: ACCCRN (2010); Sorsogon: City of Sorsogon 2000 (2011); Surat: ACCCRN (2011); Vancouver: City of Vancouver (2012).

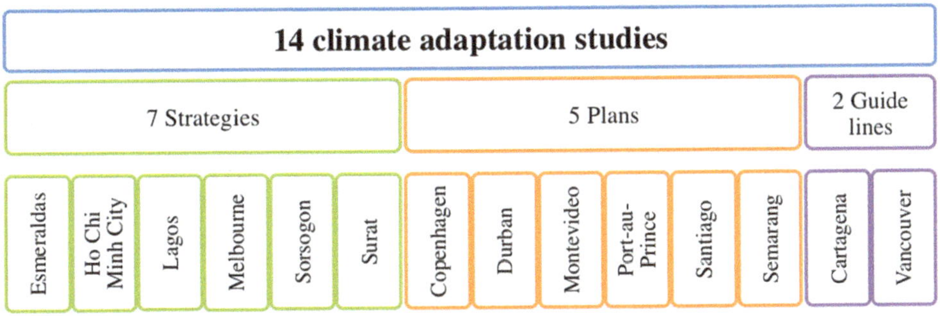

Figure 9.2: Fourteen planning tools: strategies, plans and guidelines.

- Strategies: provide information on the current situations, indicating possible future scenarios in detail enhanced by maps and data.
- Plans: complete documents containing maps (sometimes with geo-referenced information), diagrams and graphs. They do not express a strategic vision for the future.
- Guidelines: concise documents (less than 50 pages) providing information to direct future planning decisions without a time schedule and budget.

Each of these pilot projects has a couple of important features which should always be present. For example, data and maps are indispensable for the operational use of the document. If, however, they do not present possible scenarios, they risk making the utility of a document ineffectual. It is also essential to explain the aim of the document and the hazard for which the impact is intended to be reduced. This aspect is clearly indicated in guidelines only.

9.4 General View and Objectives

The general aim of these projects is to incorporate adaptation into a broader framework of risk reduction. In general, these projects contain three essential parts:
- Description of the climate (trend, most frequent extreme events)
- Presentation of the political, economic and social situations
- Measures/strategies to reduce impacts

These three components are vital for the development of the planning projects. This is especially true for the presentation of the climatic and urban context, because they lead to the generation of specific responses.

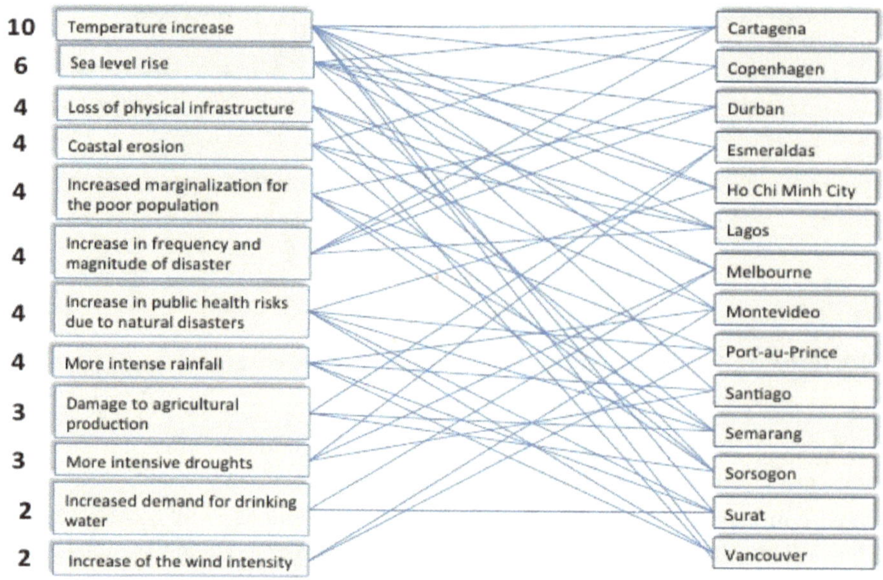

Figure 9.3: Impacts of CC according the 14 pilot projects. Numbers indicate how many times impacts recur in the projects.

9.4.1 Identification of the Local Impacts Resulting from Climate Change

Within the different pilot projects, the main impacts of CC on the 14 cities have been identified (Figure 9.3).

If we analyse the impacts in the 14 cities according to the Human Development Index (HDI) of the countries in which they are located, focusing on countries with an HDI lower than 0.65 (Figure 9.4), we have no impact on coastal erosion or loss of infrastructure.

This result can possibly be explained by modest urban expansion along the coast and also with a poor development of infrastructures, such as to render the damage due to CC as irrelevant.

In the case of countries with an HDI higher than 0.65 (Figure 9.5), the impacts on health-hygiene and on poverty disappear. However, we find confirmation that, in BRICS and in Developing countries, CC and the extreme events related to it have an impact on the poorer people, worsening their already precarious conditions.

9.4.2 Time Schedule and Budget

In the pilot projects analysed, the time schedule of the individual activities to be carried out and the budget are often developed together. The 14 cases can be divided into three groups, depending on the level of definition of these elements:

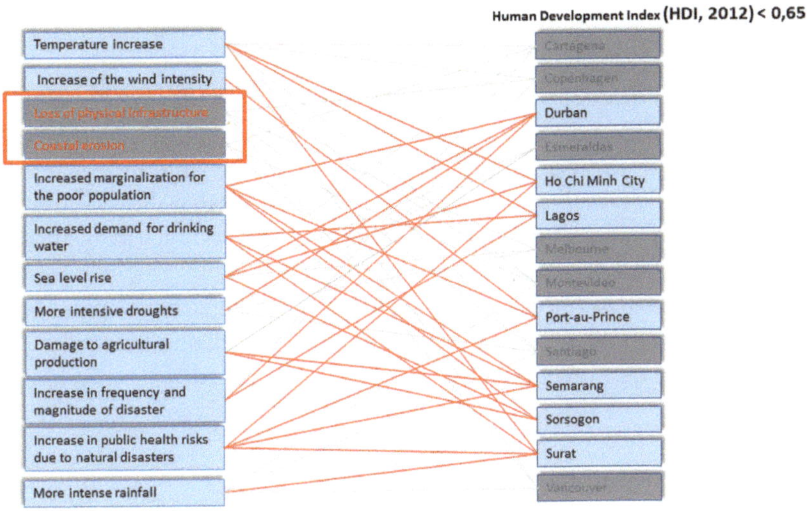

Figure 9.4: Impacts for cities with an HDI < 0.65 (blue boxes)

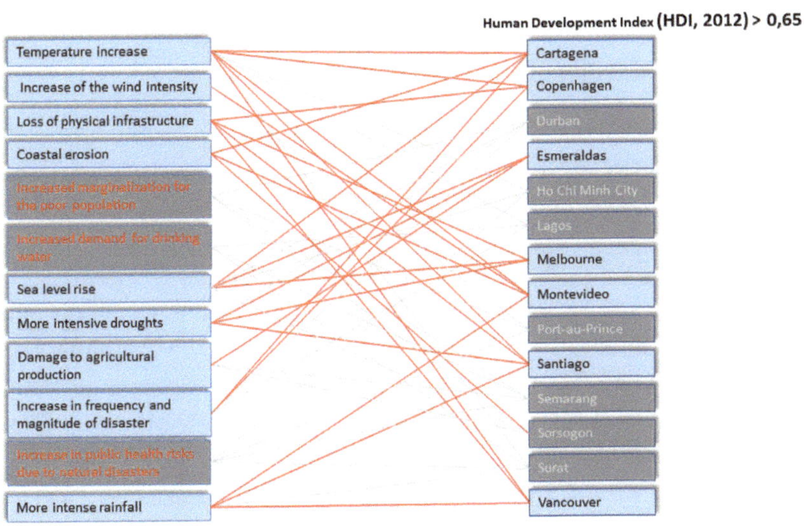

Figure 9.5: Impacts for cities with an HDI > 0.65 (blue boxes)

- High definition (2 of 14): present a detailed time schedule (monthly or four-monthly), accompanied by a cost estimate for each item (e.g.: Melbourne and Durban);
- Sufficient definition (6 of 14): rough time schedule indicating the investment required for every step; the interventions to perform actions are prioritised (e.g.: Copenhagen, Semarang, Montevideo);

- Absent (6 of 14): only the priorities are indicated, without any cost estimate (e.g.: Surat, Port-au-Prince, Sorsogon).

In some plans (e.g.: Durban and Montevideo), attention is focused on certain precise interventions and only for these are budget and time schedule given: in Developing Countries these are almost always absent. In general, we can say that the relationship between stakeholders and policy makers is always unclear in the choice for a suitable option.

We ought to remember the importance of prioritising the steps, developing those that can have the greatest impact. It is usually important to draw up a detailed risk assessment as a basis for defining the priorities of the single steps.

9.4.3 Direct and indirect Involvement of Stakeholders

The process of planning project preparation allows us to understand the main stakeholders involved:

- most of the case studies (6 out of 14) were developed directly by the municipality or other local urban authorities (e.g.: Montevideo, Copenhagen);
- in 3 cases, a team of national and international agencies was set up (e.g.: Santiago, Surat);
- in 3 more cases, civil society was involved (e.g.: Cartagena, Durban);
- in 2 cases, the initiatives were developed within international projects and private initiatives (e.g.: Port-au-Prince, Semarang).

Only in the case of plans are the local authorities directly involved.

The involvement of the local stakeholders, especially residents' organisations, cannot be taken for granted. In recent years, the experiences of community based adaptation have multiplied. The realisation of bottom-up measures is prevented, however, by the impossibility of applying this method in all the most vulnerable communities and, sometimes, by the need to make these actions consistent and cohesive with other local political, climatic and physical problems (Satterthwaite *et al.* 2007).

9.4.4 Identification of Structural and Other Measures

All the pilot projects envisage the creation of measures to reduce impacts. The strategy for Melbourne presents the highest number of measures (20). On the contrary, the plan of Port-au-Prince envisages just 7 measures (Figure 9.6).

Another key factor is the frequency of the structural and non-structural measures envisaged (Figure 9.7).

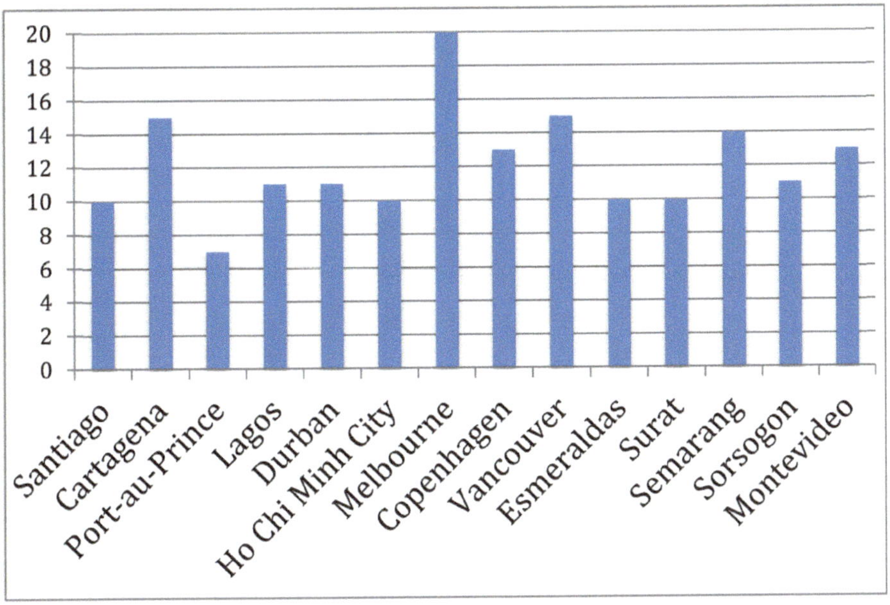

Figure 9.6: Number of measures for adaptation to CC envisaged by the 14 planning pilot projects.

By "structural measures for adaptation" we essentially mean works aimed at reducing the effects of a hazard, like resettlement, storm water drainage, dams, flood barriers, basement for buildings, seawalls, etc. In general, these measures are expensive but they have a definite impact.

By "non-structural measures" we mean those actions aimed at reducing the effects of a hazard, using intangible solutions such as preliminary risk mapping, land use planning, early warning systems, emergency kits... (FIFMTF 1992).

The measure most used is the resettlement of residents of flood prone areas to safer places. This measure is cheaper than many others but it is necessary to take into account the fact that in Developing Countries, resettlement is often poorly conceived and things deteriorate even further when it comes to putting it into practice, as verified in the case of Maputo, a recognized example of failure (Tiepolo 2014b). Moreover, it requires enforcement from the local government, to prevent new settlers from permanently occupying the areas where resettlement has taken place.

Another recurrent measure is the maintenance of storm water drains. All kinds of solid waste are often thrown into these canals, or they accumulate due to run-off. In other cases, sand accumulates or vegetation grows. When this happens, surrounding streets and properties are flooded at minimal rainfall. If maintenance is applied regularly, good results can easily be achieved.

Others are related to different forms of "planning": contingency plans, drainage networks plans and planning urban vegetation in an "adaptive way" (for example, identify the correct area where to plant trees, or mangroves).

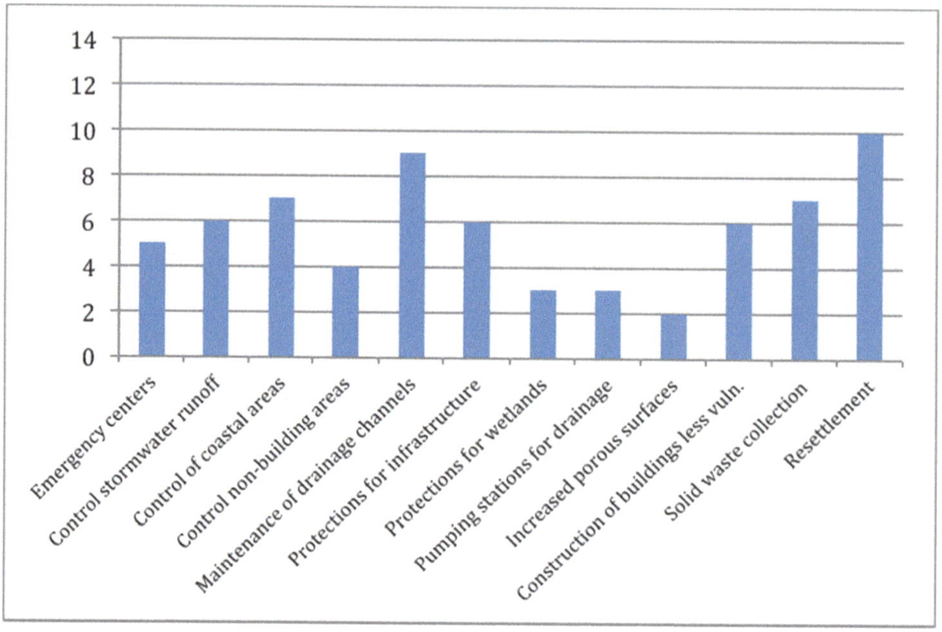

Figure 9.7: Structural measures envisaged by the 14 projects for adaptation to CC.

Among the 12 non-structural measures (Figure 9.8), those most frequently envisaged are the early warning systems and creation of awareness in the population with regard to problems linked to risks.

The early warning system is an indispensable measure as it informs the population in advance of the imminent hazard, allowing them to make preparations. The greater the advance in warning, the more positive the outcomes will be.

Creation of awareness is an important measure, as it allows the people to know what could happen in the case of extreme natural events and tells them what to do when the early warning comes. Some risk analyses (Ponte 2014) have, however, shown how important participation can also be in the creation of awareness and training. Adaptation plans conceived without much participation, involving only activities for training and the creation of awareness, seem destined to failure.

We have seen that there are more non-structural than structural measures. This is explained by the fact they cost less and are quicker to achieve. However, it is still hard to estimate the impact of non-structural measures in advance. Whether the population will be able to behave appropriately when the alarm is given is one example. Viceversa, structural measures, such as new drainage channels and their maintenance, enables us to know in advance the amount of water that they we will be capable of carrying.

Lately, information related to the costs envisaged for the different structural and non-structural measures are always missed.

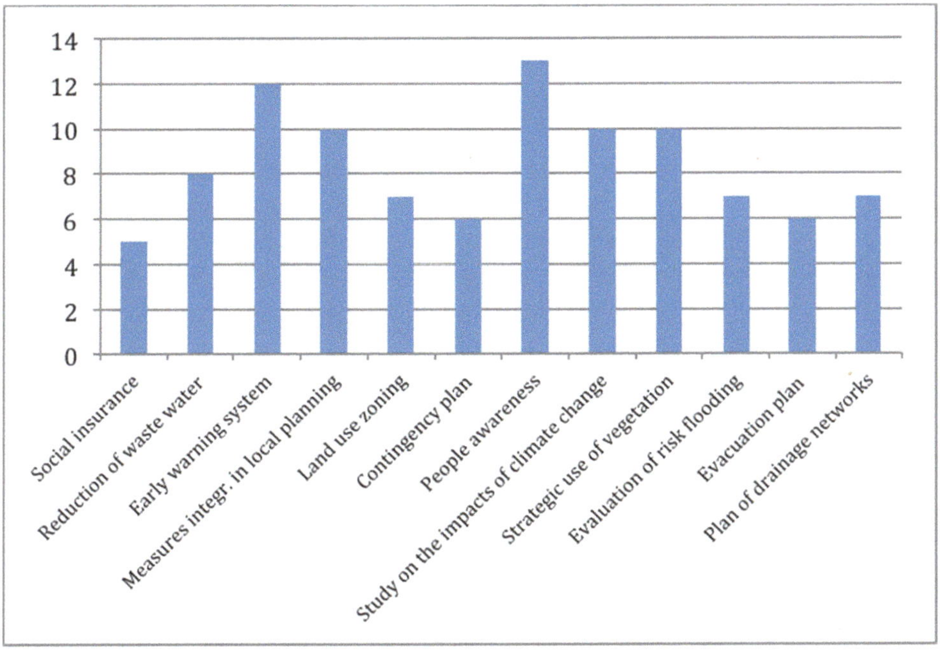

Figure 9.8: Non-structural measures envisaged by the 14 projects for adaptation to CC.

9.5 Conclusions

The analysis of 14 pilot projects recently adopted for planning city adaptation to CC in the coastal areas has highlighted methods of formulation, content, level of detail, measures and also visions that are rather different.

Obviously, the local planning capacity is different too, not to mention the capacity to implement the measures envisaged and to make sure that they continue working over time. For example, the plan of Port-au-Prince presents few adaptation measures: a decision consistent with the current situation in the Haitian capital, which still has to get back on its feet following the devastating earthquake in 2010.

Only 4 out of 14 case studies (Durban, Sorsogon, Surat, Esmeraldas) considered the guidelines suggested by HFA and CCI (the case of Durban is definitely the most important). The 14 local planning tools that we have analysed show that very few recommendations developed at an international level are adopted on a local scale, despite the long and costly action to create awareness developed so far within the framework of HFA, by UN-Habitat, the CAF and many other initiatives. An explanation for this evident detachment would require a deeper analysis of the local dynamics that have led to the preparation of the individual projects. However, it would be best to remember these results when considering the development of a post-2015 framework for disaster risk reduction.

In Melbourne and Sorsogon, it's worth noticing the effort to involve foreign financers in the implementation of particular measures. This is definitely an innovative aspect to be assessed and analysed in order to understand its possible future developments.

The 14 adaptation pilot projects examined do not provide information on the existence of an underlying geographic information system, which would seem essential for the identification of the adaptation measures, for the monitoring and for the assessment of the projects. And, at the end of the day, it is the monitoring of adaptation that is missing. It could help to identify the measures of greatest impact and those that are harder to achieve.

Despite the problems identified so far, and taking into account the considerations of Birkmann (2010), we need to acknowledge that considerable steps have been made in the adaptation to CC. The greatest progress has been recorded where there is greater social and political stability, and where the impacts of CC are well known. Despite everything, the path to obtaining planning pilot projects capable of promoting adaptation still seems very long, although we do know that there will never be an adaptation capable of eliminating the impacts of CC.

References

Adger, W.N., S. Agrawala, M.M.Q. Mirza et al. 2007. Assessment of adaptation practices, options, constraints and capacity. In *Climate change 2007: impacts, adaptation and vulnerability. Contribution of working group II to the fourth assessment report of the intergovernmental panel on climate change*, ed. M.L. Parry, O.F. Canziani, J.P. Palutikof, P.J. van der Linden, C.E. Hanson, 717–743. Cambridge: Cambridge University Press.

Asian Cities Climate Change Resilience Network. 2010. *City resilience strategy Semarang*.

Asian Cities Climate Change Resilience Network. 2011. *Surat city resilience strategy*.

Australian Government. 2009. City of Melbourne climate change adaptation strategy. Melbourne: Department of climate change.

Birkmann, J., M. Garschagen, F. Kraas et al. 2010. Adaptive urban governance: new challenges for the second generation of urban adaptation strategies to climate change. *Sustainability Science* 5(2): 185–206.

CEC-Commission of the European Communities. 2007. Adapting to climate change in Europe - options for EU action, SEC, 849. Bruxelles.

BNRCC. 2012. Lagos State climate change adaptation strategy.

CDKN. 2012. Lineamientos para la adaptación al cambio climático de Cartagena de Indias. INVEMAR-MADS-Alcaldía Mayor de Cartagena de Indias.

City of Sorsogon 2000. 2011. Sorsogon city strategy for climate change resilience.

City of Vancouver. 2012. Climate change adaptation strategy. Greenest city 2020.

COAG-Council of Australian Governments. 2007. National climate change adaptation framework. Canberra: Department of the Prime minister and cabinet.

Copenhagen Carbon Neutral. 2011. Copenhagen climate adaptation plan.

ERM. 2011. Climate change, municipal adaptation plan.

FIFMTF-Federal Interagency Floodplain Management Task Force. 1992. Floodplain management in the United States: An assessment report, volume 2: Full report.

Hegerl, G.C., Zwiers F.W., Braconnot P. et al. 2007. Understanding and attributing climate change. In *Climate change 2007: the physical science basis. Contribution of working group I to the fourth assessment report*, ed. S. Solomon, D. Qin, M. Manning, Z. Chen, M. Marquis, K.B. Averyt, M. Tignor, H.L. Miller, 663–746. Cambridge and New York: Cambridge University Press.

IPCC-Intergovernmental Panel on Climate Change. 2007. *Climate change 2007: Impacts, adaptation and vulnerability. Contribution of working group II to the fourth assessment report of the Intergovernmental Panel on Climate Change*. Cambridge: Cambridge University Press.

Meehl, G.A., T.F. Stocker, W.D. Collins et al. 2007. Global climate projections. in *Climate change 2007: the physical science basis. Contribution of working group I to the fourth assessment report of the intergovernmental panel on climate change*, ed. S. Solomon, D. Qin, M. Manning, Z. Chen, M. Marquis, K.B. Averyt, M. Tignor, H.L. Miller, 747–846. Cambridge and New York: Cambridge University Press.

MICT-Ministère de l'Intérieur et des Collectivités Territoriales. 2013. Plan de contingence national saison cyclonique 2013. Port-au-Prince.

Ministerio de Medio Ambiente Nacional. 2012. Plan de adaptacion al cambio climatico para la región metropolitana de Santiago de Chile.

NRC-National Research Council. 1989. *Improving risk communication*. Washington: National Academy Press.

PNUD. 2012. Plan climàtico de la región metropolitana de Uruguay.

Ponte E. 2014. Flood risk due to heavy rains and rising sea levels in the municipality of Maputo. In *Climate change vulnerability in southern African cities. Building knowledge for adaptation*, ed. S. Macchi and M. Tiepolo, 187–203. Springer. doi: 10.1007/978-3-31900672-7.

Romero Lankao P. 2007. Are we missing the point? Particularities of urbanization, sustainability and carbon emissions in Latin American cities. *Environment and Urbanization* 19 (1): 157–175.

Satterthwaite, D., S. Huq, M. Pelling, H. Reid and P. Romero Lankao. 2007. Adapting to climate change in urban areas. The possibilities and constraints in low- and middle-income nations. London: IIED.

Swart, R., R. Biesbroeck, S. Binnerup et al. 2009. Europe adapts to climate change: comparing national adaptation strategies, Partnership for european environmental research. Helsinki.

Tiepolo, M. 2014a. Flood risk reduction and climate change in large cities south of Sahara. In *Climate change vulnerability in southern African cities. Building knowledge for adaptation*, ed. S. Macchi, and M. Tiepolo, 19–36. Springer. doi: 10.1007/978-3-31900672-7.

Tiepolo, M. 2014b. On going and future flood adaptation measures in the municipality of Maputo. In *Climate change vulnerability in southern African cities. Building knowledge for adaptation*, ed. S. Macchi, and M. Tiepolo, 205–221. Springer. doi: 10.1007/978-3-31900672-7.

Trenberth, K.E., P.D. Jones, P. Ambenje et al. 2007. Observations: surface and atmospheric climate change, in *Climate change 2007: the physical science basis. Contribution of working group I to the fourth assessment report of the intergovernmental panel on climate change*, ed. S. Solomon, D. Qin, M. Manning, Z. Chen, M. Marquis, K.B. Averyt, M. Tignor, and H.L. Miller, 235–336. Cambridge and New York: Cambridge University Press.

UNFCCC-United Nations Framework Convention on Climate Change. 2010. Report of the conference of the parties on its sixteenth session held in Cancun from 29 November to 10 December 2010, Cancun. http://unfccc.int/resource/docs/2010/cop16/eng/07a01.pdf#page=4.

UN-Habitat. 2003. *The challenge of slums: Global report on human settlements 2003*. London: Earthscan.

UN-Habitat. 2007. *Global report on human settlements 2007*. London: Earthscan.

UN-Habitat 2010. *Planning for climate change. A strategic, values-based approach for urban planners*. London: Earthscan.

UN-Habitat. 2011. Adaptation and mitigation strategy for change climate for the canton of Esmeraldas.

UN-Habitat. 2012. *Mid-term evaluation of the cities and climate change initiative*. London: Earthscan.
UNISDR-United Nations International Strategy for Disaster Reduction. 2005. Hyogo framework for action 2005–2015: Building the resilience of nations and communities to disasters. Kobe.
UNISDR-United Nations International Strategy for Disaster Reduction. 2007. Implementation of the Hyogo framework for action, summary of reports 2007–2013. Geneva.
VCAPS-Vietnam Climate Adaptation PartnerShip. 2013. Climate adaptation strategy Ho Chi Minh City moving towards the sea with climate change adaptation.
Wamsler, C. 2004. Managing urban risk: Perceptions of housing and planning as a tool for reducing disaster risk. *Global Built Environmental Review* 4(2): 11–28.
Wheeler, S.M. 2008, State and municipal climate change plans. *Journal American Planning Association* 74(4): 481–496.
World Bank. 2010. *Natural hazards, unnatural disaster, The economics of effective prevention*. Washington: The World Bank.

Enrico Ponte[25]

10 National and Local Contingency Planning: a Comparative Analysis of Plans in Africa and Latin America

Abstract: The past twenty years have provided evidence that preventive preparation of exposed communities can reduce the impact of natural disasters, hence the creation of contingency plans. Today these plans are adopted at various scales, precisely national, regional, municipal and, at times, even village-centred. But the consistency of contingency plans and their interconnection with the various scales is rarely achieved, especially in Least Developed Countries. The contingency device should not only be organised at various territorial levels, but it should also enable the actors to take coordinated action in case of a disaster. This crucial point is still scarcely discussed in the literature. The scope of this chapter is to present the differences between national and local contingency planning, the relationship between the two scales and the weak points that can still be found in the procedure. Eight contingency plans are compared, namely 5 national (Guinea, Haiti, Madagascar, Mozambique, Senegal) and 3 local (Ecuador, Haiti, Peru). The analysis diagram proposed by Choularton (2007) has been used as a reference for the presentation of the various planning scales, phases, scenarios and considerations on climate change.

Keywords: Contingency plans, Scenarios, Least Developed Countries

10.1 Introduction

In case of a natural disaster, the intervention of humanitarian organisations can make the difference between life and death for thousands of people. These organisations have made considerable progress in preparing interventions through contingency plans that involve "making decisions in advance about the management of human and financial resources, coordination and communications procedures, and being aware of a range of technical and logistic responses" (IFRC 2015). Contingency planning analyses the potential impact of a disaster, anticipates problems that might occur during the humanitarian intervention, and organises the response in advance (CARE 2006) in order to promptly respond to the needs of the population affected by the disaster. Planning the response to the emergency entails discussing the issues and limited resources available with personnel.

25 Enrico Ponte, urban planner PhD, holds a research grant at the DIST-Politecnico and Università di Torino enrico.ponte@polito.it
The author would like to thank Maurizio Tiepolo (DIST-Politecnico and Università di Torino) for his suggestions while writing the chapter.

Experience indicates that a prompt response to a crisis reduces its impact. This depends on the expertise and plans of organisations that first intervene in affected areas (Swart *et al.* 2009).

However, the majority of contingency plans put more emphasis on the plan's formulation and on defining actions than on testing, evaluating and refining the plan (UNISDR 2012). Hence, contingency plan assessments that can help us strengthen the contingency device are still very few (IFRC 2012).

The aim of this chapter is to define the main features of contingency plans at the various territorial scales, and the critical aspects that hinder their effective implementation in case of a disaster.

This chapter presents phases, actors and scenarios of eight contingency plans, primarily designed for Least Developed Countries (LDC), namely 5 national (Guinea, Haiti, Madagascar, Mozambique, Senegal) and 3 local (Urdaneta, Ecuador; Tabarre, Haiti, Lima, Peru).

10.2 Contingency Plans

Contingency planning, which is at times defined as emergency planning, specifies how a single organisation, a whole community (from region to village) or a country must behave in case of drought, earthquake, flood, heatwave, landslide or mud flow (WFP 2010). Hence, contingency planning identifies potential disasters and establishes how to manage them (James 2008). This entails simulating the procedures to be followed in frameworks that can become problematic in order to be ready when required.

When an emergency occurs, contingency planning optimises the response (IFRC 2012), which must be rapid, appropriate and efficient (WFP 2010). A contingency plan establishes the management of human and financial resources, as well as the coordination and communication procedures in advance. The plan is a management tool for all sectors that can rapidly and effectively provide humanitarian aid to those who need it in case of a disaster (Ackermann 2013). The time spent on contingency planning is time that will be saved in case of a disaster.

Contingency planning is an ongoing process (Choularton 2007). The plans must be tested and updated to ascertain their relevance, especially in rapidly evolving frameworks (great migratory flows, devastating natural disasters). Contingency planning, the importance of which is readily acknowledged today (Mitchell 2008), must always be performed when there is a high risk of either a disaster or an emergency. In this case, the periods in which natural catastrophes are recurrent (e.g. seasonal event, such as floods, hurricanes or cyclones and drought) must also be defined.

Very often contingency planning is an informal process because it is carried out by individuals or groups of individuals who have not been assigned to do it by any authority. However, formal contingency planning too features a considerable degree of informality, which is deemed useful, though it follows a commonly agreed and defined process that is converted into an emergency plan (IASC 2007, Choularton 2007). Contingency plans should start by identifying the areas exposed to disaster by the communities concerned (Pasteur 2011).

Climate change increases the hazards to which a community is exposed and, at times, introduces new ones. Contingency planning should carefully consider the changes in progress and forecasts of their future progress when preparation for, response to and recovery from a disaster are defined (Vogel *et al.* 2012).

10.2.1 The Implementation Scale of Plans

Most contingency plans are organised on a national scale in order to ensure coordination and response to very big disasters. At times potential disasters are so vast (hurricanes, cyclones) that contingency plans even involve several countries.

Local governments play a role of primary importance in national plans because, according to the principle of subsidiarity, the administration that is closest to the population with limited access to community services will be required to reduce the gap (UNDP 2010a) (Figure 10.1). Basically, if a local government can efficiently provide a specific service, it should be assigned this responsibility (Wilson 2006). National and local governments can significantly reduce the risk generated by climate change, for instance, with a precisely planned response to disasters (rescue services, emergency services and medical assistance) and reconstruction (assistance to those who have lost their home and means of survival) designed to enhance resilience (Satterthwaite *et al.* 2007).

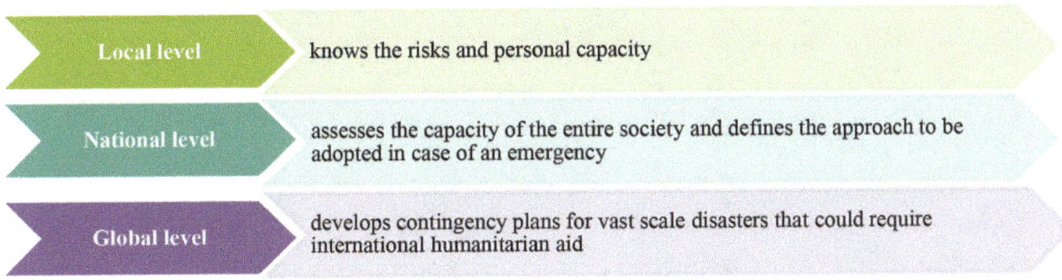

Figure 10.1: The three territorial levels of contingency plans (IFRC 2012)

10.2.2 Phases of Contingency Planning

Though the major part of humanitarian organisations follow their own guidelines for drawing up contingency plans, they present certain common phases, such as description of scenarios, strategic response, preparation for the plan, implementation of the plan, operations to support the plan, and budgeting (Table 10.1). Local contingency plans should be consistent with provincial or regional contingency plans (IASC 2007b). For instance, the databases and maps created by regional plans will be useful to locate temporary recovery facilities and evacuation routes on a local scale (UNDP 2008).

Table 10.1: Principal phases of contingency planning (Choularton 2007)

Phase	Usefulness	Notes
Scenario	Provide the foundations for planning Generate consent on the nature of the potential crisis	Simple and general, if there is no specific threat Detailed in case of an emerging crisis
Strategic response	Define the humanitarian responses	Strategy for a simple response, if there is no threat, complex in case of an emerging crisis
Implementation of the plan	Implement the response	Detailed planning in case of a potential crisis
Support for the plan	Identify human, administrative, financial and safety resources	Support is more effective in case of a potential crisis
Preparation for the plan	Identify the actions that can be carried out before a crisis	Preparation is useful when there is no specific threat
Budget	Define the cost of preparing an emergency response	The budget is more useful in case of an emerging crisis, and can be converted into a project budget The budget for preparation activities is recommended in all phases of the process

10.2.3 Defining the Scenarios

All contingency plans are based on the definition of scenarios, namely an assumption of what might occur following a potential natural disaster (IFRC 2012). A very accurate scenario will improve the organisation of the response (UNCHR 2011). Today, there are several methods for defining possible scenarios (Sikisch 1995). Usually three scenarios are developed, precisely the best, the worst and an intermediate one (Choularton 2007) (Table 10.2).

Contingency plans that include scenarios of climate change, such as the ones defined by the IPCC, are still few and are, generally, flood emergency plans. Two examples include the plan for the Colorado River Basin that was affected in recent years by drought (USBR 2014), and Durban's contingency plan in the framework of the climate adaptation plan (Cartwright *et al.* 2013).

Table 10.2: Potential damages for three scenarios in case of drought, floods and earthquake (Choularton 2007)

Scenario	Drought	Floods	Earthquake
Best	Absent	Absent	4.5 Richter's scale, minor damages
Probable	In one part of the country	100,000 disaster victims	6.5 Richter's scale, damages to rural areas and small cities
Worst	In the major part of the country	1,000,000 disaster victims	8.0 Richter's scale, epicentre in a metropolis

The scenario-based approach allows planners to work on various scales for the same potential crisis. For instance, the definition of the National Contingency Plan can resort to information provided by people from various regions. The construction of scenarios also exploits the "step by step" approach to describe the potential escalation of a crisis and of the corresponding measures required. This approach is often adopted to manage refugee flows. When the degree of aggravation of the crisis increases the number of refugees, thresholds are defined beyond which additional measures or more economic resources are demanded. For instance, every scenario of UNHCR contingency plans envisages additional measures (Table 10.3); hence, such a scenario is defined as "step by step."

Table 10.3: Measures for four scenarios based on the number of refugees (Choularton 2007)

Scenario	Refugees no.	Measures
1	5,000	Record and protect the refugees who stay with hosting families
2	15,000	Open a refugee camp Appoint an Emergency Manager Buy a pickup truck to distribute aid
3	50,000	Open a second refugee camp Increase personnel by 5 units Rent a warehouse
4	100,000	Open a third refugee camp Increase personnel by 6 units Expand the warehouse

10.3 Examples of Contingency Plans

By analysing eight plans we can better understand how they are organised. The plans considered are five national (Guinea, Mozambique, Senegal, Madagascar and Haiti), and three urban (Lima, Peru; Tabarre, Haiti; Urdaneta, Ecuador).

10.3.1 National Contingency Plans

The five plans analysed were drawn up between 2009 (Senegal) and 2013 (Haiti) (Table 4). With regard to the scenarios considered, Haiti's plan (MPCE 2013) and Senegal's plan (UNDP 2009) define scenarios based on the number of hours or days that have elapsed after the event. For every scenario, the plan specifies the activities to be carried out, the people to be mobilised, and the areas to be monitored.

Table 10.4: Comparison of 5 national contingency plans.

Phase of the plan	**Guinea (UNDP 2012)**	**Haiti (MPCE 2013)**	**Madagascar (BNGRC 2011)**	**Mozambique (UNDP 2010b)**	**Senegal (UNDP 2009)**
Scenario	Best, probable, worst	Temporal (Day 1, Day 2, ...)	2 scenarios based on the cyclone season	3 scenarios based on the probability of occurrence	Temporal (24 h, 48 h, ...)
Strategic response	For every administrative unit (cluster)	None	Entrusted to the individual clusters	Strategies for each cluster	Based on the risk
Implementation	Detailed timeline of the programme	Estimated community needs and response capacity	Presentation of clusters and logistic, administrative responses	Presentation of clusters, logistic and administrative responses	Logistic and administrative responses
Support	Humanitarian Coordinator	Identification of available local human resources	National Council for Risk and Disaster Management	The UN Coordinator appoints the Humanitarian Coordinator	Appointment of a Humanitarian Coordinator, other human resources
Preparation for the plan	Actions to be implemented for each cluster	Role of national, departmental and municipal operative centres	Actions to be implemented for each cluster in 24h, 48h, etc.	Actions to be implemented, appointment of Cluster Managers	Actions to be implemented for each cluster
Budget	Absent	Detailed	Detailed based on needs and scenarios	Absent	Absent

The plans for Guinea (UNDP 2012) and Mozambique (UNDP 2010b) define three scenarios (better, probable and worst), indicating the procedures to be followed for each of them. Madagascar's plan refers to the different periods of the year, based on the cyclone season.

The strategic response refers almost entirely to clusters, which are either precisely described (Madagascar) or barely outlined (Haiti).

The plans are specifically implemented by administrative and political actors according to the various logistic duties defined in the scenarios. In the case of Haiti, the plan requires an assessment of the needs and response capacity of the various population groups when a disaster occurs.

Some plans define specific figures, such as the Humanitarian Coordinator or the Emergency Manager (Guinea, Madagascar, Mozambique), while others recall the need to mobilise a larger number of people for emergency management (Haiti and Senegal).

As regards drawing up plans, the various clusters are assigned precise roles based on the number of hours or days that have elapsed after the event (Madagascar), or on the scenario (Senegal).

In 3 out of 5 cases, the budget is not specified (Guinea, Mozambique, Senegal), while it is detailed by items for Madagascar.

10.3.2 Local Contingency Plans

The three plans analysed were drawn up in 2011 (Lima and Urdaneta) and in 2014 (Tabarre). These administrative authorities are quite different in terms of dimensions and population, so they are representative of the various types of local plans. Lima has a surface area of 2,672 km^2 and counts 7.6 million inhabitants. Urdaneta has a surface area of 290 km^2 and has 29,000 inhabitants (2010). Tabarre is a municipality in Port-au-Prince, about 10 km from the centre of the capital city, which stretches over 25 km^2 and counts 170,000 inhabitants. It has an international airport, the American embassy and several UN operation sites. Tabarre's contingency plan can be considered an integration to the national one, considering the low quality of the departmental contingency plan. It follows the provisions laid down by the National Contingency Plan and the local scale structure to provide useful and simple tools for the local authorities (Table 10.5).

Compared to national plans, the ones for Lima and Tabarre define the scenarios (weak, moderate and catastrophic), and the objective, response and preparation for each of them (Table 10.6). Urdaneta's plan, instead, describes geographical scenarios that refer to precise territorial sectors. Climate change is not one of the scenarios considered.

Responses either implement national plans and strategies (Lima, Urdaneta), or they are carried out locally (Tabarre). In the case of local plans, the resources avail-

Table 10.5: Summary of the analysis of three local contingency plans.

Plan phase	Lima (COOPI 2011)	Tabarre (COOPI 2014)	Urdaneta (Los Rios Prefectura 2011)
Scenario	According sectors of the city	Light, moderate, catastrophic	High, medium, low
Strategic response	Refer to the Strategic development plan	According sectors and scenarios	Not specified
Implementation	Commissions for protocols planning and coordination	Identification of strength and weakness of institutions and organisations	Operative procedures by sector
Support operations	Emergency communication, coordination and control	Civil protection, local authorities, United Nations	Responsibility assigned to each actor
Preparation	Specified	Specified by sectors	Specified by sectors
Budget	Absent	Absent	Absent

Table 10.6: Tabarre (Haiti). Objectives, response and preparation according to scenario 3.

Scenario 3	Objectives	Response	Preparation
Disaster victims	Evacuate, provide accommodation and food to disaster victims within 72 hours Psychological assistance Economic aid for funerals	Hospitality sites Distribution of water, food, hygiene kits from the second week	Chemical toilets (mobile) Food stocks Agreements with dispensaries and NGOs
Community	Create access routes Repair damaged streets Drinking water	Human resources of the municipality	List of materials available Establish watering points

able are often limited. Hence, achieving the objectives inevitably depends on actors and competences situated outside the municipality. However, the coordinated action of the municipal Civil Defence Department, municipality, NGOs, committees and local organisations can reduce the impact of the disaster. Constantly updated needs assessment and a training plan can contribute to achieve the goals of the plan.

Regarding the implementation of the plan, dedicated committees that establish communications between actors involved in formal and informal planning procedures are created in Lima. Moreover, coordination protocols are defined to be implemented during emergency response phases. Instead, the actors to be involved in case of a natural disaster (national and departmental civil defence, the United Nations, the

Red Cross, etc.) are specified in Tabarre, taking into account the precarious political and administrative situation. Urdaneta's plan (Los Rios Prefectura 2011) provides a detailed description of the roles played by both public administration and society.

None of the three plans has a budget. In the case of Tabarre, preparation costs are included in the annual budget of the Civil Defence Department, while emergency funds are allocated for the specific situation.

Tabarre's plan has a map. It would be helpful if maps were always present as a useful tool for people who will have to implement the plan.

All the plans analysed neglected disaster simulations that would train the actors.

The national and local plans are complementary. However, a local plan can mirror the national one, but the reverse is never the case.

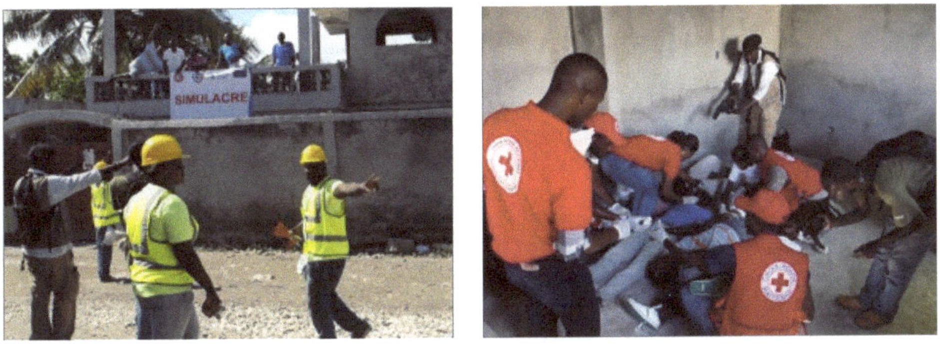

Figure 10.2: Tabarre (Haiti), 2013. Simulations envisaged by the emergency plan (photo E. Ponte)

10.4 Conclusions

The comparative analysis of 8 contingency plans, especially for LDC, underscored some weak points.

First, national contingency plans that are not associated with departmental and local contingency plans, have a low impact potential. This is often the case of LDCs.

Second, the coordination between local, national and international actors breaks down without a local contingency plan, and so does the prompt efficiency of the response (Choularton 2007).

Third, the lack of an Emergency Manager reduces the efficacy of the response.

Four, neglecting the climate change in progress and its envisaged future progress impairs the quality of contingency plans. This common deficiency is now being improved in the contingency plans of advanced economies.

Five, the lack of physical plans for territories exposed to hazards (Haiti) entails that urban expansion is not managed, natural resources are exploited in an unsustainable manner, and both factors hinder emergency management.

Six, without strengthening the municipal administration with regard to disaster management and without economic aid provided by the national government (Haiti), local governments are unable to independently draw up their own contingency plans.

Though the plans considered present several specificities, the above deficiencies are common in many other Least Developed Countries.

References

Ackermann, M. 2013. Mine closure: a contingency plan to mitigate socio-economic disasters. Magister degree in Development and management at the Potchefstroom Campus of the North-West University.

BNGRC-Bureau National de Gestion des Risques et des Catastrophes du Madagascar. 2011. Plan national de contingence sur les cyclones et les inondations.

CARE. 2006. Emergency preparedness planning guidelines. London.

Cartwright, A., J. Blignaut, M.D. Wit, K. Goldberg et al. 2013. Economics of climate change adaptation at the local scale under conditions of uncertainty and 2010–11 constraints: the case of Durban, South Africa. *Environment and Urbanization* 25(1): 1–18.

Choularton, R. 2007. Contingency planning and humanitarian action: a review of practice, HPN Network Paper 59.

COOPI-Cooperazione Internazionale. 2011. Plan de operaciones de emergencia ante sismos centro histórico de Lima. Lima.

COOPI. 2014. Plan de contingence communal de Tabarre de 2014.

IASC-Inter-Agency Standing Committee. 2007a. Inter-agency contingency planning guidelines for humanitarian assistance. Geneva: IASC working group.

IASC. 2007b. Challenges and suggestions for enhancing inter-agency contingency planning: report of the 1st Global consultation of contingency planners in humanitarian agencies.

IFRC-International Federation of Red Cross and red crescent societies. 2015. Contingency planning. http://www.ifrc.org/en/what-we-do/disaster-management/preparing-for-disaster/disaster-preparedness-tools/contingency-planning-and-disaster-response-planning/. Accessed 6 Dec 2015.

IFRC. 2012. *Contingency planning guide*. Geneva: IFRC.

James, E. 2008. *Managing humanitarian relief an operational guide for NGO's*. Intermediate Technology Publications Ltd.

Los Rios Prefectura. 2011. Plan de contingencia por inundación Cantón Urdaneta, Urdaneta, Ecuador.

Mitchell, T., and M. van Aalst 2008. *Convergence of disaster risk reduction and climate change adaptation*. London: DFID.

Pasteur, K. 2011. *From vulnerability to resilience. A framework for analysis and action to build community resilience*. Rugby: Practical Action Publishing Ltd.

Satterthwaite, D., S. Huq, M. Pelling, H. Reid, and P. Romero Lankao. 2007. Adapting to climate change in urban areas. The possibilities and constraints in low- and middle-income nations. London: IIED.

Sikisch, G. 1995. *Emergency management planning handbook*. New York: McGraw-Hill.

Swart, R., R. Biesbroeck, S. Binnerup et al. 2009. Europe adapts to climate change: comparing national adaptation. Helsinki: SYKE-Finnish Environment Institute.

USBR-United States Bureau of Reclamation. 2014. Reclamation, managing water in the west. Climate change adaptation strategy.

UNDP-United Nations Development Program. 2012. Plan de contingence multirisque de la Guinée.

UNDP. 2010a. Local governance and climate change. Discussion note.
UNDP. 2010b. Mozambique humanitarian country team inter-agency contingency plan. Maputo.
UNDP. 2009. Plan de contingence pour la République du Sénégal. Dakar.
UNDP. 2008. Mainstreaming disaster risk reduction in sub national development and land use/physical planning in the Philippines. National Economic and Development Authority.
UNHCR. 2011. Contingency planning. Learning module. Tokyo. http://www.coordinationtoolkit.org/wp-content/uploads/UNHCR-Contingency-Planning-Training-Module.pdf. Accessed 5 Dec 2015.
UNISDR. 2012. Drought contingency plans and planning in the Greater horn of Africa. Geneva.
Vogel, J., E. Gordon, J. Brislawn et al. 2012. Boulder county climate change preparedness plan. Boulder: Boulder county commissioner's sustainability office.
WFP-World Food Program. 2000. Contingency planning guidelines. Rome.
Wilson, E. 2006. Adapting to climate change at the local level: The spatial planning response. *Local Environment. The International Journal of Justice and Sustainability* 11(6): 609–625.

Rita Biconne[26]

11 Measures of Adaptation and Community-based Water Management in Mékhé, Senegal

Abstract: Water and ecological stressors that have decreased resource availability result in increased social and territorial disparities from the viewpoint of quantity and quality. The challenge increases in contexts characterized by critical environmental and climatic conditions, such as those found in many Sub-Saharan African countries, where local priorities are heavily influenced by survival needs. The ability to cope with a situation of water scarcity is contingent on the interaction among environment, economy and society. According to Ohlsson, adapting to natural resource scarcities involves the mobilization of an increased level of social resources. These social resources can be identified with the "adaptive capacity" of a given society. To make the relevant adjustments needed to cope with increased resource scarcity, a social entity could find the social tools enabling it to deal with the consequences of the water scarcity. In this framework, local communities and associations play a central role relative to institutional failures in an articulated relationship between responsibility and interest. The experience of water access and management for irrigation led by an inter-village peasant organization in Mékhé, Senegal, will be analysed to show a feasible way to implement new opportunities facing climatic challenges.

Keywords: Adaptive capabilities, Water scarcity, Climate change

11.1 Introduction

Water is the fundamental link between the climate system, human society and the environment, and it is the primary medium through which Climate Change (CC) impacts the livelihood and wellbeing of societies.

According to the Fourth Assessment Report of the Intergovernmental Panel on Climate Change (IPCC), rainfall is generally expected to become more variable, floods are expected to become more common, droughts are expected to become more intense and last longer, and sea-levels are expected to rise. "The effect of all these impacts is reduced availability of water (surface and groundwater) and reduced security of supply. This phenomenon is spatially variable, particularly against the background of

---

26 Department of Architecture, University of Florence, Italy, rita.biconne@gmail.com.
This chapter resume part of the PhD thesis "Access to water. Actors and planning in the territories of Senegal" discussed in 2013 at the University of Florence. The author thanks "Fratelli dell'Uomo" NGO, Union of peasants of Mékhé and KAYER – Kayor Energie Rurale for the support and information provided in Mékhé.

high evaporation rates and low rainfall under natural conditions as well as increased demands for water to support development, among others. This effect also depends on the water supply system and the ability of politicians, planners and managers to respond also to population growth and changes in demands, technology, economic, social and legislative conditions" (IPCC, TAR 2001).

The literature recognises the impact of climate-driven water shortages and emphasizes the vulnerable and poor communities due to their lack of resilience (Ziervogel *et al.* 2008; IPCC 2007). In particular, according to Eboh (2009), countries of sub-Saharan Africa are likely to suffer from these devastating impacts because of their geographical location, low incomes, low technological and institutional capacity to adapt to rapid changes in the environment, as well as their higher reliance on climate-sensitive renewable natural resources sectors such as water and agriculture.

This chapter will discuss some of these aspects, mainly focusing on water management connecting environmental challenges to the agricultural use of water. First, the paper proposes some concepts related to water scarcity, focusing on physical and climatic issues (as aridity, drought), as well as political and social features linked to the theory of "social scarcity" and adaptation processes. Secondly, through a case study, this essay will explore a possible path to cope with water scarcity, aiming to integrate the social, environmental, spatial, economic and political issues of water management.

The hypothesis we are going to present is the following: in the process of water management and its related responsibilities concerning monitoring, regulation and usage, especially under the condition of vulnerability resulting from CC, might give rise to ways of adaptation propelled by the skills of local actors to mobilise the available resources, both material and immaterial ones.

The selected case study takes place in the extended hinterland of the Senegalese capital city. It represents a very interesting context to study due to its location in the ever-growing urban corridor[27] between Dakar and Touba, the two biggest cities in Senegal.

Relating to CC impacts, this interest is due to the very poor scientific data and awareness of what is going on in the hinterland and in particular in the new spatial configuration of territories touched by urban corridor development. This chapter aims to contribute to this topic, focusing on water scarcity and related adaptation measures at the community level.

27 "In sub-Saharan Africa, as well as in all regions where trade between cities and their hinterlands has accelerated, urban development corridors are now arising in the wake of rapid demographic expansion and urbanisation. An urban development corridor can arise where two or more large urban centres are located along a single connection trunk line (road, rail, sea or river) that is organised so as to attract flows of people, goods and services while large and regular trade flows pass through urban or rural transit points between the larger urban centres" (UN-Habitat 2010: 129).

The selected context is characterized by a webbed and hierarchical structure of villages. Both the social and spatial implications stemming from the location and the distribution of water resources require accurate reflection.

The starting point of this discussion is recognizing that territory is not only deemed to be a physical support, but also influences the social organization and structure on the basis of history of settlement and on the ties that social groups have established with the territory itself, and with other social groups. Therefore, managing the territory and its related natural resources means not exclusively considering technical decisions as the rules to adopt. Advantages and utility are shaped by the array of rights and obligations of social actors in this resource-space where different roles, responsibilities and forms of power are taking new shapes to face current challenges.

11.2 Adapting to Water Scarcity

Climate Change is aggravating the challenges of water resources access and management in Sub-Saharan Africa, with serious implications on socio-economic development. IPCC (2001) noted that these challenges include population pressure, problems associated with land use and possible ecological consequences of land-use change on the hydrological cycle. Climate change – especially changes in climate variability through droughts and flooding – will make addressing these problems more complex. The current condition of water access worsened because of the many components exerting an impact on water scarcity. The Human Development Report *Beyond scarcity: Power, poverty and the global crisis* (UNDP 2006) considered water scarcity from two points of view: a) as a crisis arising from a lack of services that provide safe water, and b) as a crisis due by scarce water resources. Above all, it underlined that water scarcity in the World is not related to the physical availability, but to unbalanced power relations, poverty and related inequalities.

Water scarcity is a very critical phenomenon in arid and semi-arid regions in different parts of the World (Figure 11.1). Specifically, the Sub-Saharan African countries suffer from a particular type of water scarcity, an important issue that requires precise consideration.

Despite some common errors of understanding, scarcity is not a synonym of aridity; the two terms are not equivalent concepts. Aridity expresses a condition that characterizes specific ecosystems and natural environments; scarcity, which is often mistakenly associated with data exclusively related to climatic, physical and hydrogeological features, represents a problem of various kinds. In fact, water scarcity can have manifold causes, both natural and man-made, and corresponds to several regimes.

As underlined by Pereira (2005), the main natural causes for scarcity are aridity and droughts, and he offers the following definition of the two concepts.

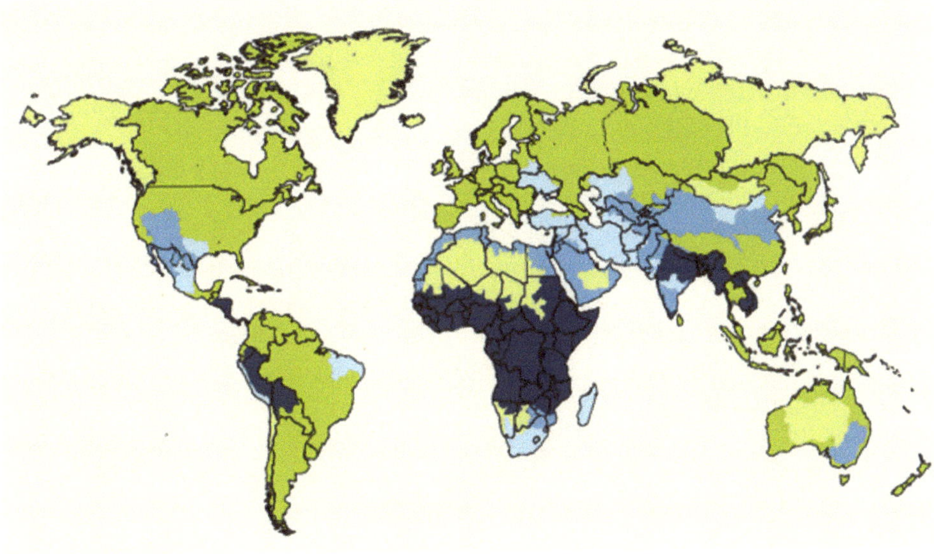

Figure 11.1: Physical water scarcity (light blue), economic water scarcity (dark blue), no water scarcity (green). Source: UN-Water, 2012.

"Aridity is a nature produced permanent imbalance in the water availability consisting in low average annual precipitation, with high spatial and temporal variability, resulting in overall low moisture and low carrying capacity of the ecosystems. [...] Drought is a nature produced but temporary imbalance of water availability, consisting of a persistent lower-than-average precipitation, of uncertain frequency, duration and severity, the occurrence of which is difficult to predict, resulting in diminished water resources availability and carrying capacity of the ecosystems" (Pereira *et al.* 2005: 2).

More recently, man-driven processes, such as man-induced desertification and water shortages[28], are aggravating natural scarcity, while population growth and the demand for water faces an increased competition among water user sectors and regions.

According to Pereira (2005), the fact is that negative socio-economic conditions in the vulnerable areas are often related to poor land resources, mainly soil and water,

28 "Desertification is a man-induced permanent imbalance in the availability of water that occurs in arid, semiarid and sub-humid climates, which is combined with damaged soil, inappropriate land use, mining of groundwater, increased flash flooding, loss of riparian ecosystems and a deterioration of the carrying capacity of the ecosystems. [...] Water shortage is also man-induced but temporary water imbalance including groundwater over exploitation, reduced reservoir capacities, disturbed and reduced land use, and consequent altered carrying capacity of the ecosystems" (Pereira *et al.* 2005: 4).

inadequacy of ecosystems to support development, low response capacity to climate forcing, and non-existence of alternative activities to agriculture and forestry. Thus, climate conditions may exacerbate the vulnerability on water scarcity, affecting the socio-economic dimension, in addition to the environmental one.

Moreover, a region is in a state of high water-related criticality (susceptibility of a region or its population to crises) if water scarcity coincides with a low problem-solving capacity of the population (WBGU 1998).

In this sense, the research carried out by Ohlsson is a crucial turning point for the scientific literature on water scarcity. He introduces a level of scarcity – different from unchangeable data linked to environmental conditions – defined as "social scarcity", which is related to the lack of institutional and economic tools.

As a historical factor subject to constant change, according to Ohlsson, society (even if subjected to multiple political and cultural limits and constraints) can mobilize developing adaptive capacity allowing it to cope with the physical scarcity of resources. Indeed, in addition to the so called 'first-order resource', (i.e. a natural resource which is relevant to a population over time, and is becoming either more scarce or abundant), a 'second-order resource' has been identified as "the set of potential 'adaptive behaviours' stemming from the broader social context that can be used by decision-making elites, either legitimately or illegitimately. 'Adaptive behaviour' is an explicit response to the changing level of water scarcity which can assume a number of forms, such as voluntary rationing schemes, changes in cropping cycles, rainwater harvesting, groundwater mining, formal policies etc." (Ohlsson and Turton 1999: 3).

Framing this theory, Ohlsson was the first to develop the interrelationship between water scarcity and adaptive capacity. By shifting the focus from the 'first-' to the 'second-order resource', an improved understanding of why and how certain societies cope with resource scarcity can be achieved.

Based on the concept of adaptive capacity, the notion of adaptive management has also been advocated by other scholars as the paradigm within which communities, planners and managers should operate, since resource management systems need to be able to adjust to sudden changes in the system (Jeffrey and Gearey 2006).

This requires a paradigm shift, where water management is flexible enough to adjust to changing socio-economic and environmental conditions. Therefore, it is necessary to take into account the environmental, technological, economic, political and cultural features of water systems (Biconne 2013a). Through the case study, the next section of the chapter will show the complementarity of these components to face water scarcity at the community level.

Indeed this paper aims to highlight that overcoming "social scarcity" can represent a necessary step in order to face the impacts of CC and also to turn this challenge into opportunity not to be missed.

11.2.1 Understanding Adaptation at Community Level

According to the Intergovernmental Panel on Climate Change, adaptation to climate changes and variability requires the system adaptation to downsize the impact of climate changes, in order to exploit new opportunities and to cope with the consequences (IPCC 2001). In this sense, adaptation involves the action that people take as a response to current or foreseen climate changes to reduce adverse impacts or exploit the opportunities offered by climate changes. In terms of CC, this last part of the definition is meaningful since climate changes also offer certain opportunities and advantages for communities to reduce their vulnerability or increase their resilience to climate shocks (Parry *et al.* 2005; Dinar *et al.* 2008).

Considering adaptation in terms of capabilities to respond to a stressor and in terms of education allows for the identification of material adaptation, individual and collective actions, and socio-political modification as valid adaptive strategies. The relational attributes of local actors, their degree of knowledge, and social implications are central to adaptive capacity, which enables responses not only to the unforeseen shocks related to climate change, but also to the corollary incertitude of economic, social, and political change (Biconne 2013b). In this framework, the array of adaptation options is very large, ranging from purely technological measures to managerial adaptation and policy reforms (Falkenmark *et al.* 2007). According to Ngigi (2009), one of the most important features of an adaptation strategy is that it should reflect the needs and aspirations of the society or community to which it is meant to bring an advantage. Thus, the most effective mechanisms are flexible and the adaptation efforts would be coordinated across sectors and between different agents, which is a great challenge in practice.

Therefore, in many cases adaptation activities are local – district, regional or supra-regional – issues rather than international. Since communities show different vulnerabilities and adaptive capabilities, they tend to react differently, thereby showing different adaptation needs. According to Paavola and Adger (2006), successful adaptation at community level will depend on:

- Establishment of appropriate social institutions and arrangements that discourage marginalization of vulnerable populations and enhance collective/participatory decision-making process;
- Diversification of income sources and livelihood systems reducing vulnerability and risks, especially for the poor people;
- Introduction of collective security arrangements such as farmers' cooperatives and community-based organizations;
- Provision of knowledge, technology, policy, institutional and financial support (e.g. credit facilities) to the vulnerable communities;
- Prioritization of local adaptation measures and provision of feedback to stakeholders.

These main elements outlined by scholars are going to be used in order to discuss the general strategy and the measures of adaptation deployed by the inter-community organization in Mekhe. In the selected case study, the focus will be mainly on the introduction of new technology for water irrigation and related new hybrid forms of management.

The use of water for agricultural purposes related to the on-going CC is a main challenge in many tropical and sub-tropical regions. More than 70 per cent of people living in African, Caribbean and Pacific (ACP) countries work in the agricultural sector: for them, understanding and responding to CC is not only a theoretical debate, it is an existential issue. Precisely, smallholder farmers are the most vulnerable to CC, and they have no alternative but to adapt their livelihood systems to changing climatic conditions.

11.3 The Case Study of Mékhé

Mékhé is a municipality in the region of Thiès, the central-northern area of Senegal's historical groundnut basin. Since the time when Senegal was a colony, this area has always been exploited for the production of monoculture for international trade. In this region of the ancient reign of Cayor such a strategic function was carried on after the Independence under the aegis of the national agricultural strategy until the year 2000, when the State marked a turning point by privatizing the groundnut production chain. Such a policy of privatization pushed local farmers to adopt new strategies and engage in growing alternative crops such as manioc, orchards and breeding.

According to the oral tradition, in ancient times the area was full of forests, rich in animal species and had high soil fertility, which eased the development of agricultural and breeding activities, and human settlements. There were not so many villages and the size of agricultural surfaces was not bigger than a hectare. These tiny fields were located at the margins of the *concession*, the traditional house, paying particular attention to the interchange of times of harvest and periods of fallow (Fall 2009). The farmer system was based on the familial or collective organization.

The development of the settlement of Mékhé was originally linked to its location in one of the main strategic axes in the country, the National Road 2 (RN2). Mékhé is currently the intersection between two main routes connecting the capital city to Saint Louis (the abovementioned RN2) and to Touba (RN3).

As mentioned at the beginning of the chapter, this case study represents a very interesting context in which to investigate because of its localization in the hinterland of the capital city, especially for the on-growing urban corridor between Dakar and Touba (Figure 11.2). This new link between the two large urban centres (the undisputed primacy of the capital city and the growing role of Touba as a religious and economic centre) is the main cause of relevant spatial transformations in the region.

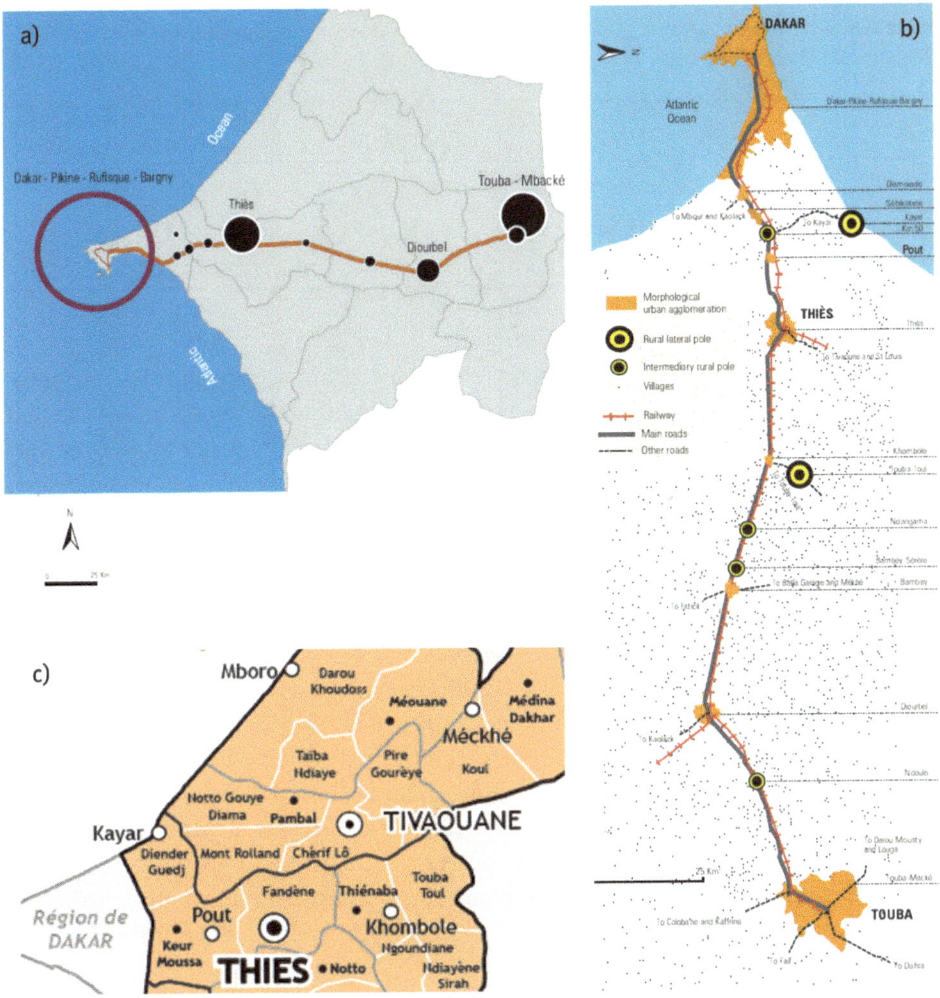

Figure 11.2: The urban corridor of Dakar-Touba in 2002 (a) and 2010 (b) Source: UN-Habitat 2010. Localization of Mékhé (c). Source: www.au-senegal.com.

The 200-km long Dakar-Touba corridor owes its dynamism to the line of towns and cities, which lie along the highway linking these two main urban centres. Thiès, the largest one, stands as a major transition point along the route (UN-Habitat 2010). A number of rural settlements hosting weekly markets lie between the main corridor cities, helping to maintain the commodity flows within the corridor and creating very active lateral links.

Thiès, an area of 6,670 km^2 that is 3.4% of the national territory, is the second most populated region after Dakar. In 2013, it shows an urbanization rate of 49% and about 14% of the national urban population (Table 11.1).

Table 11.1: Data on demographic growth of Thiès region. Source: Conseil Regional de Thies 2007; Ministère du Plan – ANSD 2014.

Year	1976	1988	2013
Total Population	654,046	941,151	1,709,112
Urban Population	195,906	331,549	837,331
Rural Population	458,140	609,602	871,781

Furthermore, this area was chosen also for the degradation problems due to persistent drought, affecting the agriculture as well as the natural resources. The expansion of Thiès, intense mining activities and overexploitation of the water table have also had a great impact on the neighbourhood. The whole area is thus undergoing deep changes in the organization of land, requiring a redefinition of land use planning (IAO 2003).

Agriculture has historically been the main activity, but now it is undergoing major changes. Up until the 1970s, agriculture flourished thanks to ideal soil, climatic conditions, and market opportunities. Later on, agricultural market outlets become poorer and poorer, due to rainfall decreases, soil erosion, reduction of soil fertility, increased salinity and vegetation degradation, mainly due to deforestation. Nowadays, after the privatization implemented by the national agricultural strategy, the farming patterns in selected areas changed slightly according to the land morphology. However, about 70% are largely dependent on rainfall and the rest on irrigation, which is fuelled by the shallow water table. Crop production, livestock husbandry and small trading are the main sources of income in this region. The widespread cultivation is rain fed, breeding is extensive and transhumance is common when food for animals is lacking or mainly during the dry season.

This framework is highly aggravated by climate change, essentially consisting of rainfall regime fluctuations and the rise of temperatures, leading to a continuous degradation of the other natural resources. The study of the precipitation index of the Thiès region reveals the contrast between two periods: a first period from 1930 to 1970 with positive rainfall indices in most years, and a second period from 1971 to 2012 with negative rainfall indices for almost all years (ANACIM), (Figure 11.3).

From the point of view of water infrastructure, the region has 75 boreholes realized by the State, with a production capacity of 18,720 m^3.

Despite this important hydraulic equipment, the demand for drinking water for human, livestock and agricultural use, estimated at 47,473 m^3/day, is covered at 39% only.

In rural areas, water consumption comes mainly from boreholes and wells. Wells in particular, which were the main sources of water supply in the villages, are no longer functional due to the lowering of the water table. Therefore, the majority

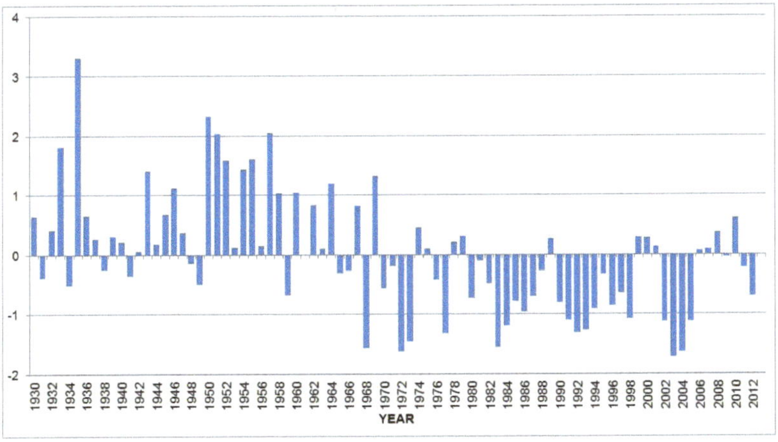

Figure 11.3: Rainfall index of climatological station of Thiès (1930–2012). Source: ANACIM-Senegal National Meteorology Agency.

of villages now rely on drilling for drinking water sold at 250 FCFA per m^3, which does not favour the agricultural activities development (Conseil Regional de Thies, 2007).

11.3.1 Methodology

The chosen methodology has added to the bibliographic survey the analysis of the case study as a research strategy to empirically deepen the issue, as the border between the examined process and the context is not always easy to identify (Yin 2003). A strategy based upon the usage of multiple sources to gather and analyse information was designed in order to identify "who" and "how" the forms of adaptation are taking shape in conditions of water scarcity.

Added to the complexity of the topic considered is the choice of combining different methods in order to integrate the different levels of analysis. By making use of the tools belonging to ethnography, it was considered expedient to make use of the *observation from a distance* and the *semi-structured interviews* in the awareness of the benefits coming from the complementarity of the two techniques. Indeed, the non-involvement of the researcher in the events, which has led us to choose this method of enquiry due to the high-quality information that it allows to collect, does not entail investigating the way things are perceived and learnt by people, which instead has been outlined through interviews. The detached viewpoint of the observer is useful to learn about interactions and behaviours in the way they usually occur, or – better – without the risk of contamination by external actors (Ronzon 2008). Nevertheless, "the limits of the observation in the distance lie in the fact that in order to provide complete findings it needs to be matched with other techniques allowing more participation, allowing to collect additional contextual information and allowing to understand social roles, functions and the meaning given to the ongoing actions" (Ronzon 2008: 21).

Two types of interviews – individual and collective – were performed according to the need to focus attention on the perception, cognitive and / or technical level of the single person or group. The interviews were conducted in French and sometimes in Wolof with the support of local interpreters. About 30 semi-structured interviews were realized and addressed to several people, as the representatives of the villages – *Chefs du village*, farmers, the UGPM president, the UGPM representative for sustainable development, the UGPM responsible of the Solar platform project, the UGPM facilitator for peasant groups, a hydraulic technician, the KAYER Laboratory members and technicians, and cooperation project representatives.

The questions were calibrated with respect to the interviewee and concerned different issues, such as: the socio-economic characteristics of the territory; water use and agricultural practices, and related difficulties; the organizational structure of UGPM and peasant groups in the villages; the technological and economic components of the new system, the method of decision-making and agricultural fields management and infrastructure; changes, benefices and limitations of the initiative.

The research was carried out from January 2011 to April 2013, in particular the fieldwork was conducted from January to March 2012.

11.4 Adaptive Capabilities of the Inter-village Organization

> *"When territory, social actors and the relations between social actors in the local systems of power are not taken into account, the resource management cannot do other than floating in a sociological void allowing bewilderments and manipulations. Moreover, the issues of water management depend on understanding the local arrangements of power." (Lavigne Delville 2005: 155)*[29].

The last three decades have featured bottom-up actions aiming to the re-appropriation and exploitation of local resources in Mékhé, especially water and land. The main agent of this transformation is the Union of peasants of Mékhé (*Union des Groupements Paysans de Mékhé* – UGPM), which is organized in groups of 90 different villages for a total of 5,000 members. It has had a great evolution if we consider that in 1985, when the *Union* was created, only 5 groups formed it.

The main choice of UGPM is to improve the living conditions of the peasant groups and to strengthen their role in the protection of natural resources, diversification of sources of income, the fight of the rural exodus and the reinforcement of social solidarity.

---

29 Translation by author.

The starting point was the awareness of local needs and potentialities to face anthropic and environmental challenges, fundamental elements recognized by some scholars (Chauveau *et al.* 1999).

The association recently strengthened this ever-growing knowledge on local conditions and on endogenous understanding and practices deployed in the various villages, by systematizing them in documents and maps. For example, the organization created a map that identifies five agro-ecological zones in the area under UGPM influence (UGPM 2009), based on the peculiarities of the main economic activities, of the nature of the soil and vegetation, of the localization of villages and their relation with near-by settlements and their type of production (manioc, cereals, horticulture, sheep-breeding, groundnuts, pearl millet, and *niébé*).

11.4.1 Provision of Knowledge and Technology for the Communities

The state-led policies aiming at privatization and to agribusiness, as well as the impacts of environmental and climatic factors, have pushed rural communities to seek alternative solutions. The network of peasants groups in Mékhé has tried to introduce new systems of water management and farming that would allow for facing the critical environmental and socio-economic conditions.

In this framework, one of the main interventions of UGPM, supported by a decentralized cooperation program, has been in some villages the realization of solar pumping platforms connected to a network of micro irrigation and water for human consumption (Figure 11.4).

The intervention was based on the reactivation of *forage*[30] in order to build a micro-irrigation system allowing the exploitation of near-by unused horticultural fields (*périmètres maraichers).* Through giving support to an integrated approach aimed at local development and food security, UGPM has favoured the introduction

Figure 11.4: The solar platform of village, Ndia (Photo by the author, 2013)

30 *Forage* is a well drilling that enables reaching deep groundwater, and through a pumping system provides water to the surface.

of renewable energies as a way to preserve the environment and to enhance the conditions of a local communities lifestyle, also giving support to the process of technological transfer.

The technological transfer of the new system and its related storage and micro-irrigation is a process entailing information activities and training of some local inhabitants responsible for the intervention, and the progressive assumption of the infrastructures management so as to ease a process of gradual appropriation. The technical sustainability of the new system is mainly guaranteed by the fact that the collective platform goals are set up by existing groups and volunteers, and by KAYER, a local cooperative company created in 2006 by UGPM and funded since 2011 thanks to a microcredit system.

However, through setting up solar energy systems that satisfy domestic and village platforms, the organization has tried to respond to the basic needs identified by the communities as time went by. The problem was finding a way to ease the water access for farming and finding an alternative solution to the usage of expensive generators. This would allow the restoration and the boosting of production for selling, but above all, the survival of the families largely dependent on this activity. The solar platform indeed allows the supply of about 20mc of water per day which is used for irrigation, and no additional costs or forms of payment apply. It is essential to outline that the introduction of technological innovations of solar energy systems is totally in line with the recent national and international guidelines on the replacement of the use of fossil fuels with renewable energy.

Last but not at least, this intervention has represented an opportunity to slow down the process of rural exodus affecting young people of the area: some of them are voluntarily involved in the KAYER laboratory.

11.4.2 Establishment of Appropriate Social Institution

Even though the recent Senegalese reform on administrative decentralization has committed the natural resources and land management to local communities for about 20 years, an effective implementation of the decentralized action is still far from being undertaken (Poteete and Ribot 2011). Owing to the lack of institutional answers and measures of environmental and climate change impacts, as well as to the already existing problems, the first reflection is about the role of inter-villages organization. Rooted in the territory and increasingly recognized by the populace, it can represent a useful process of re-appropriation of resources management in a broader vision of endogenous development.

By synthesizing spatial, cultural and social needs of various villages involved, UGPM is a catalyst of the local potentialities and the international engagement, raising alternative forms of response to the current challenges. The organization has revealed the ability to encourage and keep alive the relations within the communities, in order to promote social and cultural peculiarities, fostering the exploitation of local resources for empowerment and adaptation process.

Basically there is no conflict between the official and unofficial figures, owing maybe to the perceived distance of the population compared to legal forms of representation and intervention. It amplifies the gap between the channels of planning and territorial government where the institutional authorities seem to be the armed wing of choices distant from and poorly shared by the population. Instead, the capacity of social mobilization makes UGPM a significant experience of community self-organization that over time has been able to convey, coordinate and exploit the social, cultural and productive potentiality.

The growing demand coming from other villages for the construction of solar platforms puts on UGPM an important responsibility for the transmission of approaches and methods linked to this technology, which is undoubtedly favoured by the power and legitimacy granted by the population.

11.4.3 Collective Decision-making and Socio-territorial Arrangements

As Venot underlines, small-scale irrigation fits well within the development narrative of participation (Venot 2011). In the selected experience, UGPM was able to undertake the upgrading of the water supply for agricultural purposes thanks to its ability to mobilize endogenous and exogenous resources. The project was proposed and designed during a meeting organised by UGPM as the forerunner of the initiative, to which the members of the peasants' unions in Mékhé had been invited. The first four villages in which to begin the programme were identified on the basis of the adhesions expressed by the spokespeople of the villages during a meeting, and according to the criteria to start the intervention (among them there was the presence of disused *forage* in the village). This tiny discourse on the initial implementation phase of the programme showed the main features of the decision-making process of the peasants' unions.

As already highlighted, UGPM embodies that role of linkage able to intercept and embrace the main needs of local communities in order to promote an alternative path of development where members and representatives of each community are directly involved and responsible for the initiative. Within the village, this network of relationships can be identified also in the management of the realized hybrid system, where the community choices are mainly discussed and taken inside the meetings to which dwellers and local representatives take part. In these assemblies, a division of the responsibilities and roles related to the entire chain is set up, from water management to land management, from the choice of crops to be plant to the distribution of products and income. For instance, in these assemblies are identified those responsibles for the *périmètre maraicher*, the contact person for each field, as well as people responsible for the maintenance of *forage* and hydraulic system.

The activation of this hybrid water and land management scheme shows an adaptation process – that we could call experimental – of which two main levels of transformation can be highlighted. The first is about the introduction of new horticultural perimeters to collective management (Figure 11.5), a hybrid structure based partly on the need to foster a dynamic access to the available resources, and partly on the rediscovery of *exploitations familiales*, the traditional structures of farm. Family is indeed recognised as the foundation from which both social and settlement relations most often arise, as well as basis of the production dynamics traditionally centred on the small-scale agricultural activities (UGPM 2009). Family being the core of local activities and initiatives, it is perceived and lived as a valid alterative and – to some extent – as a change-resistant social institution, which is not affected by the dynamics of social fragmentation and environmental challenges.

Figure 11.5: Mékhé collective management of the *périmètres maraichers* (Photos by the author, 2013).

The second issue is related to the changes at a territorial scale. The rehabilitation of community water facilities favoured significant changes in the dynamics of relationships among settlements in the area. First of all, the presence of peasants' groups and their work on the territory (although it may change in the different villages) has meant that every settlement gains a significant weight in the territory.

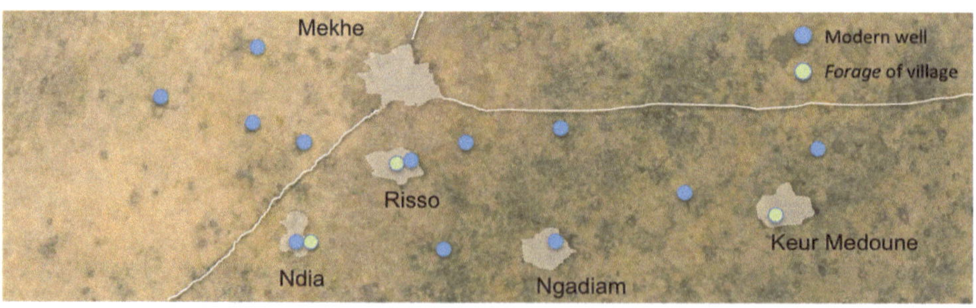

Figure 11.6: Localization of the main points of water (Elaboration of the author, 2013).

Particularly the process of *forage*, reactivation in some villages (Figure 11.6) has increased the exchanges and inter-village mobility, creating a more complex geography of relationships and hierarchies. The new system for the access and use of water raised new functions and related activities that have expanded the range of influence of the village on the surrounding environment involving – more or less directly – the bordering settlements. The increase of horticultural production, the exchange of related products, as well as the dynamics of formal and informal markets, not only attract neighbouring people, but also gradually bring the interest of other professional associations to gravitate to where more opportunities are being created.

11.5 Conclusions

The climate change discourse recognises the impact of climate-induced water shortages and stress on the vulnerable and poor communities, due to their lack of resilience.

This chapter aimed to reflect on the "social resources", which are meant to be the abilities of the community to adapt to climate change and to turn this challenge into opportunity.

The case study showed an experience in which the response to cope with water scarcity and the impacts of CC is based on a desire for change in the local community, sustained in the start-up by the technological and economic support of the international cooperation. This path shows the virtues of integration of the rehabilitation and upgrading of water infrastructure and technical components, but indissolubly combining the "rehabilitation" and enhancement of the local socio-economic patterns.

The tangible answer is therefore the increase in water availability and in agricultural production realized in a common framework of organization and management.

From this, a first criterion to be highlighted emerges. The allocation of roles and responsibilities as well as rights of access to available resources, based on socio-cultural peculiarities and on a shared choice of the organization and management forms, is a *conditio sine qua non* that initiatives or practices cannot provide an effective way of adaptation at the community level.

In this regard, the results of the interviews, used for the reconstruction of the case described, revealed some common elements in the people's opinion. The awareness of an increasingly critical vulnerability and experienced impotence of the individual to cope with it, has given rise to the perception that acting jointly, the network skills and knowledge across the territory may offer a viable alternative strategy.

Compared with the technological aspect, this element provides an answer less tangible but more intense at a procedural level. The considered experience shows a first but crucial step for the shift towards solutions based on the synergy between social and political potential – where "politics" should be understood in the broader sense as the art of governing, and therefore of leading, organizing and administering public goods.

In this sense, it should be noted that – regardless of the forms through which adaptation can take shape (voluntary rationing schemes, changes in cropping cycles, rainwater harvesting, groundwater mining, technology solutions) – it is undoubtedly the mobilization process, the decisive act at the community level for facing physical scarcity of resources and to the impacts of CC that are most impactful.

In the first part of the chapter, Ohlsson's theory described the method by which shifting the focus from "resource" to "social scarcity" can achieve an improved understanding of why and how certain societies cope with resource scarcity better than others.

But who takes or is able to take the lead in an adaptation process at community level?

An element that becomes so central in this proposal, it is the understanding of the social actors that can collect and deal with individual and collective instances, that take-charge of these, in order to direct them in a shared process of adaptation.

In the case study, the views of interviewed peasant groups emphasized the role of UGPM in helping them to gradually acquire a more detailed awareness on local issues, and to ensure, encourage and keep alive those relations of community, neighbourhood and sometimes symbolism, which keep the social and cultural patterns, by building a consensus more and more entrenched.

On the basis of this reflection, three main features would characterize the "elites", able to gather the potential of the broader social context:

- The awareness and knowledge on territorial specificity, socio-economic issues and on the related criticality/potential of the local resources;
- The quality of legitimacy by various social groups and populations, and therefore the level of authority recognition as guarantor for the entire community (that has a direct impact on the degree of responsibility that it can take);
- The ability to integrate demands and local needs that directly affect the degree of sharing of the community.

In conclusion, it is necessary to take into account that where there is no pre-existing social bias, communities can be supported by referring to their traditions, associations and indigenous systems. Self-management and popular participation are neither immediately nor easily transferable and can be full of pitfalls, however, awareness that the benefits of wide participation instead accrue over time and space is necessary.

References

Biconne, R. 2013. Discourse and practice on water management in the Southern countries. *Planum. The Journal of Urbanism* 27 (2): 36–51. ISSN: 1723-0993.

Biconne R. 2013b. Knowledge sharing for adaptation to Climate Change: a case in Malika, Senegal. In *Climate change vulnerability in southern african cities. Building knowledge for adaptation*, ed. S. Macchi, and M. Tiepolo, 125–140. Springer. doi: 10.1007/978-3-31900672-7.

Chauveau, J.P., M.C. Cormier-Salem, and E. Mollard. 1999. L'innovation en agriculture. Questions de méthodes et terrains d'observation. Paris: Éditions de l'IRD-Institut de Recherche pour le Développement. Issn: 0998-4658 Isbn: 2-7099-1442-5.

Conseil Regional de Thies. 2007. PRDI-Plan régional de développement intégré de Thies.

Dinar, A., H. Rashid, R. Mendelsohn, and J. Benhin. 2008. *Climate change and agriculture in Africa: Impact assessment and adaptation strategies*. London: Earthscan.

Eboh, E.C. 2009. Introduction: Debating Policy Options for National Development. Enugu Forum Policy Paper 10. Enugu, Nigeria: AIAE-African Institute for Applied Economics. http://www.aiaenigeria.org/Publications/Policypaper10. pdf.

Falkenmark, M., A. Berntell, A. Jägerskog, J. Lundqvist, M. Matz, and H. Tropp. 2007. On the verge of a new water scarcity: A call for good governance and human ingenuity. SIWI Policy Brief. SIWI.

Fall, A.D. 2009. Etude des impacts écologiques des stratégies économiques des exploitations agricoles familiales rurales dans la région de Mékhé au Sénégal. Mékhé: UGPM.

IPCC-Intergovernmental Panel on Climate Change. 2001. *Third Assessment Report of the Intergovernmental Panel on Climate Change. Parts 1, 2 and 3, Synthesis Report and Policy Makers Summaries*. Cambridge, UK: Cambridge University Press.

IPCC. 2007. *Impact, adaptation and vulnerability contribution of working group I of the IPCC to the third assessment report of IPCC*. London: Cambridge University Press.

IAO-Istituto Agronomico per l'Oltremare, Ministero degli Affari Esteri. 2003. Land evaluation in the Province of Thies, Senegal. Professional master "Geomatic and Natural Resources Evaluation".

Jeffrey, P., and M. Gearey. 2006. Integrated water resources management: lost on the road from ambition to realisation? *Water Science & Technology* 53 (1): 1–8.

Lavigne Delville, P. 2005. Conditions pour une gestion décentralisée des ressources naturelles. In *L'État et la gestion locale durable des forêts en Afrique francophone et à Madagascar*, ed. A. Bertrand and P. Montagne. Cirad.

Ministère du Plan, ANSD-Agence Nationale de la Statistique et de la Démographie. 2014. Recensement Général de la Population et de l'Habitat, de l'Agriculture et de l'Elevage (RGPHAE). Rapport Mars 2014.

Ngigi, S.N. 2009. Climate change adaptation strategies: Water resources management options for smallholder farming systems in Sub-Saharan Africa. New York: The MDG Centre for East and Southern Africa, The Earth Institute at Columbia University.

Ohlsson, L., and A. Turton. 1999. Water scarcity and social stability: towards a deeper understanding of key concepts needed to manage water scarcity in developing countries. Occasional paper. London: SOAS-KCL Water Issues Group, University of London.

Paavola, J., T.W. Adger. 2006. Analysis of fair adaptation to climate change. *Ecological Economics* 56: 594–609.

Parry, J., A. Hammill, and J. Drexhage. 2005. Climate change and adaptation. Manitoba, Canada. http://www.iisd.org/.

Pereira, L.S. 2005. Water and agriculture: Facing water scarcity and environmental challenges. *Agricultural Engineering International- The CIGR Journal of Scientific Research and Development* 7.

Poteete, A.R., J.C. Ribot. 2011. Repertoires of domination: Decentralization as process in Botswana and Senegal. *World Development* 39 (3): 439–449. doi:10.1016/j.worlddev.2010.09.013.

Ronzon, F. 2008. Sul campo. Breve guida pratica alla ricerca etnografica. Roma: Meltemi editore.

UGPM-Union des Groupements Paysans de Mékhé. 2009. Pour un développement "social" de la région de Meckhé. Document d'orientation stratégique de l'UGPM.

UNDP-United Nations Development Programme. 2006. Human Development Report "Beyond Scarcity: Power, Poverty and the Global Water Crisis". New York: UNDP-Human Development Report Office.

UN-Habitat. 2010. *The state of African cities 2010, Governance, inequality and urban land markets*. Nairobi

UN-Water. 2006. Coping with water scarcity. A strategic issue and priority for system-wide action. UN-Water thematic initiatives.

UN-Water, 2012. World water development report 4. World water assessment programme.

Venot, J.P. 2011. What commons? Rethinking participation in the sub-saharan african water sector. 13th Biennial Conference of the International association for the study of the commons, Hyderabad-India, Jan. 10–14, 2011.

Yin, R.K. 2003. *Case study research: design and methods*. SAGE Publications.

WBGU-German Advisory Council on Global Change. 1998. World in transition: Ways towards sustainable management of freshwater resources. Report 1997. Berlin, Heidelberg, New York: Springer.

Ziervogel, G., A. Catwright, A. Taas, J. Adejuwon, F. Zermoglio, M. Shale, and B. Smith. 2008. Climate change and adaptation in african agriculture. Research report for Rockefeller foundation prepared by the Stockholm Environment Institute. Stockholm.

Maurizio Tiepolo[31] and Sarah Braccio[32]

12 Flood Risk Preliminary Mapping in Niamey, Niger[33]

Abstract: Flood mapping is still rare in the large cities South of Sahara. The lack of information on the characteristics of floods, the orography of the sites and the receptors hampers its production. However, even with scant information, it is possible to create preliminary risk mapping. This tool can be used by local administrations in decision making on emergency plans or on climate change (CC) action plans. From 2010 onwards, the River Niger at Niamey (1.1 million inhabitants, 123 km^2 in 2014) swelled at unseasonal times. This new river flood pattern can be linked to CC. Each flooding event affected thousands of people and homes. The steady development of areas that did not appear to be flood prone in the past is the main cause of these impacts. These areas require special measures if further impact is to be avoided in the future. This chapter presents the preliminary flood risk map of Niamey 1:20,000. The map was built up using an historic approach (flooded area derived from satellite images) and considering risk (R) as the result of hazard (H) and damage (D), R = H * D. Risk was measured according to two scenarios: medium and high probability of flooding. The inverse of the return period of river and pluvial flooding (H) and the potential damage to buildings and crops according the water depth were used. Information to measure risk components was sourced by daily rainfall and daily discharge of the River Niger from 1946 to 2014, and from high-resolution satellite images (2014). The risk map id entifies hot spots for emergency and CC action planning. The fifth district alone contains 52% of the potential damage. Ninety-nine percent of the potential damage is concentrated within 225 hectars. Reinforcing existing embankments and constructing new ones to protect these areas seems more appropriate than resettle-

31 Maurizio Tiepolo, associate professor of urban and regional planning at DIST-Politecnico and University of Turin, is the author of paragraphs 12.1, 12.2, 12.5, 12.6 and 12.7, maurizio.tiepolo@polito.it
32 Sarah Braccio, planner PhD and research grant holder at DIST-Politecnico and University of Turin, is the author of paragraphs 12.3, 12.4 and 12.5, sarah.braccio@polito.it
33 This chapter is created with information taken from the Atlas of local resources (Braccio and Tiepolo 2014) produced by the EuropeAid 127763/C/ACT/TPS project, supplemented by geomorphological, meteorological and hydrological data collected as part of the ANADIA-Niger project (Italian Ministry for Foreign Affairs, Directorate General for Development Cooperation). The Authors would like to thank Katiellou Gaptia Lawan and Moussa Moujaimini (National Directorate for Meteorology of Niger), Abdourhamane Daouda (Hydraulics Ministry), Harouna Mato (IGNN) for the information, Abdou Adam and Ibrahim Moussa Maiga (Niamey Municipality) for the information and field work, Alessandro Sotrios d'Acquisto and Elena Ferro for the photo-interpretation of the fields and built-up areas in the flood prone areas.

ment. The cost of the works would equal the potential damage if it remains within 2,580 euro/ml.

Keywords: Preliminary flood risk map, Hazard, Pluvial flood, River flood, Climate change, Damages, GIS, Niamey, Niger

12.1 Introduction

In recent years, the large cities of South of Sahara have been struck increasingly frequently by floods (Tiepolo 2014a), events that are believed to be attributable to climate change (CC) (UN-Habitat 2014, 2011; Douglas *et al* 2008; Satterthwaite *et al* 2007). The initiatives to support local governments in adopting climate emergency and action plans are multiplying, such as the *Cities and Climate Change Initiative* (UN-Habitat 2012). Localising the areas at highest risk (hot spots) and estimating the potential damage are the first steps for planning the adaptation to CC. This information should be provided by flood risk maps. However, urban flood mapping in the South of Sahara is still sporadic and is mostly limited to identifying flood prone areas (Ould Sidi Cheikh *et al* 2007), main classes of receptors as built-up, farm land, etc. (Mbaye *et al* 2004; Bello *et al* 2014), occasionally estimating the amount of affected people (Mutanga *et al.* 2013; Oriola *et al* 2012; Mendelshon *et al* 2010). Few large cities have a fully-fledged flood risk map (Ponte 2014; Credel 2010; Diallo 2009) and this tool rarely estimates the potential damage. Flood mapping of a single hazard is more frequent. The main reasons for these gaps without a doubt include a deficit of information: fragmented databases or databases that are too brief to characterise CC (Tiepolo 2014a), lack of recordings of flood depth, submersion time, damage to the receptors. Another reason is the propensity to use hydrological or hydraulic models in flood mapping: an approach that requires lengthy timescales and rarely generates products that are appropriated by the local South of Sahara administration.

This chapter demonstrates that it is possible to create a preliminary flood risk map even with scant information and simplified methods functional to decision-making. The chosen case was Niamey (1.1 million inhabitants, 123 km^2). From 1998 to date, Niamey has been flooded 4 times by the Niger River. Thousands of homes have been destroyed and a large part of urban and peri-urban agricultural production has been lost. According Sighoumou *et al* (2013), the main cause of flooding would be the increased run-off resulting in the soil denudation process and consequent soil crusting after agricultural area increases and abandonment of the practice of fallow. This interpretation, however, is not based on a sound analysis of rainfall. In this regard, Panthou *et al* (2014) have found a sharp increase in extreme rainfall (by number and contribution to yearly precipitations) between 2001 and 2010 compared to previous decades. The phenomenon reached a peak in the northeast of Burkina Faso, at the head of the watersheds of the right bank of River Niger tributaries: Gorouol, Dargol and Sirba.

Following these disasters, various attempts at flood mapping were made. Bechler-Carmaux *et al* (2000) opened the way with a river flood vulnerability map which estimated the population and the major equipment exposed to flooding. At least three river flood event maps followed, which identified the flooded area (SERTIT 2012a, 2012b; OCHA 2012) and the flood depth (ABN/CRA 2007). However, none of these tools is functional to decision-making.

This chapter proposes instead a preliminary flood risk map that stands out from previous maps due to its purpose, approach and content. The purpose is to produce a tool for decision-making on flood adaptation measures. The data, managed in a GIS and displayed on the map, should help to identify the adaptation measures. It is a case of estimating the potential damage in monetary terms relative to affordability of adaptation measures.

In the following pages, we will introduce (i) the methodology, (ii) the hazard, (iii) the flood prone areas, (iv) the damage, (v) the risk, (vi) the use of a preliminary risk map and (vii) the conclusions.

12.2 Methodology

The determination of the boundaries of the flood prone areas was done using the historic approach (WMO 2013: 18), based upon the perimeter of the flooded areas on the occasion of previous floods. As a result, the preliminary flood risk map allows for the pluvial and fluvial flood risks to be calculated and compared. Having identified the flood prone area according to different scenarios of probability of the event, the risk (R) is determined by two components:

- Hazard (H), "a dangerous phenomenon that may cause loss of life, injury, property damage, loss of livelihoods, social and economic disruption or environmental damage" (United Nations 2009)
- Damage (D), "direct damage to buildings or motor vehicles and their contents, which requires the repair, reconstruction or replacement of the property, losses of use for industrial and commercial activities, indirect losses such as rental, resettlement costs, supplementary costs, etc." (MEDD, ONRN 2012)
  according to the equation $R = H * D$ (UNISDR 2011; UNDP, BCPR 2010; Marzocchi *et al* 2009).

The river flood prone area was determined according to a scenario of high and medium probability of occurrence. The high probability scenario corresponded to the flood event map produced by SERTIT (2012a and 2012b) by the interpretation of data from the satellite image radar TerraSAR-X 29 August 2012, taken when the discharge was 2,105 m^3/s and the height of the water was 580 cm above zero of the flood gauge (corresponding to 175 m altitude). This is a plausible flooding scenario with an estimated 34% annual chance of occurrence. The flooded surface was 1,847 Ha. The key assumption was that the embarkments remained intact. The scenario of medium probability corresponded to the peak of 617 cm (discharge of 2,477 m^3/s) reached on 22 August 2012. This was the flood peak of the last 69 years. The flooded area was 3,019 Ha. The key assumption was that the embarkment on the River's left banks broken, as happened during the 2012 flood. This should have generated an increase of 759 Ha in the flood prone area. For the purposes of determining the boundaries of the flood prone area, we considered that the water line was slightly sloping: 16 cm/km upstream and 7 cm/km downstream of the Kennedy Bridge (ABN, CRA 2007: 2). Since the built-up area in the floodplain today extends along the axis of the river for 6 km to the North and to the South of the Kennedy Bridge, we calculated a variance of 138 cm between the two extremes of the flood prone area and an average value of the water level of + 69 cm to be added to 617 cm, i.e. + 106 cm compared to 580 cm altitude. Close to the Kennedy Bridge, this operation generates a water level that exceeds 617 cm but which should be considered as the precautionary principle with respect to the error between the Digital elevation model (DEM) and the real profile of the land. This operation required the production of a DEM based upon the contour lines of the IGN maps 1:20,000 and 1:50,000 (IGN 1978 and 1980) with 2.5 m interval. The two perimeters allowed for the identification of 2 areas of flood depth: 0-1 m and higher than 1 m.

The pluvial flood prone area is determinate according a high probable scenario. Using a methodology already tested in Maputo (Braccio 2014), which calculates the MNDVI index of the Landsat images, water bodies and wet soil after the rainfall on 13 June 2002 (68.5 mm), were identified on a Landsat image taken on 17 June 2002. This event is a pluvial flood with a high probability of return. Within the flood prone area, the main receptors (buildings and crops) that may be damaged were identified.

The hazard was expressed as the probability of a pluvial or river flood occurring once in the next year. The probability is the inverse of the return period: $RP = (n+1)/m$ where n is the number of years on record and m is the number of recorded occurrences of the flood/rain being considered. The discharge/rainfall annual peak was provided by the Hydraulic Ministry and by the National Meteorological Directorate, respectively. The period observed was 1946–2014, as the set of data on discharge of the river in the previous 18 years is incomplete (Abrate 2007).

Flood damage depends upon the flood depth and duration, the flood velocity and the size of the sediments in the water. In Niamey, as in other tropical cities, the damage assessment can only take account of the flood depth, since there is a dearth of information on the other components. Also, when considering only flood depth, we do not know the damage (% on the monetary value of the building) with respect to the water depth. In this case, we used the average between 2 curves stage-damage produced for buildings with one floor and no basement and for buildings with several floors and no basement similar to those that were found in Niamey (NFIP 2013; Davis and Skaggs 1992). The average values were 27% and 45% for water 1m and 2m deep for buildings, respectively, with one floor and no basement. Those values dropped to 23% and 34% for buildings with a number of floors. In the case of adobe housing, curves stage-damage were not used because the building collapses as soon as the water enters it.

The damage was expressed as a percentage of the cost of construction. The value of the buildings could not be acquired from the Cadastre, as it is possible to do in other countries, as informal constructions, common in the flood prone area, are not registered. As a result, the buildings were identified by way of photo-interpretation of the satellite image taken on 16 October 2014 available on Google Earth, i.e. just over two years after the great flood. Instead of the value, the cost of construction of a one-floor 4 x 6 m module in perforated concrete blocks and one in adobe plus bathroom and kitchen were estimated. In the flood prone areas, homes can consist of one, two or four modules. The costs of construction were deducted from the price list created by Zaneidou (2013) for the International Fund for Agricultural Development (IFAD/FIDA) which, in turn, is also based upon the national reference price established by the government on January 15, 2012.

For crops, the damage (D) was calculated based upon average yields per hectare and prices per kg (2008–2012) of millet and paddy in Niamey (RN, MF, INS 2013a) assuming that those crops behave like wheat, which, as is known (Förster *et al* 2008: 314), does not withstand submersion for longer than one day.

Contrary to the definition of damage provided above, we did not estimate the potential damage to building contents, urban infrastructure, vehicles, clean-up costs, losses of sales, production, or intangible damages.

In order to facilitate interpretation of the flood risk map, only the 5th *arrondissement* (fifth district) will be shown. That decision is justified by the fact that on August 22, 2012, 52% of the collapsed houses and 57% of those damaged in the whole city were located in this jurisdiction, followed at a distance by the fourth district with 36% and 37% respectively (RN, VN 2012: 4).

The receptors and construction costs are constantly evolving. In just two years in the flooded area, 20% of the collapsed homes have been rebuilt. The prices of construction materials increase significantly from one year to the next due to the strong demand for cement and iron rods to implement the national program of public works: infrastructure in Niamey (including the second bridge on the River Niger and the railway Niamey-Cotonou), the refinery in the region of Zinder, the pipeline, the construction of three cement factories in the region Tahoua, etc. Since Niger is not a producer of modern construction materials and has a limited truck fleet, the high demand of cement and iron rods increases costs. Storage costs in Niamey contribute increasing costs of construction materials. For these reasons the risk map should be regularly updated.

12.3 Hazard

Niamey is exposed to pluvial and river floods. The inverse of the return period (probability of the flood occurring once in the next year) of the annual maximum discharge (Figure 12.1) will be used to express the hazard in the equation of the risk.

This chapter considers a scenario for pluvial floods (high probability) and two scenarios for river floods (high and medium probability). The thresholds of the scenarios are less than 3% of probability to occur in the year for the medium scenario, and more than 3% to occur for the high probability scenario. There are no scenarios with low probability (less than 1%) as the records are limited to 69 years. These ranges are used in various countries (EXCIMAP 2007).

13 June 2002 is an example of a pluvial flood (68.5 mm) with a high probability (34%) of occurring every year.

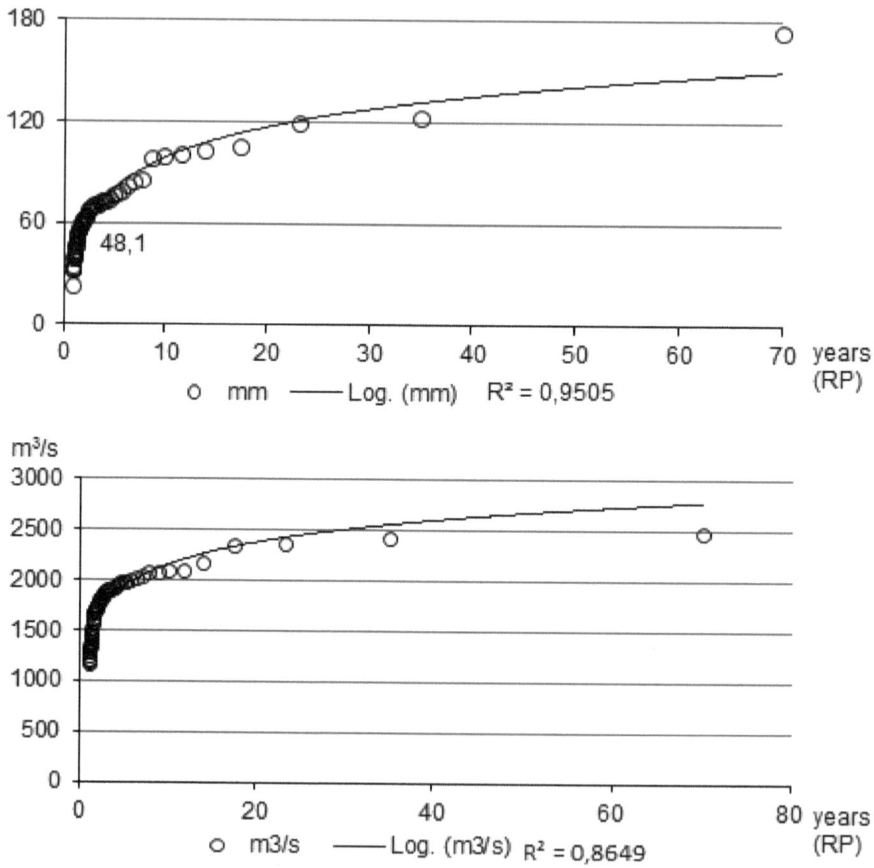

Figure 12.1: Niamey, 1946–2014. Annual maximum rainy days at the Airport (top) and annual maxium discharge (m³/s) of the River Niger (bottom) organised according mm and return period (years)

For river floods, the reference was the annual maximum discharge (m^3/s) of the Niger River as recorded in Niamey from 1945 to 2014. The flood stage of 22 August 2012 (2,477 m^3/s, 617 cm at the river gauge) was one scenario of medium probability (1.4%) and the one flooded on 29 August of the same year (2,105 m^3/s) was a scenario with high probability (10%) of occurring every year.

More probable scenarios do not threaten buildings or crops and therefore do not generate potential damages.

The values of expected frequency for pluvial floods and for river floods to be considered in the hazard were respectively 0.34 in the scenario of pluvial flood, 0.01 and 0.1 in the two scenarios of medium and high probability of river flood.

Regarding the exposure to disastrous natural events, Niamey has, therefore, a greater probability of pluvial than fluvial flood. In the former case, the duration rarely exceeds one day, while in the latter case it can last up to two weeks.

In the first decade of the 21st century, the heavy floods resumed but with a completely different pattern compared to the past. The river is characterized by two floods in Niamey. First of all, a local flood, which occurs between August and September, and is fuelled by rainfall on the drainage basins of the right bank tributaries upstream of the capital, known as "red flood" due to the suspended matter transported by the run-off on laterite soils. Secondly, a "black flood" originating from the upper River Niger, which comes between December and January (dry season). In the last 9 years, the local red flood exceeded the discharge of the black flood. The change was striking in 2006, 2010, 2012 and 2013 with floods in rainy season far in excess of those in dry season (Figure 12.2).

The disastrous floods of recent years have been attributed (Sighomnou *et al* 2013) to the run-off on increasingly extended denuded surfaces. Any attempt at understanding the dynamics of the phenomenon should begin with the pluviometric characterisation in the watersheds of Sirba, of Gorouol, of Dargol (in order of average annual

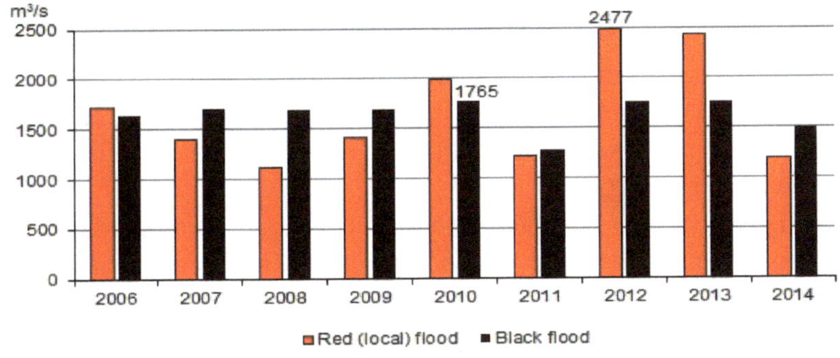

Figure 12.2: Niamey, 2006–2014. Black flood peak (black) and local flood peak (red) (source: Min. Hydraulique Niger)

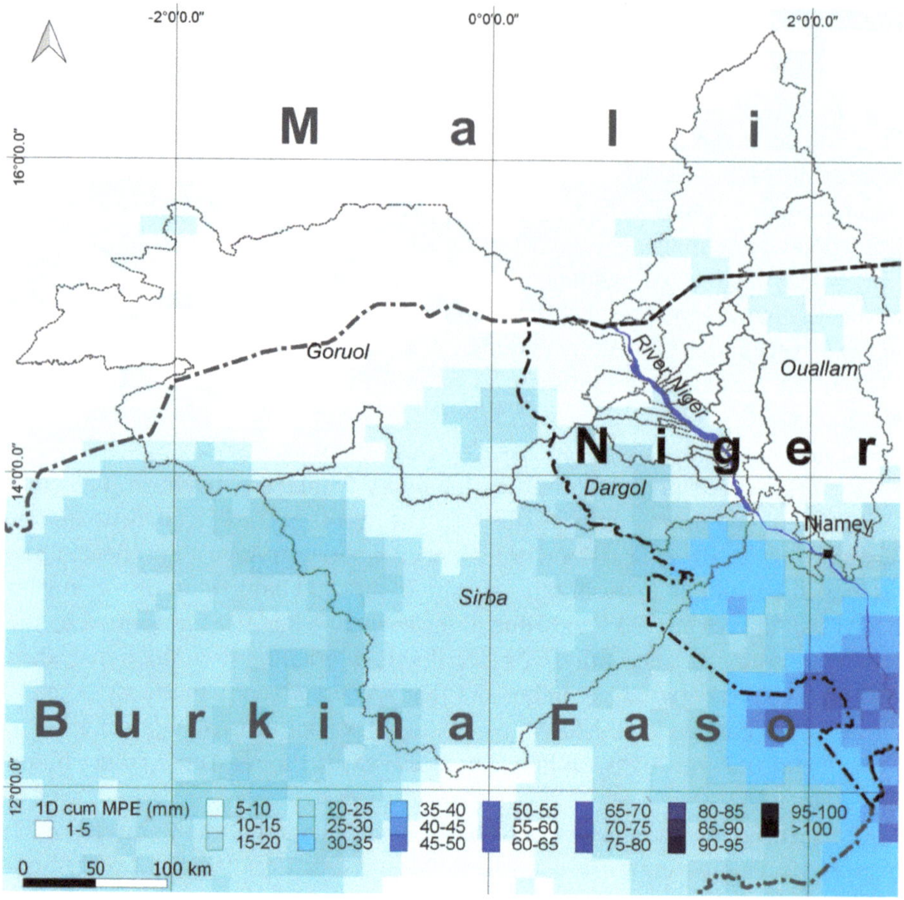

Figure 12.3: Niamey, 21 August 2012. Daily precipitation in the main catchments upstream of the capital city on the eve of the great flood of 22 August 2014 (EMMA 2012, by S. Braccio)

discharge) and the numerous intermittent creeks, i.e. on approximately 90,540 km^2, which the 5 rain gauges considered by Sighomnou *et al* do not represent adequately. In order to increase the significance of the analysis, the total daily precipitation from EMMA (2012) can be used. This analysis shows that the Niamey flood of 2012 was preceded by intense rainfall in the watersheds of the right bank upstream of the capital on the eve of the event (Figure 12.3).

August's flood peak coincides with the maturation of rain-fed crops and with the transplant of the seedlings in the paddy fields (Sido 2011). The flood, at this critical time, destroys the crops.

At this point, the event of coincidental flooding (from river and rain source on the same days) should be evaluated. If we consider the past three major river floods, it is noted that in 2012 and 2013 the flood peak was reached in the rainy season, lasted only one day, and was reached quickly. In the 4 days preceding the flood peak,

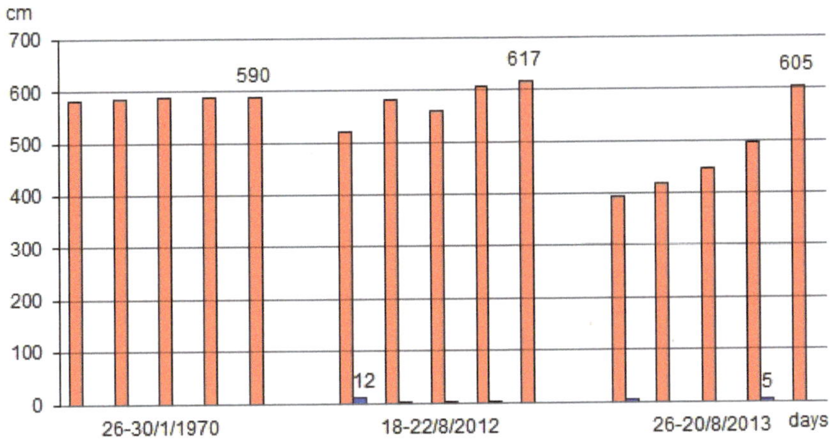

Figure 12.4: Niamey, 1970, 2012, 2013. Coincidental flooding analysis: cm of river peak (red) and mm f precipitation (blue) in the 4 days before the three main river floods of the last 69 years.

there had been a significant rainfall in Niamey only in 2012. But this event happens 4 days before the peak: too a long period to be considered responsible for the flood. In 1970 the flood peak was reached very slowly in the dry season and lasted 6 days (Figure 12.4). Up until now, coincidental flooding has not occurred, but in the future, if the river regime remains the same as the last years, it could happen. Therefore, it is worth considering both hazards.

12.4 Flood Prone Areas

The flood-prone areas were determined in the five districts according to hazard (pluvial or river flood), flood depth (0–1, > 1 m) and probability scenario (high, medium) (Figures 12.5 and 12.6). These areas are located in 4 different positions: in the flood plain of the Niger River (river flood), at the confluence of the intermittent creeks with the river, along the intermittent creeks and in the depressions, particularly on the left bank (pluvial flood). These are therefore fragmented areas, especially on the left bank.

The river and pluvial flood prone areas together extend for 28.4 km² (Table 12.1). However, in this enormous space, only 2 km² are now urbanised: 1.6% of the capital's built-up area (Tiepolo and Braccio 2015). This is therefore a very limited surface area compared to the other large cities South of Sahara provided by flood mapping, like Cotonou.

Moreover, in the fifth district of Niamey, the urbanised river flood prone area is distributed over just 4 main sites. This was a particularly favourable condition for implementing the adaptation measures as it allowed for efforts to be concentrated

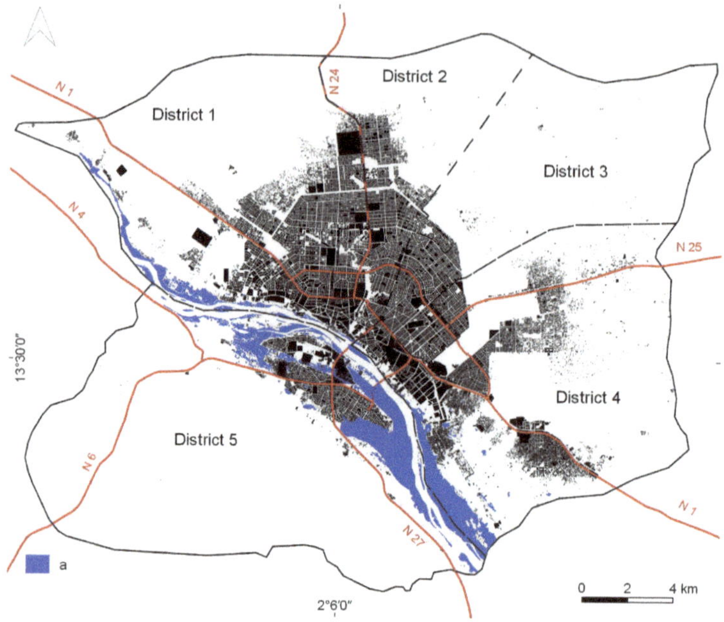

Figure 12.5: Niamey, 2014. Areas prone to (a) highly probable pluvial flood (by S. Braccio)

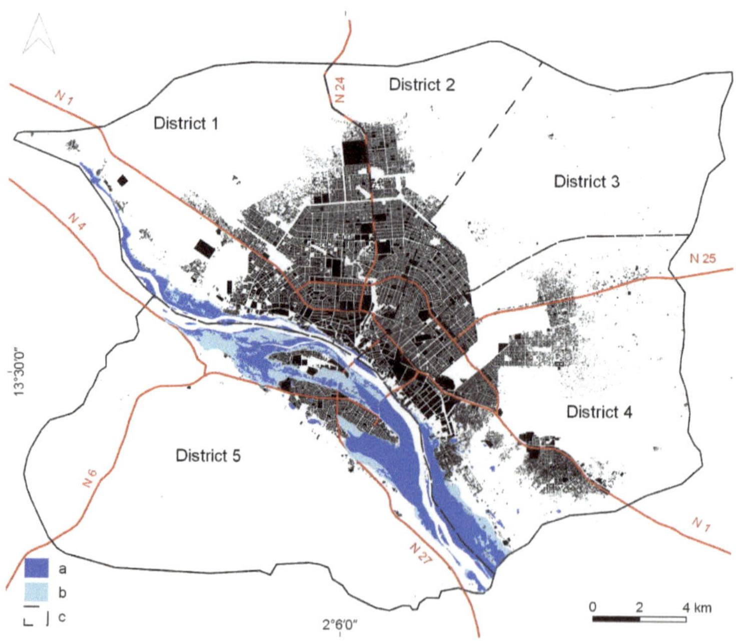

Figure 12.6: Niamey, 2014. Areas prone to river flood with (a) over 1 m water depth, (b) 0–1 m water depth, (c) districts limits, built up areas at 2014 (black) and national roads (red) (by S. Braccio)

Table 12.1: Niamey, Flood prone areas (by S. Braccio)

Jurisdiction	Flood prone areas due to	Scenario	
		Medium probability of occurrence Ha	High probability of occurrence Ha
Niamey	River flooding > 1 m	1,847	0
	River flooding 0–1 m	1,172	1,847
	River flooding total	3,019	1,847
	Pluvial flooding	–	1,010
Fifth district	River flooding > 1 m	951	0
	River flooding 0–1 m	907	951
	River flooding total	1,858	951
	Pluvial flooding	–	485

rather than distributing them across many different sites, as would be the case in Luanda (15 sites), Maputo (11 sites), Dar es Salaam (7 sites), Nouakchott (5 sites).

In the fifth district, 4.8 km^2 have a high probability of being flooded after heavy rains. Regarding the river flood, 18.6 km^2 have a medium probability of being flooded and 9.5 km^2 have a high probability of being flooded (Table 12.1).

12.5 Damages

Today, the river flood prone areas in the fifth district are mainly cultivated (49%), unused or little used (38%), and to a lesser extent, built-up (13%). The cultivated parts include irrigated crops (paddy, gardens) and rain-fed crops (millet).

The unused or little used parts include barren land, road side slopes, ponds, or receptors for which the damage is difficult to estimate (groves, pastures).

The built-up part of the fifth district includes plots not yet constructed, which, in the event of pluvial flood, are exposed to erosion and damage to the enclosure wall. Then, there are over 8,000 one floor houses, 67% in durable materials and 33% in adobe. There are also schools which, if flooded, risk losing their function as the initial reception for victims, or simply remain unusable if the classrooms consist of wooden shacks (Figure 12.7).

Although the area was devastated by the flood of 22 August 2012 and by that of 30 August 2013, the reconstruction was rapid. Occasionally, the homes were rebuilt in adobe, an inexpensive material not resistant to water, sometimes on a reinforced

Figure 12.7: Niamey, fifth district, 2013. A selection of receptors: gardens (top left), classrooms in wooden shacks (top right), incremental adobe housing (bottom left), river banks (bottom right) (pictures by S. Braccio and M. Tiepolo)

concrete slab. The stubbornness to remain in the same flood prone site is explained by the lack of alternatives at such a short distance from the downtown (just 4–6 kilometres from the Big market).

Moving to the suburbs, in search of a new plot, would involve huge expenditure and a distance of over 10 kilometres from the downtown.

In the fifth district, according to the high probability scenario, the river flood prone area was 951 hectares and is essentially occupied by fields. The potential damage was in the order of 1.2 million Euros.

According to the medium probability scenario, there were total hectares of 1,858 including 1,832 which were fields and 26 which were built-up where there were just over 8,000 buildings (Table 12.2). The potential damage was 1.4 million Euros for the crops and 40 million Euros for the buildings making a total of 41.4 million Euros.

According to the high probability scenario, the pluvial flood covers a much smaller surface area (485 hectares) and mainly affects the fields, while only two hectares were covered by buildings (Table 12.2). The damage was concentrated only on buildings in adobe and not on those in durable materials or the crops. The poten-

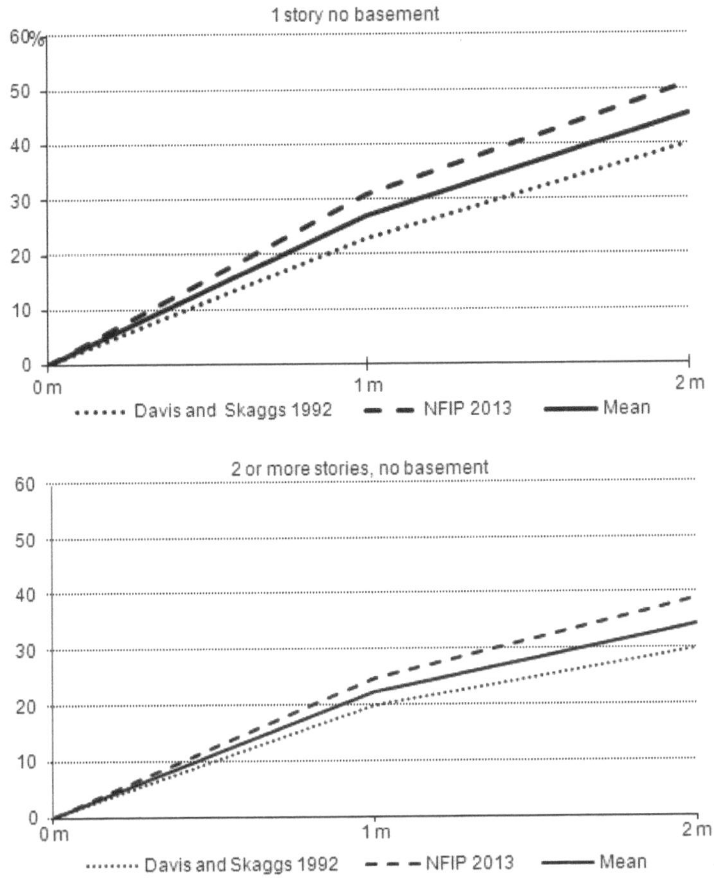

Figure 12.8: Curve stage-damage for buildings with one floor and no basement (a) and two or more floors and no basement (b) in durable materials and the mean considered for this study (source: Davis and Skaggs 1992; NFIP 2013)

tial damage due to collapse or flood of residences in adobe could be estimated at 1.1 million Euros (Table 12.2).

12.6 Risk

The risk is based upon the hazard and the damage: $R = H * D$; where $H = 1/RP$ (inverse of the return period), D = damage based upon flood depth, building materials and number of floors.

The hazard (1/RP) varies from 0.01 (medium probability of river flood) to 0.34 (high probability of pluvial flood) and influences the value of the risk less the damage, which varies from 0.23 to 1. The housing stock in adobe greatly affects the value of the

Table 12.2: Niamey, fifth district, 2015. River flood damage according the 2 scenarios.

Scenario Receptor	m²	€/m²	Total M €	D %	D M €	H	R
Pluvial probable							
Housing concrete	12,049	519	6.25	0	0	0.34	0
Housing adobe	5,934	191	1.13	100	1.13	0.34	0.38
Paddy	3,320,167	0.23	0.76	0	0	0.34	0
Millet	1,013,579	0.03	0.03	0	0	0.34	0
Barren, pasture		0		0	0	0.34	0
Total pluvial probable	4,492,772				1.13	0.34	0.038
River probable 0–1 m							
Housing concrete	648	519	0.34	27	0.09	0.1	0.01
Housing adobe	0	191	0	100	0	0.1	0
Other buildings	376	519	0.20	23	0.04	0.1	0
Plots and streets	39,272	0	0	0	0	0.1	0
Paddy, gardens	4,529,298	0.23	1.04	100	1.04	0.1	0.1
Millet	1,429,565	0.03	0.04	100	0.04	0.1	0
Barren, pasture	3,511,936	0	0	100	0	0.1	0
Total river probable	9,511,095		1.61		1.21	0.1	0.12
River middle 0–1 m							
Housing concrete	150,622	519	78.17	27	21.11	0.01	0.21
Housing adobe	74,894	191	14.30	100	14.30	0.01	0.14
Other buildings	38,396	519	19.93	23	4.58	0.01	0.05
Plots, streets	1,987,117	0	0	100	0	0.01	0
Paddy	265,862	0.23	0.06	100	0.06	0.01	0
Millet	2,977,458	0.03	0.08	100	0	0.01	0
Barren, pasture	3,573,159	0	0	100	0	0.01	0
Total	9,067,510		112.47		40.06	0.01	0.40
River middle > 1 m							
Housing concrete	648	519	0.34	46	0.15	0.01	0
Housing adobe	0	191	0	100	0	0.01	0
Other buildings	376	519	0.20	35	0.07	0.01	0
Plots and streets	39,272	0	0	0	0	0.01	0
Paddy	4,529,298	0.23	1.04	100	1.04	0.01	0.01
Millet	1,429,565	0.03	0.04	100	0.04	0.01	0
Barren, pasture	3,511,936	0	0	100	0	0.01	0
Total	9,511,095		1.61		1.30	0.01	0.01
Total river middle			114.08		41.36		0.41

risk, despite constituting only one third of the total stock and being worth just 37% of the equivalent stock (by surface area and number of floors) in durable materials. The fact is that buildings in adobe, if flooded, collapse within a few minutes. This is what leads to the potential damage of 100% (1) while for buildings in durable materials, the potential damage never exceeds 47%. Therefore, the pluvial floods seem hazardous only for buildings in adobe and not for other constructions or for crops.

Conversely, the river flood is always disastrous for buildings in adobe and for crops.

Areas with a medium probability of river flood presented the highest overall risk value, followed by those with high probability of river flood and, finally, those with high probability of pluvial flood (Table 12.2).

In the fifth district, the biggest risk (hot spots) was concentrated in four areas (Figure 12.9) which overall measure just twenty-six hectares and present a potential damage of 40 million Euros.

In Niamey the building process is hectic due to rural-urban land use conversion, the increase of density in built-up areas, and continuous increase of prices of construction materials. Consequently, the value of recipients of damage change, even though buildings will always be worth more than crops by unit of surface area and on the total of the exposed receptors. As receptors change over time, the preliminary flood risk map changes. Maps should therefore be understood as a graphic expression of an information system to be kept regularly updated.

12.7 Use of the Preliminary Flood Risk Map

The preliminary flood risk map is based upon the estimate of the construction costs (buildings) or on the sale price (crops) of the receptors and the flood probability: information that is essential in decision making.

The flood adaptation measures are positioned between two extremes: resettlement of inhabitants and activities and the protection of buildings. The receptors in the fifth district alone have (2013) an estimated cost of 114 million Euros including 113 million for the building cost.

The resettlement of inhabitants cannot be limited to finding 8,000 plots in a non-peripheral area. It is also necessary to provide a home (or a core of a home), drinking water, storm water drainage and electricity mains, otherwise people will not move or, if they do, it will only be temporarily. The costs of resettlement thus amount to a minimum of 113 million Euros, plus infrastructures and the purchase of land. This is a significant sum for a city that in 2012 had revenues for investments of just 9 million Euros (RN, VN 2013). It would also then be necessary for there to be a condition that the local government currently does not possess: an iron grip on illegal buildings, to prevent them reoccupying flood prone areas after resettlement.

On the other hand, it is worth constructing and strenghtening embankments for 15.5 km to protect the hot spots in flood prone areas, leaving only the fields exposed to flood, provided that this costs less than the presumed damages to the existing buildings (40 million Euros, in other words 2,581 euro/ml).

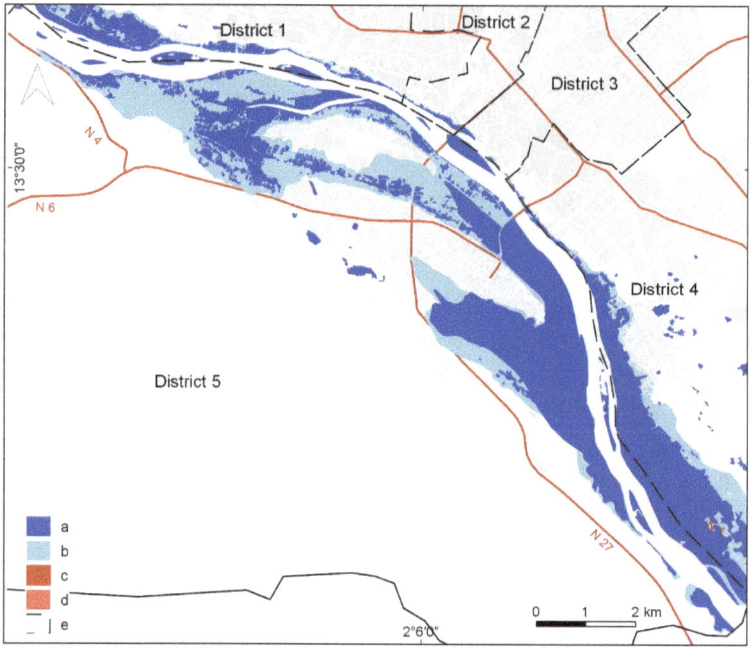

Figure 12.9: Niamey, fifth district, 2014. Preliminary river flood risk map: (a) over 1 m water depth, (b) 0-1 m water depth, (c) flooded houses in over 1 m waters, (d) flooded houses in 0-1 m waters, (e) district limits and built up areas (grey) and national roads (red) (by S. Braccio)

A number of intermediate measures are, of course, possible. They range from the early warning, to emergency plans, to preparedness, to raised entrances, to elevated latrines, etc. (Tiepolo 2014b).

The estimate proposed here is, at present, the best possible within the snapshot tools: the flood prone area, the receptors, their value can be estimated more accurately by way of the following, in order of importance: a more sophisticated Digital Elevation Model (DEM), the production of a curve stage-damage referring to the case of Niamey, the estimate of the impact of runoff-damage, updates of the potential damage based upon the Information system on agricultural markets and the prices index. It remains true that, in any information system, it is necessary to start with an embryo to be expanded later. This was the objective of this chapter.

12.8 Conclusions

From 2008 to date, the regime of the River Niger in Niamey has changed. The alternation of local (red) floods (less important) in August to black floods (annual peak) between January and February, as we have known since the end of the century, inverted in 2010, 2012 (flood peak of the last 69 years) and 2013.

The red flood is generated by particularly intense rainfall in a vast region upstream of the capital which includes south-east Burkina Faso, crossed by the major tributaries of the River Niger (Sirba, Gorouol, Dargol).

In the 34 years preceding 22 August 2012, the river at Niamey had been silent: it had never exceeded 1,940 m^3/s of discharge (544 cm at the river gauge). In the meantime, the built-up area had grown from 43 to 123 km^2, also occupying the flood plain on the right bank. The flood of 2012 (617 cm at the river gauge) "broke up" 5,533 adobe houses and flooded a further 3,324 in durable materials, producing 44,713 displaced persons on the right bank alone (RN, VN 2012).

The new regime of the River Niger at Niamey also affects the rain-fed cultures (millet) and irrigated cultures (paddy, gardens). At the time, in August, the floodplain on the right bank was dry, the rain-fed crops were maturing and the rice seedlings were being transplanted in the paddy fields of the ONAHA protected by embankments. The peak of 22 August 2012 flooded these areas and the water remained enough to destroying the crops.

Pluvial floods are responsible for lesser damages: there are few areas in which pluvial water stagnates and, for the most part, they are not inhabited.

However, Niamey has reached a critical threshold: almost all flat land, free from risks, is built-up or parcelled out. If the built-up surface area of the capital continues to grow by 50% even in the coming decade – and this will depend upon the demographic dynamic and the need to invest the oil revenue – it will move even further into the flood prone areas, being those costing less and the nearest to the downtown. The first signs are already visible on the sites flooded on 22 August 2012, where at least 20% of the houses seemed to be rebuilt.

Today, in the fifth district (where over half of all the capital's receptors are present) there are 26 hectares of buildings with a medium probability of being flooded with presumed damage of 40 million Euros. The maximum concentration of receptors is in the districts of Lamordé, Karadjie, and Kirkissoye (hot spots). One-third of homes are made from adobe. The areas are delimited: they could be protected with 15.5 km of embankments. They are also close to the downtown, which could make the protection a priority over resettlement, if the intention is to ensure that most of the people should live close to the downtown, reducing commuting and GHG emissions. In addition, protection costs a third of resettlement.

The preliminary flood risk map is the graphic expression of a GIS. Three improvements would be recommended to the work performed up until now:

- More accurate damage assessment. DEM with a high geometric resolution, less than or equal to 1 m and curve stage-damage ascertained for the specific context of Niamey.
- Enhance the pluvial flood scenario. By selecting and analysing satellite imagery in correspondence with the rains with medium probability of occurring, a more accurate perimeter of the flood prone areas could be defined.

– A study on run-off prone areas. The extension of those areas is significant, albeit not yet very built-up. However, these areas are now parcelled up: today's city and that of tomorrow is being extended to them. These are therefore the hot spots of the coming years on which Niamey municipality should focus its preventive actions to reduce the pluvial flood risk.

The preliminary flood risk map presents two useful cases:

First, it demonstrates the potential of flood risk mapping by way of free sources of information and a GIS that can be used by the local administration;

Second, it is an aid to decision-making, not so much to block the extension of construction in the risk sectors, as intended by Bechler-Carmaux *et al* (2000), but, rather, to localise the adaptation priority measures in the short and medium-term: early warning, public awareness and preparedness, local physical planning, emergency planning, raised entrances, elevated latrines, resettlement or building and strengthening river embankments.

References

ABN-Autorité du Bassin du Niger, CRA-Centre Régional Agrhymet. 2007. Prevention du risque inindation à Niamey. Note explicative 19/01/07.

Abrate, T. 2007. Essai de segmentation des séries annuelles des débits du Niger aux stations de Koulikoro, Niamey et Lokoja. Paris: Univ. P. et M. Curie, École des mines de Paris et École nat. génie rural des eaux et des forêts, thesis.

Bechler-Carmaux, N., M. Mietton, and M. Lamotte. 2000. Le risque d'inondation fluviale à Niamey (Niger). Aléa, vulnerabilité et cartographie. *Annales de Géographie* 612: 176–187.

Bello, M.N., D.N. Jeb. 2014. Analysis of flood risk inundation hazard in Birnin Kebbi town, Kebbi State, Nigeria, *International Journal of Geomatics and Geosciences* 5(1): 119–131.

Braccio, S. 2014. Flood prone areas due to heavy rains and sea level rise in the municipality of Maputo. In *Climate change vulnerability in southern african cities, Building knowledge for adapation*, ed. S. Macchi and M. Tiepolo, 171–186. Springer. doi: 10.1007/978-3-31900672-7.

Braccio, S., and M. Tiepolo. 2014. Atlas des ressources locales : Kébémer, Louga (Sénégal), Niamey (Niger), INS Project, http://cooptriangulaireins.net/wp-content/uploads/Annexe-2-Atlas-des-ressources-locales.pdf.

CREDEL. 2010. Changements climatiques et inondations dans le grand Cotonou: situation de base et analyse prospective, Projet de protection de la communauté urbaine du Grand Cotonou face aux changements climatiques.

Davis, S.A., L.L. Skaggs. 1992. Catalog of residential-damage functions used by the Army corps of engineers in flood damage estimations, Ft. Belvoir, U.S. Army corps of engineers Water resources support center Institute for water resources.

Diallo, B. 2009. Élaboration d'une méthodologie de cartographie des risques d'inondation et application à l'échelle regionale à la région de Dakar. Rapport analytique provisioire. UNDP-Ministère de l'intérieur-Direction de la protection civile.

Douglas, I., K. Alam, M. Maghenda, Y. McDonnel, L. McLean, and J. Campbell. 2008. Unjust waters: climate change, flooding and the urban poor in Africa. *Environment and Urbanization* (20)1: 187–205.

EMMA. 2012. estation AMSD multisensor precipitation estimate 21 and 22 august 2012.
EXCIMAP. 2007. Handbook of good practices for flood mapping in Europe.
Förster, S., B. Kuhkman, K.-E. Lindenschmidt, and A. Bronstert. 2008. Assessing flood risk for a rural detention area. *Natural Hazards and Earth System Sciences* 8: 311–322.
IGN. 1978. République du Niger Niamey et ses environs 1:20.000.
IGN, Service de la topographie et du cadastre. 1980. Niamey 1c, 3° ND-31-IX carte 1:50.000.
Marzocchi, W., M.L. Mastellone, A. Di Ruocco et al. 2009. Principles of multi-risk assessment. Interactions among natural and man-induced risks. EUDGR.http://cordis.europa.eu/docs/publications/1060/106097581-6_en.pdf Accessed 11 Jan. 2015.
Mbaye, M., and R. Laganier. 2004. Apport des systèmes d'information géographique et de la télédétection à l'analyse du risqué d'inondation dans la ville de Saint-Louis du Sénégal, Xèmes journées scientifiques du réseau télédétection de l'AUF, Ottawa, 24–28 mai 2004: 139–141.
MEDD-Ministère Environnement Développement Durable (France), ONRN-Osbervatoire National des Risques Naturels. 2012. *Glossaire*
Mendelsohn, J., S. Mendelsohn, and N. Nakanyete. 2010. An expanded assessment of risks from flooding in Luanda.
Mutanga, S.S., and B. Mwiruki. 2013. Managing vulnerability to climate change extremes in Africa: some policy insights into frequent flooding in Dar es Salaam, Tanzania. AISA Policy Brief, 82.
NFIP-National Flood Insurance Program. 2013. Technical documentation of NFIP actuarial assumptions and methods. Supporting rates effective october 1, 2013.
OCHA-Office for the Coordination of the Humanitarian Affairs. 2012. Niamey: crue du fleuve Niger, 5 december.
Oriola, E., and S. Bolaji. 2012. Urban flood risk information on a river catchment in a part of Ilorin metropolis, Kwara State. Nigeria. *Information and Knowledge Management* 2(8): 70–84.
Ould Sidi Cheikh, M.A., P. Ozer, and A. Ozer. 2007. Risque d'inondations dans la ville de Nouakchott (Mauritanie). *Geo-Eco-Trop* 31: 19–42.
Panthou, G., T. Viscel, and T. Lebel. 2014. Recent trends in the regime of extreme rainfall in the Central Sahel. *International Journal of Climatology* 34: 3998–4006.
Ponte E. 2014. Flood risk due to heavy rains and rising sea levels in the Municipality of Maputo. In *Climate change vulnerability in southern african cities. Building knowledge for adapation*, ed. S. Macchi, and M.Tiepolo, 187–203. Springer. doi: 10.1007/978-3-31900672-7.
RN, MF-Ministère de Finances, INS-Institut national de la statistique. 2013. Annuaire statistique 2008–2012. Edition 2013, novembre.
RN, VN-Ville de Niamey. 2012. Rapport de la Commission des affaires sociales, culturelles et sportives, 25 août.
RN, RN, VN. 2013. Comptes de gestion 2012.
Satterthwaite, D., S. Huq, M. Pelling, H. Reid, P. Romero Lanako. 2007. Adapting to climate change in urban areas. The possibilities and constraints in low- and middle income nations. IIED Human Settlements Discussion Papers Series.
SERTIT-Service Régional de Traitement d'Image et de Télédétection. 2012. Niger-Niamey. Flood extent map – Detail. Observation the 29/08/2012. http://sertit.u-strasbg.fr.
SERTIT. 2012b. Niger-Downstream off Niamey. Flood impact map – Detail Niger river – Situation the 29/08/2012. http://sertit.u-strasbg.fr.
Sido, A. et al. 2011. État des lieux de la riziculture au Niger. Rome: FAO.
Sighomnou D., L. Descroix et al. 2013. La crue de 2012 à Niamey: un paroxysme du paradoxe du Sahel? *Sécheresse* 24: 3–13. Issn: 1777–5930.
Tarchiani V., and M. Tiepolo. 2016. *Risque et adaptation climatique dans la région Tillabéri, Niger*. Paris: L'Harmattan. Isbn: 978-2-343-08493-0.

Tiepolo, M., and S. Braccio. 2016. Urban form dynamics in three sahelian cities: consequences on climate change vulnerability and sustainable measures. In *Adaptation planning in mutable environment: from observed changes to desired futures*, ed. S. Macchi, and L. Ricci. Springer.

Tiepolo, M. 2014. Flood risk reduction and climate change in large cities south of the Sahara. In *Climate change vulnerability in southern african cities. Building knowledge for adapation*, ed. S. Macchi, and M. Tiepolo, 19–36. Springer. doi: 10.1007/978-3-31900672-7.

Tiepolo, M. 2014b. Ongoing and future flood adaptation measures in the municipality of Maputo. In *Climate change vulnerability in southern african cities. Building knowledge for adapation*, ed. S. Macchi, and M. Tiepolo, 205-221. Springer. doi: 10.1007/978-3-31900672-7.

United Nations. 2009. *UNISDR terminology on disaster risk* reduction. Genecva: UNISDR.

UNDP-United Nations Development Program, BCPR. 2010. Disaster risk assessment.

UN-Habitat. 2014. *The state of african cities 2014. Re-imagining sustainable urban transitions.* Nairobi: United Nations Human Settlement Programme.

UN-Habitat. 2012. Mid-term evaluation of the cities and climate change initiative. London: Earthscan.

UN-Habitat. 2011. *Cities and climate change. Global report on Human settlements 2011.* London-Washington DC: Earthscan.

UNISDR 2011. *Global assessment report on disaster risk reduciton.* Geneva: UNISDR.

WMO-World Meteorological Organization (2013) Flood mapping.

Zaneidou, D. 2013. Bordereaux des prix unitaires des principaux investissements realisés par les projets FIDA au Niger – Capitalisation. Volume 1. FIDA.

Maurizio Tiepolo[34] and Sarah Braccio[35]

13 Flood Risk Assessment at Municipal Level in the Tillabéri Region, Niger[36]

Abstract*:* The Tillabéri region (population 2.7 million, 97,250 km^2) is the hinterland of the Niger's capital city and the second most susceptible region to flooding in the country, with 416 settlements hit from 2008 to 2013. This chapter aims to present the potential benefits of flood risk assessment at the municipal scale, a tool that can help local authorities in disaster risk reduction. Risk (R) is considered a function of Hazard (H), Exposure (E) and Damages (D) according the equation R = H * E * D. Risk is measured using six indicators. The probability of rain causing settlement flooding each year is measured for each municipality using daily rainfall from meteorological stations (1981-2010) and three-hourly Tropical Rainfall Measuring Mission (TRMM), datasets by NOAA (1998-2011). Settlements flooded (E), people affected, homes destroyed, fields flooded and livestock killed (D) are sourced from Niger's early warning system and disaster prevention unit (EWS DP), all errors corrected and units of measurement standardised. From the results, it emerged that 765 settlements were flooded between 1998 and 2013. Contrary to what one might expect, the floods caused by the swelling of the River Niger hit few settlements. Most of the areas susceptible to flooding were located in the vast Bosso and Maouri dallols, two fossil rivers that run from Mali towards Niger for over 300 km. The right-bank tributaries of the Niger and along the minor hydrographic network were the next most affected areas. Ninety-five settlements were hit more than once and 19 flooded in two or more consecutive years. Seven municipalities out of 41 were at very high or high risk of being flooded. These are crossed by the River Niger or by its main tributaries on the right bank, by the Ouallam intermittent creek or the Bosso dallol. Seven municipalities showed damage in three areas (people, dwellings, fields).

Keywords: Damages, Exposure, Hazard, Local development plan, Risk.

---

34 Maurizio Tiepolo is an associate professor of urban and regional planning at DIST–Politecnico and University of Turin. He is author of paragraphs 13.1, 13.5, 13.6, 13.7, 13.8, maurizio.tiepolo@polito.it

35 Sarah Braccio is a lecturer at the same department. She is author of paragraphs 13.2, 13.3, 13.4 and of all maps, sarah.braccio@polito.it

36 This chapter was produced within the ANADIA-Niger project, cofounded by the Directorate General for Development Cooperation of the Italian Ministry of Foreign Affairs, Ibimet CNR, DIST-Politecnico di Torino and National Directorate of Meteorology (DMN) of Niger. The authors would thank Issa Hassimou (EWS DP) and Mamoudou Idrissa (CNEDD) for data providing, Maurizio Bacci for hazard assessment, and the DMN staff for first outcomes discussion.

13.1 Introduction

The Tillabéri region (2.7 million inhabitants, 97,250 km^2) is the vast rural hinterland and main migration basin of the Niger capital city, Niamey (1.1 million inhabitants in 2012). Between 2008 and 2013, 416 settlements of the Tillabéri region were flooded, ranking the region amongst those most affected by the disaster in the country. The range of the impact had many causes: climate change over recent years, the presence of a vast hydrographic basin (Figure 13.1), and insufficient disaster risk reduction (DRR) measures developed by municipalities. Prevention and adaptation require coordinated commitment by both central and local structures – the latter being nearest to the affected populations. If the State were to fund municipalities for DRR, a municipal level flood risk assessment could become a useful tool in decision-making.

This chapter aims to test, on the Tillabéri region, the potential of a snap shot flood risk assessment at the municipal level, updatable within the national early warning system and disaster prevention unit (EWS DP).

The Tillabéri region is divided, according the 2006 administrative subdivision (RN, MEF INS 2006), into 44 municipalities (Figure 13.2). These are vast (often more than 2,200 km^2) administrative units which often contain more than one hundred of settlements. Three municipalities were lacking of some information thus were not considered in our assessment. Our flood risk assessment did not take into account the specifics of the various settlements within each municipality.

Risk (R) is «the combination of the probability of an event and its negative consequences» (UNISDR 2009: 25). This may be calculated in various ways (Tiepolo 2014): for example, using the equation R = (H * E * V) / A (Gotangco *et al* 2010) where hazard (H) is the probability of the event at the source of flooding (rain); exposure (E) is the property that may be damaged or lost following flooding; vulnerability (V) is the weakness that may amplify the effect of the flooding and adaptation (A) covers all the measures used to reduce impact. With the exception of hazard, the other three flood risk components can be calculated in two ways:

a) by measuring the factors that are presumed to determine the value of each component (Ponte 2014), for example poverty, population density, adaptation measures, etc.
b) by measuring the effect of the components in terms of damages and losses presumed (ECHO, UNDP 2008; CDEMA 2009; Marzocchi *et al* 2009; Sayers *et al* 2010) or having occurred in the past, being damage (D) the «total or partial destruction of physical assets existing in the affected area» (World Bank 2010: 2).

Which of the two ways is chosen depends on the aims of the risk assessment, on the extension of the territory on which it is calculated and on the information, methods and time available.

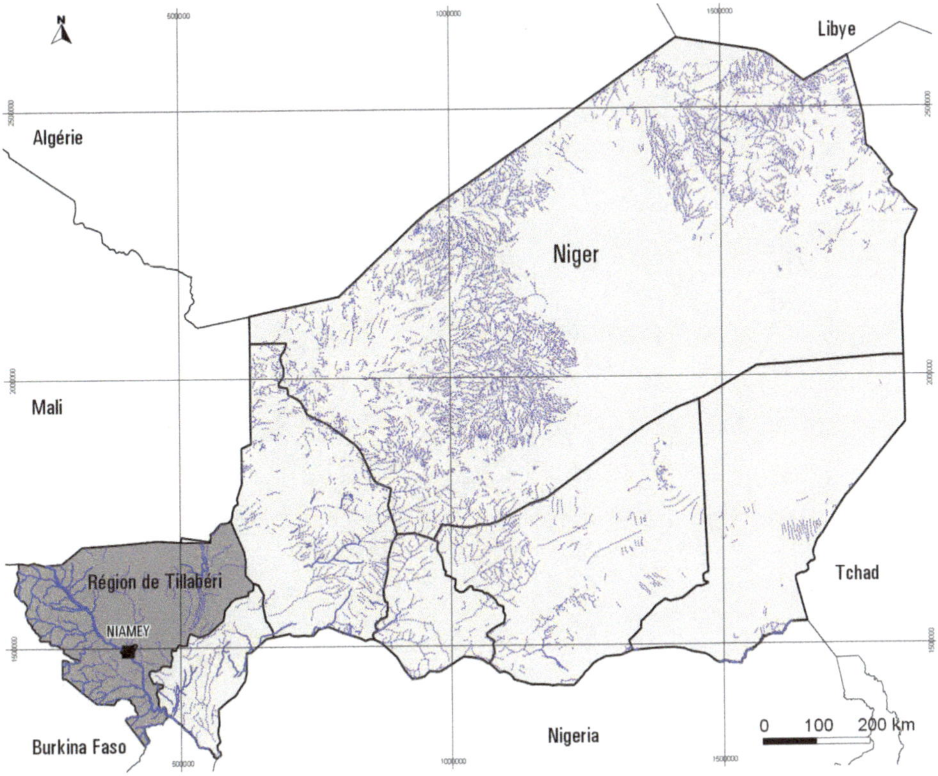

Figure 13.1: Tillabéri (dark grey) and the other regions of Niger (thick line) with the hydrographic network (blue)

The first way requires detailed information that allows the value of the E, V and A components to be appraised. Usually this approach is used for territories of limited extension, like a city (Holloway *et al* 2010; Ponte 2014) or a district, or more specifically over a few hundred or few thousand km^2.

In Niger, there is currently no information available to calculate E, V and A on a municipal level. Therefore, we chose the second way. We should discard any specific collection of information if we wish to build a sustainable risk assessment. Instead, we should rely on regular collections of information that have already been active and tested. We can appraise a hazard by using the data collected regularly from the National Directorate of Meteorology (DMN according to the French acronym) through the pluviometric station network and that recorded by satellite. The EWS DP database (DB) in the prime minister's office has held records of flooding dates since 1998, settlements flooded, the number of people affected, the number of houses destroyed, fields flooded and livestock lost as well as other occasional or heterogeneously-quantified damage to equipment and infrastructure.

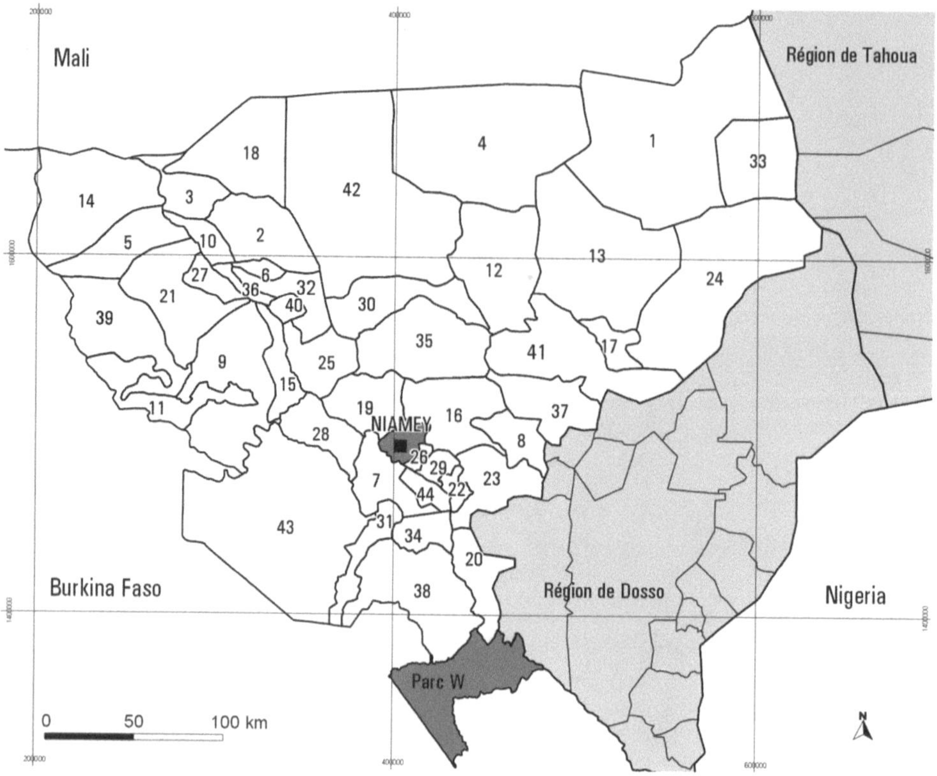

Figure 13.2: Tillabéri region, 2006. Subdivision into 44 municipalities: 1 Abala, 2 Anzourou, 3 Ayerou, 4 Banibangou, 5 Bankilaré, 6 Bibiyergou, 7 Bitinkodji, 8 Dantchandou, 9 Dargol, 10 Dessa, 11 Diagourou, 12 Dingazi, 13 Filingué, 14 Goroual, 15 Gothèye, 16 Hamdallaye, 17 Imanan, 18 Inates, 19 Karma, 20 Kirtachi, 21 Kokorou, 22 Kollo, 23 Kouré, 24 Kourfeye, 25 Kourteye, 26 Liboré, 27 Mehana, 28 Namaro, 29 N'Dounga, 30 Ouallam, 31 Ouro Gueledjo, 32 Sakoira, 33 Sanam, 34 Say, 35 Simiri, 36 Sinder, 37 Tagazar, 38 Tamou, 39 Téra, 40 Tillabéri, 41 Tondikandia, 42 Tondikiwindi, 43 Torodi, 44 Youri.

Consequently, hazard (H) in each municipality may be calculated as the inverse return period of the (critical) minimal rainfall presumed responsible for disastrous flooding over the past 16 years. The daily pluviometric values recorded at 14 meteorological stations by DMN (1981–2012) and the three-hourly data sets (1988–2011) as available with Ncview allowed Bacci (2016, chapter 5) to identify the rainfall responsible for flooding and how often it recurred.

Exposure (E) was expressed through flooded settlements (having registered some damage). Using a 16-year long series of data, we can assume that each municipality has already been subject to many types of rain (very heavy, heavy, out of season) and, therefore, that many flood-prone settlements have been flooded–including those resulting from the recent conversion into built-up sites.

Damage (D) was expressed through victims, dwellings collapsed, fields flooded, livestock killed during the last 16 years as registered by the EWS DP DB.

Each risk component was measured through indicators that are not weighed:

$$R = H_{1981\text{–}2011} * (E_{1998\text{–}2013} * Damages_{1998\text{–}2013})$$
$$= 1/Return\ period * Flooded\ settlements * (Victims + Houses + Fields + Livestock)$$

The following equation can be applied to municipality x:

$$R\ municipality\ x = 1/RP * (s/p * P/S) * [(v/p * P/V) + (h/p * P/H) + (f/p * P/F) + (b/p * P/B)]$$

where:
l = dead livestock$_{1998\text{–}2013}$ in municipality x
L = dead livestock$_{1998\text{–}2013}$ in the Tillabéri region
f = flooded fields$_{1998\text{–}2013}$ in municipality x
F = flooded fields$_{1998\text{–}2013}$ in the Tillabéri region
s = flooded settlements$_{1998\text{–}2013}$ in municipality x
S = flooded settlements$_{1998\text{–}2013}$ in the Tillabéri region
h = collapsed houses$_{1998\text{–}2013}$ in municipality x
H = collapsed houses$_{1998\text{–}2013}$ in the Tillabéri region
p = population $_{2001}$ of municipality x
P = population $_{2001}$ of the Tillabéri region
RP = return period of critic rain
v = victims$_{1998\text{–}2013}$ in municipality x
V = victims$_{1998\text{–}2013}$ in the Tillabéri region

In order to calculate damage to fields recorded with different units of measurement (number and surface) and to appraise the significance of the loss of dwellings and livestock of which no consistency is known on the municipal scale, the rate of such damage in each municipality was calculated in relation to damage of the same type within the region. Therefore, the municipality population rate was calculated with respect to the regional one of 2001[37]. Then, the first rate was compared to the second.

The severity of damage was, therefore, calculated compared to the municipality's demographic importance within its region.

There is also a practical reason for this choice. The State transfers funds to municipalities according their demographic size. Local taxation yield in rural municipalities is coming mainly from commercial activities that are proportionate to the number of inhabitants. When damage greatly exceeds the demographic weight of a municipality, local governments will have fewer means for DRR and post-flooding interventions.

37 The 2012 census data on municipal population are not published yet.

The method measures each component of risk consistently and thereby shows the H, E and D contribution to the risk value.

The EWS DP DB cannot be directly used for our purposes. First, checking the name of each flooded settlement is necessary: the slightest change in how the name is written would impede the flooded settlements being localised. To this end, we have referred to settlement names as supplied by the national repository of municipalities (RN, MEF, INS-Niger 2006).

Second, dividing the damage where the EWS DP DB attributes it to more than one settlement. In this case, we simply divided the damage by the number of settlements to which it was assigned. The approach may seem too simple, but it is justified by the often modest quantity of said damage. Other methods, for example based on the demographic size of the settlements, would have caused fractions of damage (e.g., half a cow, half a dwelling, etc.) to be attributed to the smallest villages.

Thirdly, we found very high damage values that may have come from transcription errors. In these cases, the damage (e.g., number of collapsed houses that largely encompass the number of flooded settlements households) was not considered.

Fourth, the flooded fields registered were sometimes reported as a number, and sometimes as hectares. We converted all registrations in hectares assuming that in Tillabéri region a field is on average 3.43 Ha, as found in the municipality of Téra (RN, RT, DT, CUT 2009: 21). Should a variety of animals be lost, we used the unit of tropical livestock (UBT following the French acronym): camels 1 UBT, horses 0.8, bovines 0.7, donkeys 0.5, sheep and goats 0.1, poultry and swine 0.01 (Jahnke 1982).

In the following pages, we will present (ii) hazard, (iii) exposure, (iv) damage, (v) risk, (vi) use of flood risk assessment in local development planning, (vii) conclusions.

13.2 Hazard

Hazard (H) is «a dangerous phenomenon, substance, human activity or condition that may cause loss of life, injury or other health impacts, property damage, loss of livelihood and services, social and economic disruption, or environmental damage» (UNISDR 2009: 17).

The municipalities of the Tillabéri Region can be characterised by a dominant watershed: River Niger and primary tributaries, River Niger, primary tributaries, secondary tributaries, Dallols (fossil rivers). For each dominant watershed, the critical rainfall generating the flood can be identified considering that the concentration time (the time needed for water to flow from the most remote point in the watershed to the watershed outlet) is hours for secondary tributaries of fossil rivers (dallols), and days for primary tributaries.

EWS DB reports the dates of 110 floods from 2006 to 2012.

Two sets of data were used by Bacci (2016, chapter 5) for rainfall appreciation: daily rainfall for primary tributaries floods, as registered at 14 meteorological stations by DMN, and 3-hourly rainfall for secondary tributaries or fossil rivers (dallols) as from TRMM (1998–2011) at the grid point on the centroid of the municipality. Rainfall was collected in the 5 days around each flood date. Only the maximum value was retained. Once the rain was associated with the flood, it was time to determine the critical rain. From 2006 to 2012, some municipalities were flooded only once, while others were flooded several times (up to 13 times). In the case of multiple floods, the lowest value of rain was selected as the critical rain (Table 13.1).

Then the probability of having a similar rain was calculated per Bacci (2016) as the inverse of the return period on the series of rainfall 1981–2010 in 14 meteorological stations. Municipalities with no meteorological station were considered to have received the same rain as the nearest upriver municipality with the same hydrographical conditions (Table 13.1).

Figure 13.3: Niger, June 2014. The River Niger around Namaro (left) and the Sirba river (right) close to the confluence in the River Niger (photo M. Tiepolo)

13.3 Exposure

Exposure refers to «people, property, systems, or other elements present in hazard zones that are thereby subject to potential losses» (UNISDR 2009: 15).

Firstly, we should remember that three characteristics refer to the Tillabéri region regarding the flooding events of the last six years:

1. it is the second most affected region in Niger after Dosso (Table 13.1);
2. it ranks second (after Agadez and Dosso) for magnitude of flooded settlements compared to its demographic proportion of the national total (Table 13.2);
3. every year the number of flooded settlements is relevant, while Dosso, Zinder and Agadez have had years with peaks followed by years without disaster (Figure 13.4).

Table 13.1: Tillabéri region. Critical rain for each municipality (Bacci 2016, chapter 5).

Municipality	Type of critical rain D = daily, 3-h = 3-hours	Critical rain mm
Abala	3h	79.6
Anzourou	3h	73.4
Ayerou	D	73.4
Bani Bangou	3h	79.6
Bankilaré	D	61.8
Bibiyergou	3h	75.2
Bitinkodji	D	61.7
Dargol	D	75.2
Dessa	D	73.4
Diagorou	D	75.2
Diantchandou	3h	51
Dingazi Banda	3h	51
Filingué	3h	42.5
Gorouol	D	66
Gothèye	D	66
Hamdallaye	D	30
Imanan	3h	45
Inates	D	73.4
Karma	D	54
Kirtachi	D	75.5
Kokorou	D	61.8
Kollo	D	75.5
Kouré	D	75.5
Kourfey	3h	40
Kourtey	D	72.8
Liboré	D	61.7
Méhana	D	72.8
Namaro	D	58
N'Dounga	D	61.7
Ouallam	3h	36
Ouro Gueladjio	D	75.2
Sakoira	D	86.3
Sanam	3h	113.9
Say	D	75.5
Simiri	3h	70
Sinder	D	72.8
Tagazar	3h	51
Tamou	D	75.5
Téra	D	52
Tillabéri	D	75.2
Tondikandia	3h	59
Tondikiwindi	3h	36
Torodi	D	75.2
Youri	D	75.5

Table 13.2: Regions of the Niger, 2008–2013. Settlements flooded, demographic weight (sources: RN, SAP 2008–2013; RN, MF, INS 2012)

Region	Flooded settlements No.	P_{2012} No.	Flooded settlements %	P_{2012} %	Δ L-P Times
Agadez	163	481,982	8	3	2.67
Diffa	47	591,788	2	3	−0.67
Dosso	617	2,040,699	32	12	2.67
Maradi	157	3,404,645	8	20	−0.40
Niamey	39	1,011,277	2	8	−0.25
Tahoua	136	1,972,907	7	11	−0.64
Tillabéri	416	2,715,186	21	16	1.31
Zinder	374	3,556,239	19	21	−0.90
Niger	1,949	17,129,076	100	100	0.00

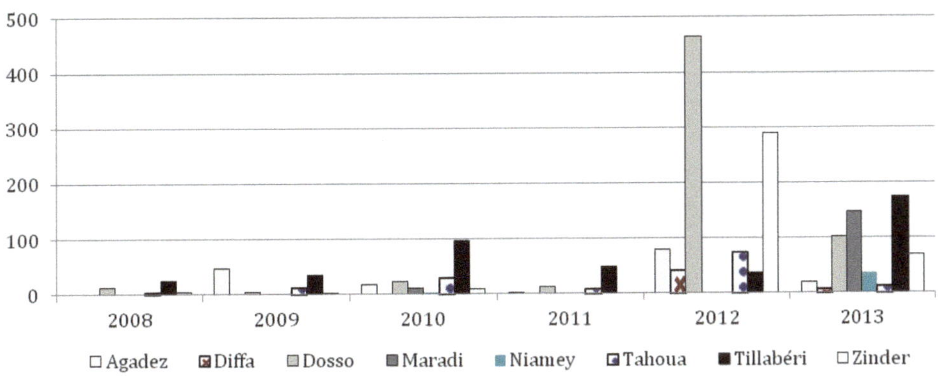

Figure 13.4: Niger, 2008–2013. Number of flooded settlements per region and per year (source: RN, SAP 2009–2013)

Secondly, considering that the smallest unit for analysis is the municipality and that we operated in a vast regional context, exposure was given by the settlements that were effectively flooded over the past 16 years.

The EWS DP's flooding DB (1998–2013) allowed us to identify 765 flooded settlements in the region of which 472 we localised with geographical coordinates (62%).

Over the course of the last eight years, there were peaks in 2006, 2010 and 2013 for number of settlements flooded. The size of these peaks increased over time. Nonetheless, we should warn the reader that this trend does not always correspond with reality – especially for the period prior to 2008, when the floods were not registered systematically.

The largest number of flooded settlements was found in the area over a kilometre away from the banks of the River Niger (27%), therefore on sites protected from flooding. These were probably flash floods of secondary tributaries of the great river. The next largest were the Bosso and Maouri dallols, followed by the main tributaries of the right bank (23%) (Table 13.3).

Table 13.3: Tillabéri region, 1998–2012. Settlements flooded by zone.

Zone	Flooded settlements	
	No.	**%**
River Niger, < 1 km	77	16
River Niger, not coastal	126	27
Right bank		
Main tributaries	107	23
Left bank		
Bosso and Maouri dallols	116	24
Minor tributaries	46	10
Total	472	100

Over the 16 years observed, 95 settlements flooded more than once: or rather, 12% of the settlements flooded in the region (Table 13.4, Figure 13.5). In some municipalities, such as Liboré and Kourfeye, this rate was closer to 30%. Thankfully, most of the repeated flooding affected 10 municipalities (which had three or more settlements affected). These were found along the Bosso dallol, the River Niger, the Sirba and the great Ouallam intermittent creek. The phenomenon is particularly serious when it happens over two consecutive years, because the populations affected don't have the time to bounce back from the first event when a new one hits again. Nineteen settlements were flooded for two or more years, an event that has mainly been found along the Niger River. In half of the cases, it was the capital of the municipality that was hit.

The amount of exposure contributing to risk was calculated by comparing the number of flooded settlements in the municipality to the total of flooded settlements in the region, and comparing this number with that of the demographic weight of the municipality relative to the region, as explained in the equation:

$E\ municipality\ x = s/p * P/S$

where:
p = population of municipality x
P = population of the region
s = flooded settlements in municipality x
S = flooded settlements in the region

Table 13.4: Tillabéri region, 1998–2013. Settlements flooded more than once.

Municipality	Settlements flooded			
	a	b	c	d
	From 2 to 4 years No.	2 consecutive years No.	More years No.	a/c %
Liboré	16	1	17	31
Gotheye	11	1	12	26
Abala	9	0	9	19
Simiri	6	1	7	27
Imanan	5	1	6	23
Tondikiwindi	2	3	5	10
Say	3	1	4	13
Kourfeye	1	2	3	29
Kourteye	3	0	3	9
Tamou	2	1	3	5
Other 18 municipalities	16	10	26	–
Total	75	20	95	12

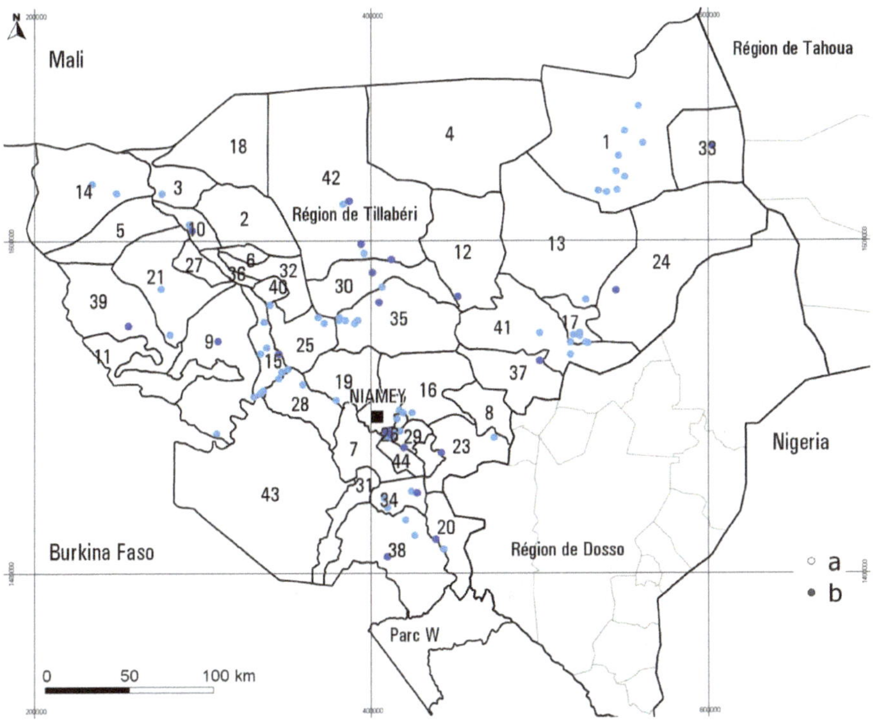

Figure 13.5: Tillabéri region, 1998–2013. Settlements flooded more than once (lift blue), for two consecutive years (blue)

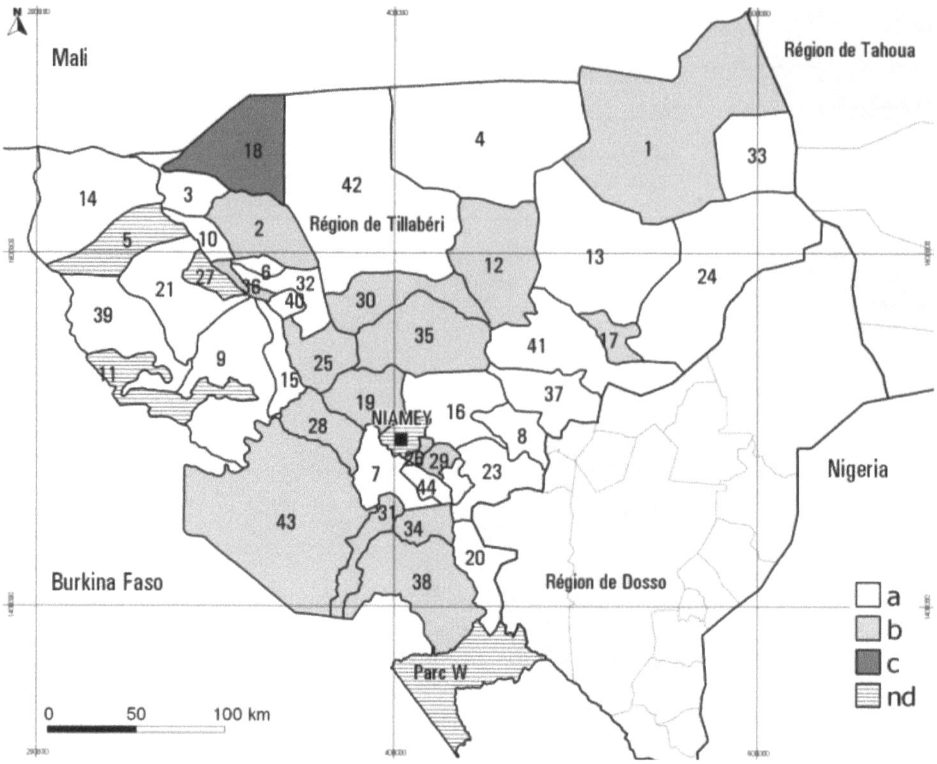

Figure 13.6: Tillabéri region. Exposure: low (a), moderate (b), high (c) (S. Braccio)

The result was that almost half of Tillabéri's municipalities had an exposure of five times their demographic weight relative to that of the whole region (maximum value for Inates). An exposure value equal to the demographic weight was performed by Tillabéri municipality. A little over half of the municipalities showed exposure values less than the respective demographic weights in the region, with Bitinkondji and Youri municipalities ranking last (Table 13.5, Figure 13.6).

13.4 Damage

In Niger, there was insufficient information to measure the factors that determined vulnerability and adaptation on the municipal level, and it was impossible to collect the information on a regular basis. This limit may be overcome if we move our attention onto the effects of vulnerability and adaptation. If a community is poor, badly organised, settled in the valley, located where fields are not protected with dykes, and livestock is not sheltered in protected areas, it is more likely that any flooding

Table 13.5: Tillabéri region. Municipalities ordered by importance of the flood risk.

Municipality	Hazard	Exposure	Damage				Risk
	Indicator 1	Indicator 2	Indicator 3	Indicator 4	Indicator 5	Indicator 6	
	1/Return period	Settlements	Victims	Houses	Fields	Livestock	Time
18 Inates	0.5	5.09	4.8	0.00	0.00	267.7	693.4
29 N'Dounga	1	2.60	5.6	2.4	12.00	0.00	51.9
26 Liboré	1	4.56	3.1	2.4	4.1	0.00	43.9
31 Ouro Gueledjo	1	2.31	0.91	10.01	1.4	0.12	29.0
28 Namaro	1	1.67	5.2	1.3	7.1	0.01	22.7
34 Say	0.22	2.93	2.50	20.9	2.6	0.15	16.9
30 Ouallam	1	2.34	1.50	0.34	2.4	0.63	11.5
42 Tondikiwindi	1	2.56	0.53	0.25	0.62	0.75	5.5
38 Tamou	0.22	2.48	1.5	5.6	1.3	0.10	4.6
12 Dingazi	1	1.86	0.4	0.07	0.88	0.95	4.3
17 Imanan	1	1.32	1.6	0.02	1.4	0.04	4.0
19 Karma	1	1.09	1.3	0.10	0.95	0.00	2.6
14 Goroual	1	0.97	0.88	0.23	1.2	0.01	2.3
22 Kollo	0.22	0.51	7.7	2.0	9.5	0.20	2.2
15 Gothèye	1	0.90	1.3	0.40	0.45	0.00	1.9
16 Hamdallaye	1	0.96	0.76	0.24	0.16	0.00	1.1
36 Sinder	0.33	1.11	1.9	0.00	0.82	0.00	1.0
10 Dessa	0.5	0.97	0.53	0.03	1.4	0.04	1.0
43 Torodi	1	0.81	0.60	0.10	0.46	0.00	0.9
35 Simiri	0.5	1.20	0.59	0.10	0.62	0.17	0.9
4 Banibangou	0.14	0.27	0.09	0.07	0.23	20.2	0.8
32 Sakoira	0.33	0.75	1.3	0.02	1.7	0.00	0.8
8 Dantchandou	1	0.26	2.0	0.23	0.38	0.00	0.7
1 Abala	0.08	1.36	0.72	0.11	0.05	5.2	0.7
41 Tondikandia	1	0.77	0.28	0.02	0.51	0.00	0.6
25 Kourtheye	0.33	1.33	0.81	0.06	0.15	0.01	0.5
7 Bitinkodji	1	0.15	1.5	0.15	1.2	0.00	0.4
20 Kirtachi	0.22	0.18	3.2	2.3	4.3	0.00	0.4
2 Anzourou	0.4	1.08	0.47	0.13	0.12	0.07	0.3
40 Tillabéri	0.33	1.02	0.40	0.17	0.38	0.00	0.3
13 Filingué	1	0.49	0.24	0.08	0.27	0.01	0.3
24 Kourfeye	1	0.24	0.10	0.07	0.73	0.00	0.2
23 Koure	0.22	0.58	1.1	0.13	0.25	0.05	0.2
9 Dargol	1	0.39	0.24	0.14	0.05	0.02	0.2
21 Kokorou	0.5	0.36	0.38	0.11	0.21	0.24	0.2
6 Bibiyergou	0.13	0.98	0.20	0.15	0.37	0.00	0.1
37 Tagazar	1	0.38	0.10	0.00	0.07	0.00	0.1
39 Téra	0.5	0.22	0.00	0.23	0.06	0.15	0.0
44 Youri	0.22	0.15	0.77	0.13	0.49	0.00	0.0
3 Ayerou	0.5	0.17	0.00	0.02	0.13	0.00	0.0
33 Sanam	0.08	0.30	0.19	0.03	0.00	0.00	0.0

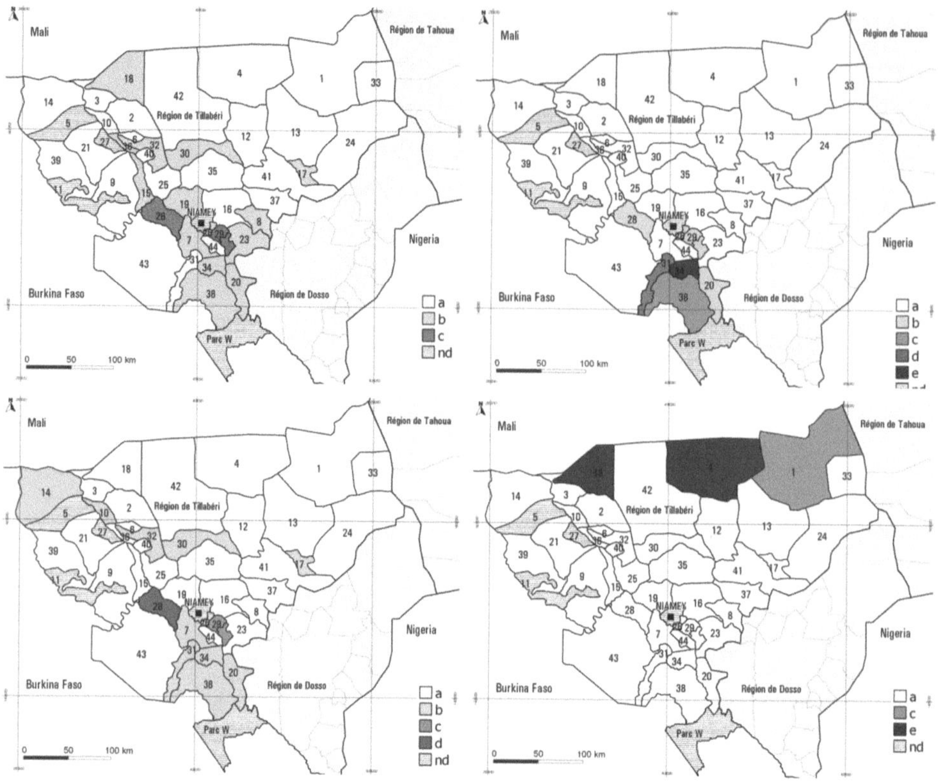

Figure 13.7: Tillabéri region, 1998–2013. Indicators of damage on a municipal scale: victims (high, left), dwellings (high, right), fields (low, left), livestock (low, right). The value classes are: < 0 (a). 0–5 (b), 6–10 (c) higher than 10 (d), data not available (nd) (elaborated by S. Braccio).

will have more widespread and significant effects on the population, dwellings, fields and livestock.

The EWS DP's DB holds four types of information on damage: the amount of population affected (victims), dwellings destroyed, fields flooded and livestock killed. We used the same procedure to obtain the value of each indicator as that for assessing exposure. For example, for indicator 3 (victims) the equation is the following:

Indicator 3 municipality $x = v/p * P/V$

where:
p = population 2001 of municipality x
P = population 2001 of the Tillabéri region
v = victims 1998-2013 in municipality x
V = victims 1998-2013 in the Tillabéri region

As for the results (Table 13.5, Figure 13.7), four elements were considered:

1. The variation field for the values of damage indicators was much higher than we had seen for the exposure indicator: 1,438 to 0 (livestock), 20 to 0.1 (houses), 7 to 0.15 (victims), 11.9 to 0 (fields). Therefore, some municipalities were badly hit by the floodings compared to their demographic weight in the region.
2. Those municipalities with indicator values higher than 1 were in the minority: 19 for victims, 14 for flooded fields, 8 for houses destroyed, 3 for lost livestock over a total of 41 municipalities on which we had data. Luckily, few municipalities were vulnerable or poorly adapted.
3. There were 12 municipalities that had more than one indicator higher than 1: Say, N'Dounga, Kollo, Namaro, Tamou, Kirtachi, Liboré (victims, houses, fields), Say and Tamou (houses, fields), Abala (fields, livestock). This means that large scale vulnerability or non-adaptation only affected a few municipalities.
4. Only three municipalities had a positive livestock indicator (Inates, Abala, Banibangou) and were all in range areas.

13.5 Flood Risk

According to the methodology used, the numeric value of a municipality's risk expresses the number of times damage was recorded between 1998 and 2013 in a given municipality compared to the total damage in the region which may exceed or be less than the municipality's demographic weight in the region in 2001, possibly reduced by hazard, and further reduced or increased by exposure (Table 13.5).

If the R value exceeds 20, the risk is considered very high. If the value is between 11 and 20 the risk is high; between 1 and 10, risk is moderate and if it is less than 1 it is low. Firstly, it should be made clear that the EWS DP DB did not contain any data on the municipalities of Bankilaré, Diagourou and Méhana. For the other 41 municipalities, five were at very high risk of flooding, two were at high risk, 11 at moderate risk and the remaining 23 municipalities were at low risk.

It should be noted that, in the event of flooding, a low risk level does not necessarily mean a municipality is immune to damage. In fact, the risk map (Figure 13.8) only shows which municipalities have suffered less damage than their demographic weight within the region, and in which the damage has been higher or much higher.

The municipalities that are at very high or high risk are, in order: Inates, N'Dounga, Liboré, Ouro-Gueledjio, Namaro, Say and Ouallam.

Luckily, this represents just 17% of the region's municipalities. With the exception of Inates and Ouallam, these municipalities are located along the River Niger or are crossed by its main tributaries. In these cases, risk value was determined by high values of five indicators out of six (Table 13.5) with the sole exception of Inates, which had three extremely high indicators out of six, especially regarding loss of livestock.

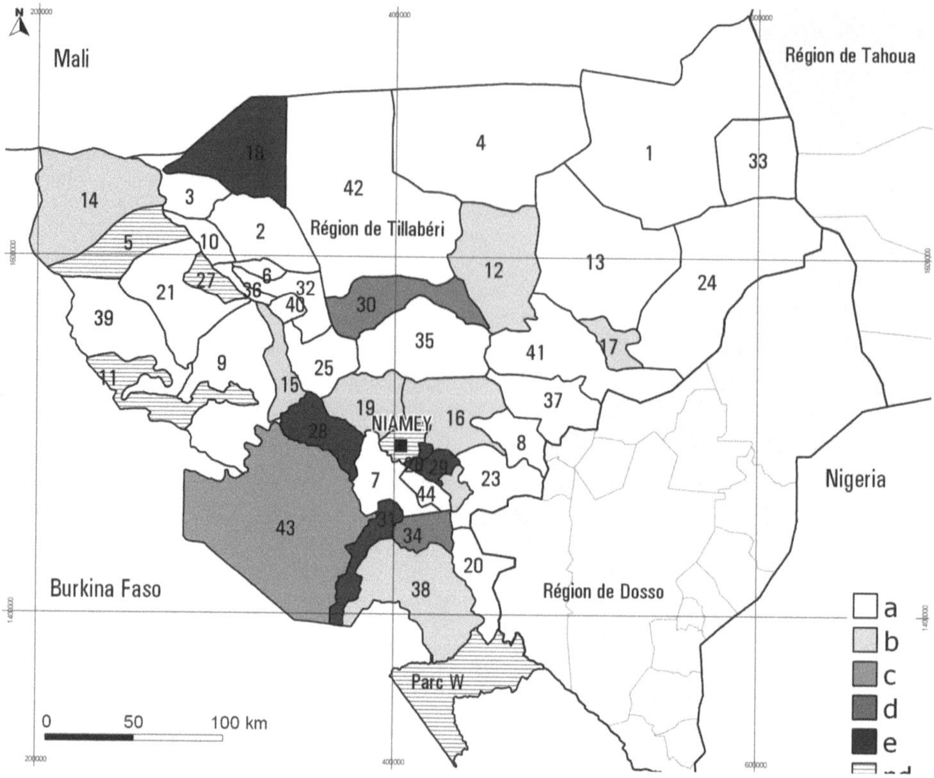

Figure 13.8: Tillabéri region. Flooding risk on the municipal level: low < 1 (a), moderate 1–10 (b), high 11–20 (c), very high > 20 (d)

Municipalities at very high and high risk covered 36% of the region's population. These were mainly medium size municipalities (40,000–80,000 inhabitants) whose chief town is within a 60 km radius of Niamey, with the sole exceptions of Ouallam (93 km) and Inates (more than 200 km). This means we are in the presence of municipalities that are linked to Niamey through asphalted roads, making the materials and the machinery for DRR available in the capital city easy accessible.

The risk components (H, E, D) analysis allowed us to make suggestions to the municipalities in identifying adaptation measures.

Nonetheless, the relationship between risk and indicators was closer for exposure, or rather, how many settlements were flooded in a municipality compared with the total number of settlements flooded in the region. All those at high or very risk were among the nine most exposed municipalities (Table 13.6).

The method used to assess the risk on the municipal level allowed this activity to be conceived as monitoring instead of a one-off measure. It is enough to update hazard, to introduce the new data on damage and exposure. Following monitoring, the position of the municipalities on the risk scale may change.

Table 13.6: Tillabéri Region. Comparison between exposure and risk.

Municipality	Exposure 1 indicator	Risk 6 indicators
Inates	5.09	693.4
Liboré	4.56	43.9
Say	2.93	16.9
N'Dounga	2.60	27.0
Tondikiwindi	2.56	51.9
Tamou	2.48	4.6
Ouallam	2.34	11.5
Ouro Gueledjo	2.31	29.0
Dingazi	1.86	4.3
Namaro	1.67	22.7
Abala	1.36	0.7
Kourtheye	1.33	0.5
Imanan	1.32	4.0
Simiri	1.20	0.9
Sinder	1.11	1.0
Karma	1.09	2.6
Anzourou	1.08	0.3
Tillabéri	1.02	0.3
Bibiyergou	0.98	0.1
Goroual	0.97	2.3
Dessa	0.97	1.0
Hamdallaye	0.96	1.1
Gothèye	0.90	1.9
Torodi	0.81	0.9
Tondikandia	0.77	0.6
Sakoira	0.75	0.8
Koure	0.58	0.2
Kollo	0.51	2.2
Filingué	0.49	0.3
Dargol	0.39	0.2
Tagazar	0.38	0.1
Kokorou	0.36	0.2
Sanam	0.30	0.0
Banibangou	0.27	0.8
Dantchandou	0.26	0.7
Kourfeye	0.24	0.2
Téra	0.22	0.0
Kirtachi	0.18	0.4
Ayerou	0.17	0.0
Bitinkodji	0.15	0.4
Youri	0.15	0.0

13.6 Use of Flood Risk Assessment for Local Planning

In 2012, the National council for the environment and sustainable development (CNEDD according the French acronym) started to promote adaptation to climate change in the Municipal development plans (MDPs). To this end, we should remember that only nine of the 44 municipalities in the Tillabéri region had started to or completed the DRR integration in their MDP so far.

Adaptation to climate change involves analysing the main hazards to which the municipality is exposed and planning DRR measures. Hazard analysis has been rather concise in the MDPs adapted so far: the trend of annual precipitations and how much time passes between rainfalls were analysed, but the recurrence of intense rainfall was not considered, nor was the discharge of the River Niger for the municipalities it crosses. Drought was not analysed through the SPI, nor was it analysed considering how much water is required for the main cultivations. Finally, flood prone settlements were not identified.

The risk assessment developed in this chapter may contribute to improving the MDPs' adaptation to climate change. In fact, settlements frequently flooded over the past 16 years (Figure 13.5) may be compared with measures provided by the MDP. Based on the measures provided for each settlement by MDPs, we can confirm that, still today, just one sixth of those settlements that flooded between 1998 and 2013 are subject to some adaptation measures. These measures are, in order of frequency, mini dams, stone cords, trapezoidal bunds, contour stone bunds, semi circular bunds, weirs, zai, restoration of wells, treatment of water course banks, protection dykes, sensitisation, planting, etc. (Critchely *et al* 1991).

Therefore, the vast majority of the flooded settlements are not subjected to any DRR measures by the MDP.

Starting from these sites and the hydrographical network, a preliminary map of the flood prone areas can be prepared for the most exposed river sectors. It could aid adaptation decision making.

13.7 Uncertainties in Input Data and Resulting Outcomes

The main sources on which the risk assessment at municipal scale was based are the EWS DP DB (damage), the meteorological stations records and the TRMM records (rain), and the national population census (population). There were no gaps in time regarding the daily precipitation (1981–2010), the 3-hourly precipitation (1998–2011) or the damage (2005–2013). Gaps in space existed for daily precipitation, which were collected by 14 meteorological stations for 44 municipalities.

Errors were limited to EWS DP DB and regarded the names of the settlements flooded and the amount of damage. These errors were often due to transcription and

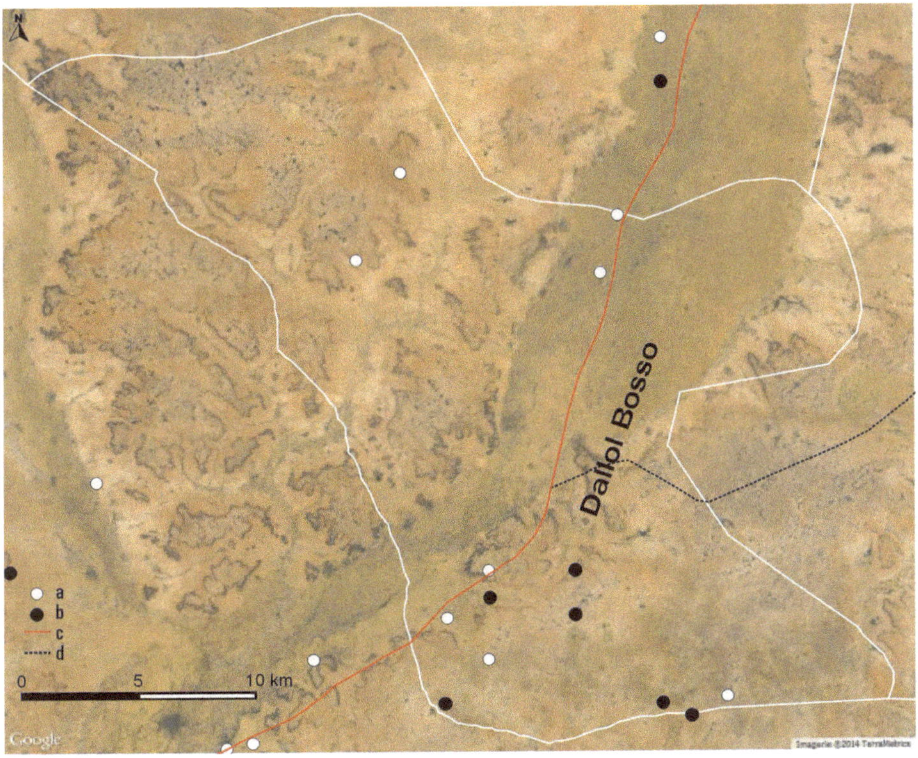

Figure 13.9: Imanan municipality, 2013. Settlements flooded once (a), twice or more (b), paved road (c), dirty road (d). Some flooded localities are just in the middle of the dallol (by E. Braccio).

were eliminated by discarding all dubious values (damage greater than the settlement population) before risk calculation.

As for critical rain probability, Bacci (2016) considered only the meteorological stations that had continuous recording from 1981 to 2010. In the case of municipalities with no meteorological station, he used the registrations from the closest meteorological station upriver with the same hydrographical characteristics. Errors should have been eliminated at this point and the impact on results concerned only approximations we made assimilating critical rain of the upriver rainfall in the case of municipalities without rain stations or no significant precipitation (less than 20 mm) in the five days around the flood date.

It is useful to remember that, at present, there are no alternatives based on regularly collected data to provide a flood risk assessment at the municipal level more accurate than the one presented in this chapter. Daily river discharge could not achieve better results since only 13 hydrographical stations are operating (out of 20), and they are located on the river Niger and on its right bank tributaries only. Consequently, using this data 28 municipalities out of 44 (64%) would have no hydrographical measurements, thus flood data.

13.8 Conclusion

The flood risk can be assessed in each municipality of the Tillabéri region (Niger) starting from the daily rainfall (DMN), from the three-hourly data, and from the damage recorded by the EWS DP DB since 1998.

Almost half the municipalities have a yearly recurrent flooding hazard, meaning that large parts of the territory are continually threatened by flooding.

The EWS DP DB provided three relevant types of information.

Firstly, the identification of those settlements that were flooded more than once was provided, especially those hit two or more consecutive years, a catastrophic condition for the interested populations. Eleven per cent of the 765 flooded settlements were hit more than once and 19 settlements were hit for two or more consecutive years. Half of these cases affected the same municipal capitals.

Second, the contribution of an isolated damage event in determining the risk was provided thanks to the adoption of indicators built with the same methodology. This allowed us to consequently identify the priority sectors for DRR (people, dwellings, fields, and livestock).

Third, the identification of the municipalities that were at high risk of flooding was made possible. Despite the fact that there were only 7 municipalities at very high, high risk and 9 at and moderate risk of flooding, these were, in any case, administrative bodies that held 39% of the region's population. Ten out of the sixteen were municipalities on the edge of the River Niger and which were sometimes also crossed by the river's many tributaries such as Gorouol, Dargol, Sirba, Goroubi, Tapoa, Mékrou.

The cases of consecutive flooding exposure that we identified between 2010 and 2013 do not seem to be explained just by the occupation of exposed areas, but by the effects of climate change (higher intensity of flooding, out of season flooding). This would explain the lack of flooding adaptation measures in MDPs before 2012.

However, a lot remains to be done in order to adapt MDPs to climate change.

The awareness of the flooding risk on a municipal level is useful for:

1. Central structures: prime minister, CNEDD, ministries:
   - organising aid in advance to municipalities that are regularly flooded
   - supporting the preparation and adoption of municipal emergency plans
   - supporting the adaptation of MDPs to climate change for municipalities at very high, high and moderate risk
   - supporting the carrying out of adaptation measures provided by the MDPs through the constitution of a special fund (for example, to be created with the help of the development partners)
2. The municipalities, in order to practice evacuation in the most exposed settlements
3. The development partners, to support the adaptation of those settlements flooded more than once and those municipalities at very high, high and moderate risk of flooding

4. Professionals, in order to:
   - address their assistance during and after disaster
   - assist young people in case of disaster
   - teach their students the relationship between meteorology, flooding and development
   - incorporate the risk concept into the approach and activities of the NGOs

This chapter closes with some recommendations to the CNEDD and the EWS DP
- Adaptation should be monitored by settlement, especially in those frequently exposed to flooding
- The MDPs of the 16 municipalities most at risk of flooding in the Tillabéri region should be adapted to climate change
- Attention should be paid to the definition of risk, which is not synonym of exposure (see RN, DNPGCCA unk)
- The annual information sheet on flooding should include the cause of flooding (intense rain, the overflowing of watercourses that cross or border the flooded settlements, formation of water stagnation, etc.)
- Indicators should be weighed according to pastoralist, agro pastoralist or farmer areas
- The possibility of interrogating the EWS DB as web GIS should be considered

To mayors
- Depending on how high the municipality ranks regarding at risk of flooding, check those indicators that have the highest values and consequently identify the DRR measures and include them in the MPD, especially for settlements flooded more than once.

References

Bacci, M. 2015. Characterization of flood hazard at municipal level. A case study in Tillabéri region, In *Planning to cope with tropical and subtropical climate change*, ed. M. Tiepolo, E. Ponte, E. Cristofori, 89–106. De Gruyter Open.

CDEMA. 2009. The regional disaster reduction management for sustainable tourism in the Caribbean project. CROSQ.

Critchley, W., and K. Siegert. 1991. Water harvesting. A manual for the design and construction of water harvesting schemes for plant production. Rome: FAO-Natural and resources management and environmental department.

ECHO-European Commission's Humanitarian aid and Civil Protection department, UNDP. 2008. Mainstreaming disaster risk reduction in sub national development land use/physical planning in the Philippines. Guidelines.

Gotangco, K., and R. Perez. 2010. Understanding vulnerability and risk: the CCA-DRM nexus. Manila: Klima Climate Change Center.

Holloway, A., et al. 2010. Radar Western Cape 2010. Risk and development annual review. Cape Town: University of Cape Town.

Jahnke, H.E. 1982. Resource for livestock production. In *Livestock development in tropical Africa*, H.E. Jahnke, 9–30. Kiel: Wissenschaftsverlag Vauk.

Mazzocchi, W. et al. 2009. Principles of multi-risk assessment. Interaction amongst natural and man-induced risks. European commission, Directorate-general for research. Environment directorate. EUDGR.http://cordis.europa.eu/docs/publications/1060/106097581-6_en.pdf. Accessed 11 Jan. 2015.

Ponte E. 2014. Flood risk assessment due to heavy rains and sea level rise in the Municipality of Maputo. In *Climate change vulnerability in Southern African cities. Building knowledge for adapation*, ed. S. Macchi, and M. Tiepolo, 187–203. Springer. doi: 10.1007/978-3-31900672-7.

RN-République du Niger, DNPGCCA. -. Carte de risque d'inondation région de Tillabéri. UNDP, Union Européenne.

RN, MEF, ISN-Niger. 2006. Niger. Répertoire national des communes du Niger (RENACOM), Niamey: INS.

RN, MF, INS. 2012. Présentation des résultats préliminaires du quatrième (4ème) recensement général de la population et de l'habitat (RGPH 2012). Niamey.

RN, RT, DT, CUT. 2009. Enquête publique dans la perspective du schéma d'aménagement foncier Commune urbaine de Téra. Tome I : consolidation des données. SPCR, CoFoDép de Téra, CoFoCom de Téra.

RN, SAP-Système d'Alérte Précoce. 1998–2013. Situation détaillée des dégâts des inondations, campagne hivernale.

Sayers, P., et al. 2013. Flood risk management. A strategic approach. Paris: UNESCO.

Tarchiani, V., and M. Tiepolo eds. 2016. *Risque et adaptation climatique dans la région Tillabéri, Niger*. Paris: L'Harmattan. Isbn: 978-2-343-08493-0.

Tiepolo, M. 2014. Flood risk reduction and climate change in large cities south of the Sahara. In *Climate change vulnerability in Southern African cities. Building knowledge for adapation*, ed. S. Macchi, and M. Tiepolo, 19–36. Springer. doi: 10.1007/978-3-31900672-7.

UNISDR. 2009. *2009 UNISDR terminology on disaster risk reduction*. Geneva: UNISDR.

World Bank, GFDRR, DALA. 2010. *Damage, loss and needs assessment. Guidance notes*. Washington: The World Bank.

Vieri Tarchiani, Maurizio Bacci[38]

14 Drought Risk in the Tillabery Region, Niger

Abstract: Understanding, monitoring and mitigating drought is a very difficult task due to the intrinsic nature of the phenomenon. In addition, the evaluation of drought impact on ecosystems and societies is a complex task because the same severity of drought can have different consequences in different regions and systems due to specific underlying vulnerabilities. Climatic analysis based on observed data and agrometeorological modeling based on geospatial information can be successfully used for drought risk assessment if combined with measured quantitative data to characterize the crop production system under analysis. Such field data are primarily used to determine the vulnerability of a system to drought and evolve meteorological drought assessment into agriculture drought assessment. In this chapter, a drought risk assessment methodology is proposed. Its main goal is to support decision making and sectoral planning for integrating Disaster Risk Reduction into development strategies. Drought risk assessment is a primary step to improve drought preparedness and mitigation. We propose a case study on the Tillabery Region in Niger (97,250 km^2, population of 2,715,186 inhabitants in 2012), which is characterized by Sahelian climate, extensive rain-fed crop production systems and poor capacities in terms of resilience to climatic stresses. The analysis of the past 10 years shows that recent food crises, even if triggered by drought, have been caused by a concomitance of different drivers. Indeed, during the last decade the region didn't experience any generalized drought during the cropping season, even in years of food shortage. The main impacts on crops have been caused by a shift in the start of the growing season or by sowing failures due to false starts of the rainy season. Climatic and agroclimatic analysis over last 30 years shows clear shifts in the beginning of the rainy season, while the end remains quite unchanged. The reduced length of the growing season is the main concern for crop production in the region.

Keywords: Climate change, Drought hazard indices, Drought risk indicators

14.1 Introduction

Climate change, climate variability and increasing extreme hydrometeorological events are a threat to the Sahel's people because they affect crop production and exacerbate food insecurity in an area where most livelihoods rely on small-farm agriculture. In Niger, crop production systems are particularly fragile due to desertification, soil degradation, low soil fertility, high levels of crop, and livestock diseases.

---

38 IBIMET-National Research Council of Italy, v.tarchiani@ibimet.cnr.it

The vulnerability of production systems to climate risks is worsened by population pressure, poverty, and food insecurity. Despite this uncomfortable situation, most smallholder farmers, who are the majority of the population, rely on annual rainfed crops for satisfying their basic food needs and are unable to implement effective strategies for minimizing actual and future climatic risks.

Drought is the main climatic phenomena affecting food production in Niger. During the '70s and '80s, drought periods triggered catastrophic food crises, while during the last two decades no generalized disaster occurred. Moreover, recent years highlight that non-biophysical factors are also affecting the food security context, showing new dynamics in the region (Tarchiani *et al.* 2006). An example was the Niger crisis in 2005, which likely is the first complex food crisis that has affected the region. Besides the concurrence of different factors, its underlying causes are more than simply drought-based, but rather the incapacity of the regional cereal trading system to supply the demand and to move goods (particularly cereals such as millet, sorghum and maize) with the same efficiency as during the '90s. For this region the crisis has been called a "Free market famine" (Mousseau and Mittal 2006). The 2012 crisis differs significantly from that of 2005. It was triggered by drought, coupled with high food prices, but other particular circumstances contributed to overwhelming regional coping strategies, including the conflict in Mali and reduced international remittances (Garvelink and Tahir 2012). Nevertheless, periodic and localized droughts regularly affect crop production in Niger, even during years considered climatologically above normal in terms of precipitation. Poorer farmers do not produce enough food to meet their family's nutritional needs for more than three to six months (Rinaudo and Yaou 2009). Depending on the context, different factors contribute to low productivity, generally associated with drought conditions, but also with socioeconomic conditions such as reduced land area for farming and poor production practices. In some zones of Niger, up to 50% of the landmass is totally unproductive because land degradation and erosion have resulted in hardpan formation hampering infiltration and boosting run-off (Amogu *et al.* 2013). Elsewhere, land mismanagement (including inappropriate crop and natural resources management practices) led not only to land degradation, but also to increased flood risks. Indeed, the whole Sahelian region (and the northern part of Sudanian region) is experiencing a change in hydrological dynamics, described as the Sahelian paradox: since the first Sahelian drought, runoff coefficients of the Sahelian basins have been increasing (Mahé *et al.* 2013). There is a consensus to interpret that this increase in runoff (which cannot be due only to climate, as climatic patterns don't show relevant increase of rainfall amounts) is driven by land use changes.

The interaction of such physical and socio-economic drivers reduces the resilience and the ability to cope with, and adapt to, climatic changes and variability (Bruce 1994; Nicholson *et al.* 1998). A negative loop is thus engaged and episodic drought triggers a shift from an existing chronic food deficit into an acute phase.

Current development policies in the Sahel do not adequately take into account the special needs of people most prone to drought risk. The common thinking is that food aid and humanitarian assistance are an appropriate response when food crises arise. However, it is becoming evident that the massive humanitarian efforts required for addressing the increased frequency and scope of food crises are no longer sustainable. There is growing recognition that a major paradigm shift in development polices is needed to foster resilience (Gubbels 2012). Thus drought risk assessment is a real priority for governments at national and sub national level in order to:

- Develop strategic policies at national/sub national level
- Define mitigation and adaptation measures
- Tune Early Warning System to local needs
- Identify research priorities

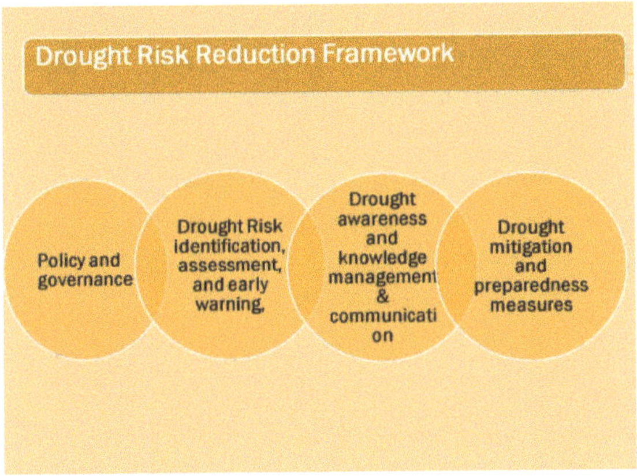

Figure 14.1: The drought risk reduction framework.

International best practices show that sustainable development requires a holistic approach that integrates disaster risk reduction with climate change adaptation in strengthening productive capacities. WMO (World Meteorological Organization) and UNISDR (International Strategy for Disaster Reduction of the United Nations) jointly encourage a culture of resilience and prevention through capacity building for better integration of weather, water and climate information, products and services for disaster risk reduction in all socio-economic sectors (WMO 2011; UNISDR 2007).

This chapter proposes a practical approach for the analysis of agricultural drought risk, through a case study in the Tillabery region of Niger.

The study builds on the experience done by the "Adaptation to climate change, disaster prevention and Agricultural Development for Food Security" project (ANADIA Niger) financed by the Italian Cooperation and implemented by the Institute of Biometeorology of the National Research Council of Italy, the DIST-Politecnico and University of Turin and the National Directorate of Meteorology of Niger, and is realized in collaboration with the Coordination Unit of the Early Warning System (SAP) and the National Environmental Council for Sustainable Development (CNEDD) of Niger. ANADIA Niger aims to develop methodologies and tools to assess drought and flood risk, to support planning at different decision making levels, to increase the resilience of local communities, and to develop a greater capacity for forecasting and response.

14.2 Methodology

Drought is a natural phenomenon that occurs when water availability is significantly below normal levels over a long period and the supply cannot meet the existing demand (Redmond 2002).

Drought is defined by UNISDR as an extensive risk, "associated with the exposure of dispersed populations to repeated or persistent hazard conditions of low or moderate intensity, often of a highly localized nature, which can lead to debilitating cumulative disaster impacts" (UNISDR 2009). Assessing the impact of drought on agro-ecosystems and societies is a complex task, because the same drought conditions may have different impacts on different systems due to their specific vulnerability. Drought vulnerability is the limited ability of a system to cope with drought, and is determined by its sensibility and resilience to water scarcity. Sensibility refers to the capacity to resist or slow the drought impact, and resilience refers to the capacity to recover after a drought. A system well adapted to climate variability will be able to cope with severe drought episodes without suffering irreversible degradation.

Risk Assessment is defined by UNISDR as a "methodology to determine the nature and extent of risk by analyzing potential hazards and evaluating existing conditions of vulnerability that together could potentially harm exposed people, property, services, livelihoods and the environment on which they depend" (UNISDR 2009). According with the generally accepted definition of risk:

Risk = f (Hazard, Exposure, Vulnerability)

Risk assessment includes an analysis of the characteristics of the hazard (such as location, intensity, frequency and probability); the analysis of exposure (such as people, property, systems, or other goods that are subject to potential damages) and vulnerability (including the physical, social, economic and environmental characteristics that make the system susceptible to the damaging effects of a hazard).

Drought can be analyzed from a meteorological, hydrological, agricultural or socio-economic perspective.

Agronomic drought can be defined as "A protracted period of deficient precipitation resulting in extensive damage to crop/pasture growth and production".

Despite the apparent simplicity of this definition, drought risk assessment is a very difficult task. Due to its long-term development and duration, the progressive character of its impacts and wide spatial extent, drought is the most complex natural

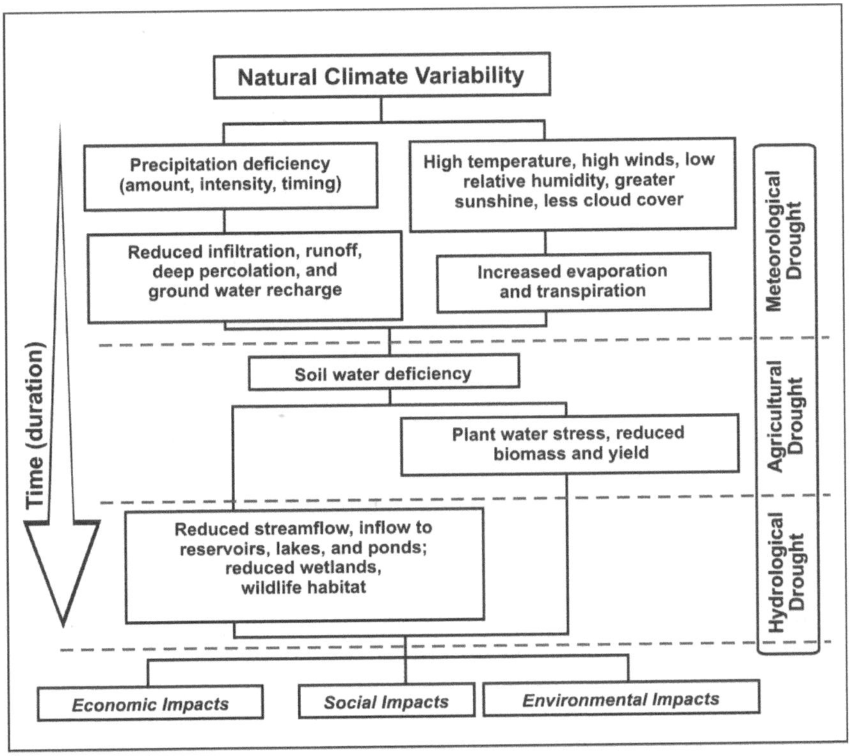

Figure 14.2: Relationships among meteorological, hydrological, and agricultural drought. (Source: National Drought Mitigation Center, University of Nebraska-Lincoln, USA)

hazard to identify, analyze, monitor, and manage (Wilhite 1993). Moreover, it is very difficult to objectively quantify drought severity, as it is a combination of the duration, magnitude and spatial extent of the drought and because the responses of biological systems are variable (Vicente-Serrano and López-Moreno 2005). Often in risk assessment, losses are used as a cumulative indicator of exposure and vulnerability. In the case of drought, losses refer to the expected damage as a consequence of a specific drought episode, which can be expressed in terms of the costs inflicted on a system, such as crops production losses, livestock mortality, etc.

But this is not the case in Sub-Saharan Africa, because an incomplete picture of drought losses and impacts doesn't allow any quantitative estimation. The main sources of drought impacts are:

- EM-DAT (The OFDA/CRED International Disaster Database – www.emdat.be, Université Catholique de Louvain, Brussels (Belgium): provides public domain coverage of large-scale mortality with a weak coverage of smaller disasters. It is inconsistent in reporting economic losses;
- NatCatSERVICE (MunichRe 2011) and SIGMA (SwissRe): are re-insurance industry databases. They cover only insured losses in developed markets.

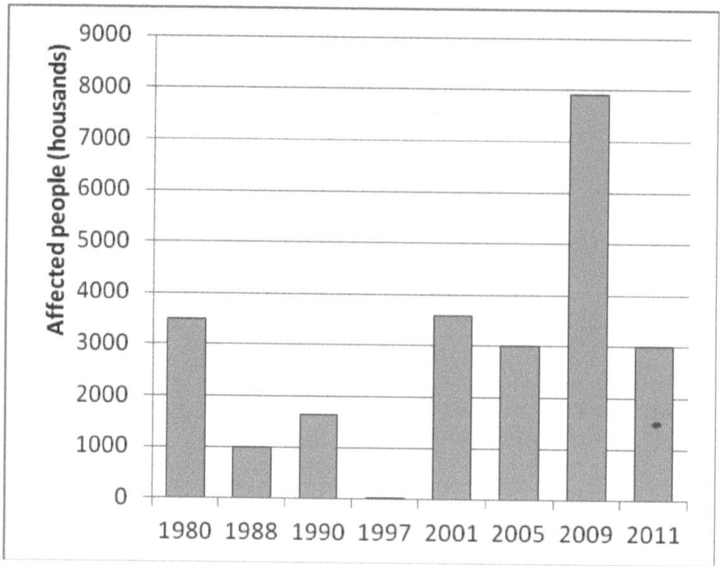

Figure 14.3: Drought events and affected people in Niger registered (Source: EM-DAT database)

- DESINVENTAR (Corporacion OSSO): DesInventar is a database of loss, damage, or effects caused by emergencies or disasters. It covers only two countries in Africa.
- National statistical data: are heterogeneous, dispersed and inaccessible data held by governments, Non Governmental Organizations, universities and others.

The proposed methodology simplifies the approach of drought risk assessment in order to face the scarce availability of impact and vulnerability data using a set of indicators.

Drought exposure is analyzed on the basis of the crop production system for the target area. National agricultural statistics provide crop production data for the main crops aggregated at the 3rd administrative level (Departments). Crop production data from 2007 to 2011 have been used to estimate the exposure to drought risk.

Drought hazard is expressed in terms of the probability or its inverse, the expected return period of a drought event occurring in a given place during a given period of time. Using data from the national climate observation network, it has been possible to calculate a set of indicators such as dry spells, SPI (Standardized Precipitation Index), lgp (length of growing period), sgp (start of growing period) and to develop spatial models of drought hazard over the target Region. Such models aid in assessing the spatial distribution of the hazard and thus in identifying the prone areas (Beguería *et al.* 2009).

Monthly SPI (McKee *et al.* 1993) has been calculated for climatic stations over the Region, for estimating trends of wet or dry conditions based on the precipitation variability for the months of the rainy period over the series of climatic data.

Dry spells are calculated as the number of consecutive days when rainfall was less than or equal to 1.8 mm. 1.8 mm/day is an empirical threshold commonly accepted for Sahelian countries. It represents the minimum daily water requirement for major crops in this area (millet and sorghum). This threshold is used by AGRHYMET Regional Center and many National Meteorological Services for calculating dry spells (Diallo 2001).

Based on historical rainfall data for the period 2003–2012, dry spells during the season were classified into five categories, which are < 5 days, 5–10 days 10–15 days 15–20 days > 20 days. The frequency distribution of the five categories of dry spells were then calculated for each station having at least 7 years of rainfall data over a period of 10 years.

Drought hazard analysis needs to be targeted on specific crop systems, considering the phenological stages, and most sensitive periods to water stress during the cropping season. Drought vulnerability of crop production systems can be expressed in terms of sensibility of the crops to water stresses and of resilience of the rural society to recover from a shock. In our approach, we considered only the physical aspect of vulnerability, the susceptibility of crops to be affected by drought, not only in terms of physiological effects, but also in terms of most sensitive periods during the phenological cycle. Such critical periods are identified for the main crops (pearl millet and sorghum) on the basis of Ministry of Agriculture decadal and mid-season bulletins, reporting the average sowing date and further phenological stages per Department (administrative units of third level). Bulletins from 2007 to 2011 have been used in order to identify the sowing period and the flowering stage, considered the most sensible to water stresses for pearl millet and sorghum.

Moreover, water balance models, such as DHC (Samba 1998) and ZAR (Bacci *et al.* 2009), were used to identify specific indexes of water stress, which are proxy indicators of drought risk for a specific crop during a season.

ZAR and DHC are simple bucket water balance models, where the control volume is considered as a bucket that is filled up by rainfall and emptied by evapotranspiration. When the bucket is full, extra water is assumed to become deep drainage. The only input data required by these models are rainfall, actual evapotranspiration estimated from potential evapotranspiration and actual soil water content, estimated from the water storage capacity of the soil. DHC uses daily rainfall from meteorological stations as input, while ZAR uses cumulated 5 or 10 day rainfall estimations by satellite.

DHC was used to calculate the water needs satisfaction index for millet for each year from the first 10-day period of June till the first 10-day period of October, to cover the whole rainy season. Then for extended periods of dry spells (more than 15 days), the behavior of these indices was analyzed by decade: decadal and cumulated indexes of water needs satisfaction for meteorological stations that recorded dry spells longer than 15 days were calculated. Logically, the index calculated with DHC assumes the interaction of the hazard (dry spell) and vulnerability (crop sensibility to stress according to the phenological phase) components.

The ZAR model was used to calculate the sowing dates for the period 2003–2012 and the sowing failures, based on satellite rainfall estimation RFE 2.0 Data produced by FEWS-ADDS. A sowing failure is defined as a condition when the cumulated rainfall of the 20 days from the sowing date is less than the minimal water needs of the crop (1/2 Maximum Evapotranspiration) during the rising phase. Sowing failures are used as an indicator of false start of the cropping season. Then, from the sowing dates, the water balance for pearl millet was calculated for each decade from the first of June to the third of September. Phenological stages were also calculated by year, and the delay relative to the average crop development calendar was detected as well.

Thus, the drought risk is described through the exposed crop production and the probability of hazard indicators related to the most critical phases of crops.

14.3 The Study Area

Tillabery region is located in the extreme West of Niger. The climate is Sahelo-Sudanese in the South and Sahelo-Saharan in the North. The average yearly rainfall varies from 700 mm in the extreme South of the Say department, to 250 mm in the North of the Filingue department (Figure 14.4).

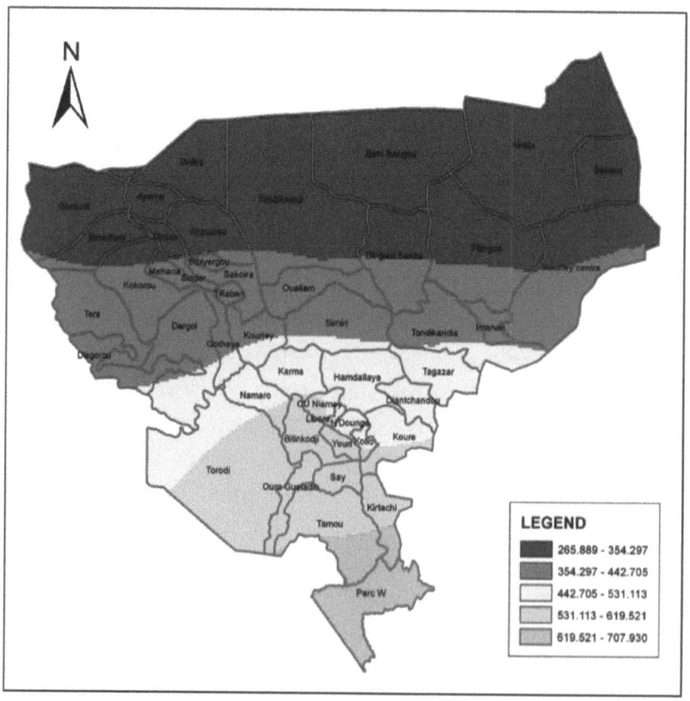

Figure 14.4: Yearly average rainfall (1980–2012).

Rainfall is characterized by strong spatial and temporal variability. Figure 14.5 shows the interannual variability over the series 1980-2012. While from 1980 till 1990 annual rainfall was clearly lower than the average, since 1998 a moderate recovery can be observed.

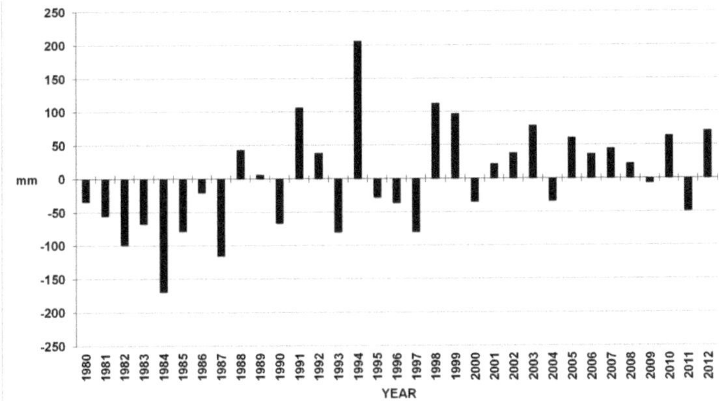

Figure 14.5: Rainfall variability 1980 – 2012 in the Tillabery region.

The population of the Region is estimated at 2,715,186 inhabitants in 2012 (RGPH 2012) with a population growth rate (3.2%) among the highest in the country. The average population density is 19 inhabitants per km², but it reaches 150 inhabitants in the valley areas.

Agriculture occupies 90% of the active population and it is divided into two major types:

- Rainfed agriculture with millet, sorghum, cowpeas and groundnuts as the main crops;
- Irrigated agriculture, particularly for rice throughout the Niger River (especially in hydro – agricultural perimeters).

The cultivable area is estimated at 4,506,122 ha, of which about 100,000 ha are irrigable and only one third is operated.

The rainfed production system is characterized by extensive monocultures of millet and cowpea-millet intercropping, with yields that are generally very low (416 kg/ha is the millet average yield of the region over the last 5 years). Production units are family-owned and have an area from 4 to 9 hectares.

The performance of the agricultural sector remains weak, being unable to satisfy population cereal needs. Indeed, the Region is still locked in a cycle of chronic food insecurity with a food-deficit situation every three years on average.

Very low yields combined with the continuous growth of cultivated surfaces show the weaknesses of rainfed agriculture in the Region, even though the availability of cultivable lands is still important (Figure 14.7).

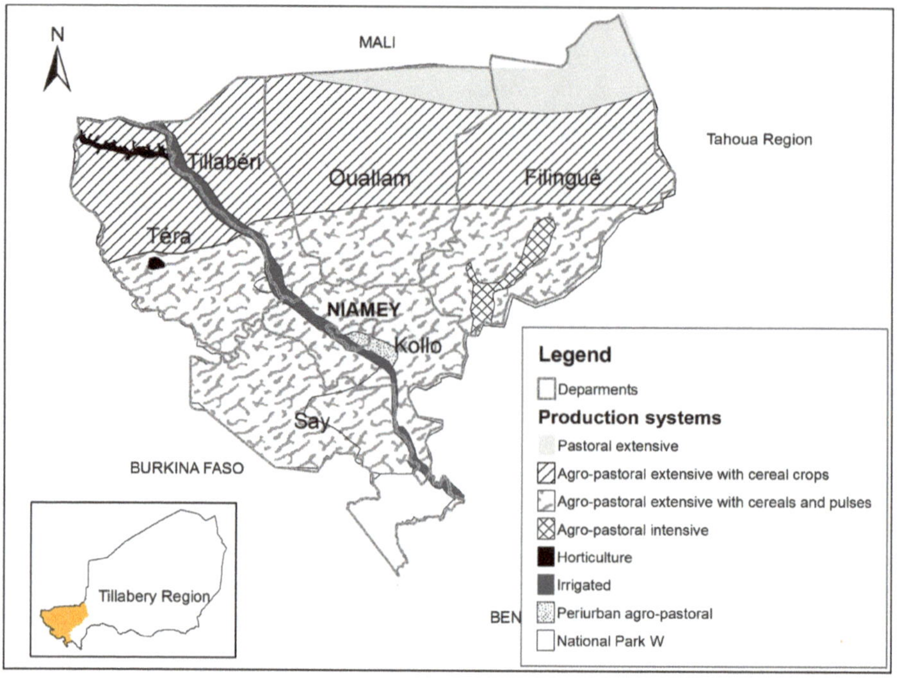

Figure 14.6: Production systems of the Tillabery region (adapted from Pini and Tarchiani 2007)

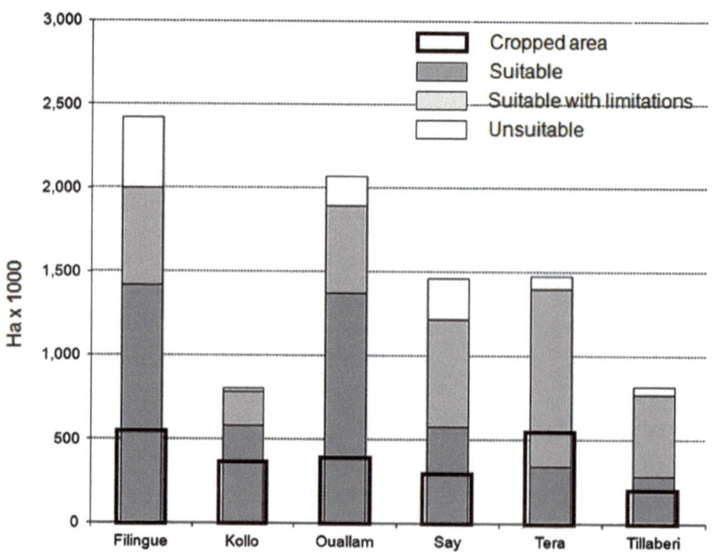

Figure 14.7: Land suitability for cropping compared to cropped areas per Department (average 2007–2011)

In this context, climate is the main limiting factor for the development of rainfed agriculture. Indeed, the main constraints can be summarized in:

- The rainfall variability which is characterized by an uneven spatial and temporal distribution;
- A high parasite pressure;
- Poor organic matter and more generally low soil fertility even in suitable soils;
- Reduction of farm size due to strong demographic pressure;
- The lack of inputs supply, especially fertilizers, and agricultural mechanization outside the river system circuits.

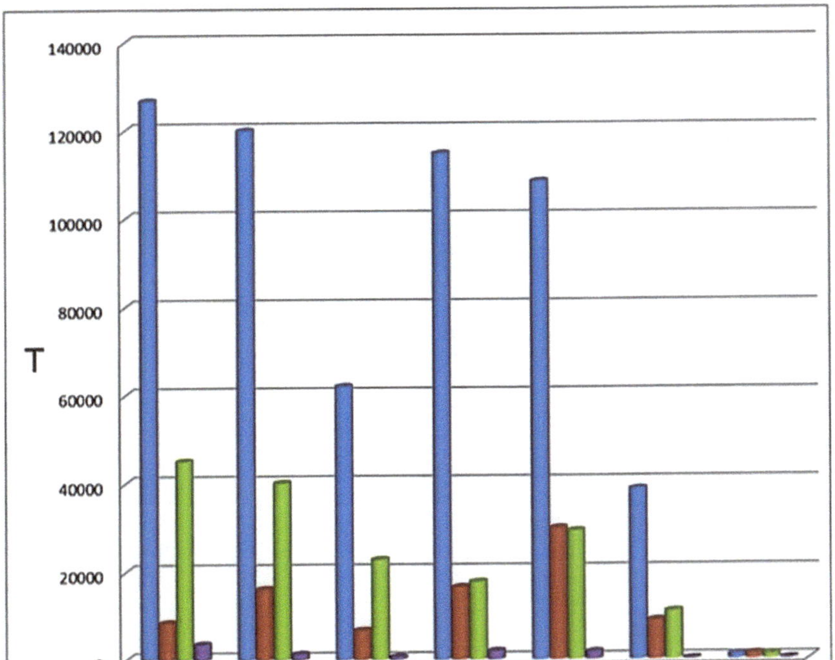

	Kollo	Fillingué	Ouallam	Say	Téra	Tillabéry department	Tillabéry Commun e
Millet	126751	119926	61916	114695	108309	38519	869
Sorgum	8333	16033	6578	16391	29712	8690	1050
Cowpea	45048	40023	22642	17565	29070	10813	963
G.nut	3473	1284	576	1824	1613	1	18

Figure 14.8: Rainfed crop production in the Tillabery region (average 2007–2011)

Due to the climatic variability, the sowing period of millet shows a strong unpredictability. As shown by Table 14.1, during last 5 years, the third decade of June is the mode of 3 departments over 6. Nevertheless, the sowing period embraces 5 decades for 5 Departments over 6.

Table 14.1: Variability (frequency) of sowing decades per Department (Average 2007–2011)

Department/Decade	16	17	18	19	20	21
Filingué	0.2	0.4	0.0	0.0	0.4	0.0
Kollo	0.0	0.2	0.6	0.0	0.2	0.0
Ouallam	0.0	0.2	0.2	0.2	0.2	0.2
Say	0.2	0.4	0.2	0.0	0.2	0.0
Tera	0.0	0.2	0.4	0.0	0.2	0.2
Tillabery	0.2	0.2	0.4	0.0	0.2	0.0

Conversely, the most critical period of water stress of millet (Table 14.2), is concentrated between the third decade of August and the second of September for the whole Region (Ouallam and Tillabery are shifted of 1 decade compared to the other departments).

Table 14.2: Flowering decades per Department (Average 2007–2011)

Department/Decade	24	25	26
Filingué	x	x	
Kollo	x	x	
Ouallam		x	x
Say	x	x	
Tera	x	x	
Tillabery		x	x

14.4 Drought Hazard Indices

The average monthly SPI (Figure 14.9) for the whole Region shows a great variability for the 4 months of the cropping season. June and September show the highest variability, but the month of July seems to experience significant and continuous cumulative deficits, especially in the late 2000s, delaying the beginning of the cropping season. September, reflecting the end of the season, shows a smaller rainfall deficit, although in the late 2000s the variability increases.

The decadal evolution of seasonal parameters shows a trend to earlier season starts until the mid-2000s, while during later years the start of the season is slightly late. So far, the length of the season has been decreasing for the whole region since 2010.

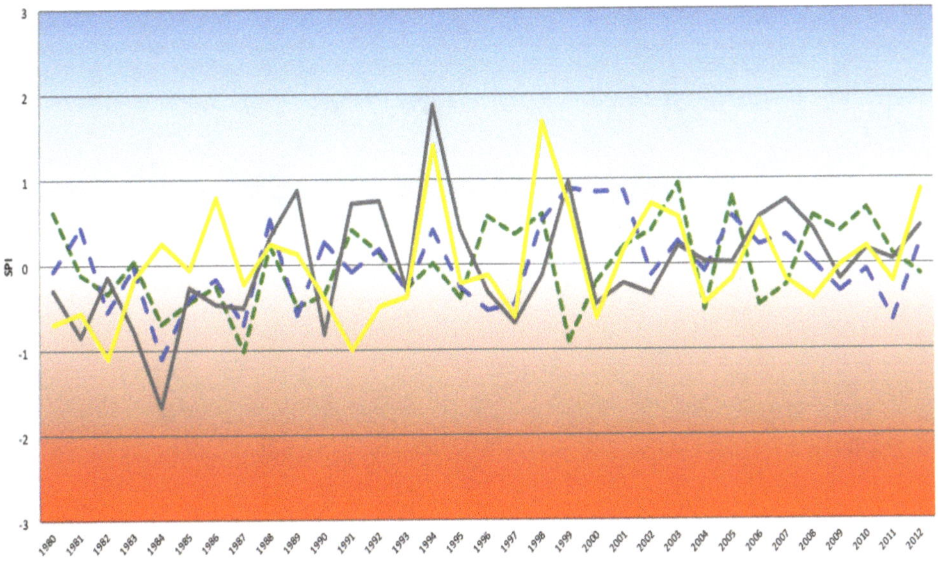

Figure 14.9: Monthly average SPI for the whole Tillabery region 1980–2012 (June: gray dashed, July: black dashed, August: gray, September light gray).

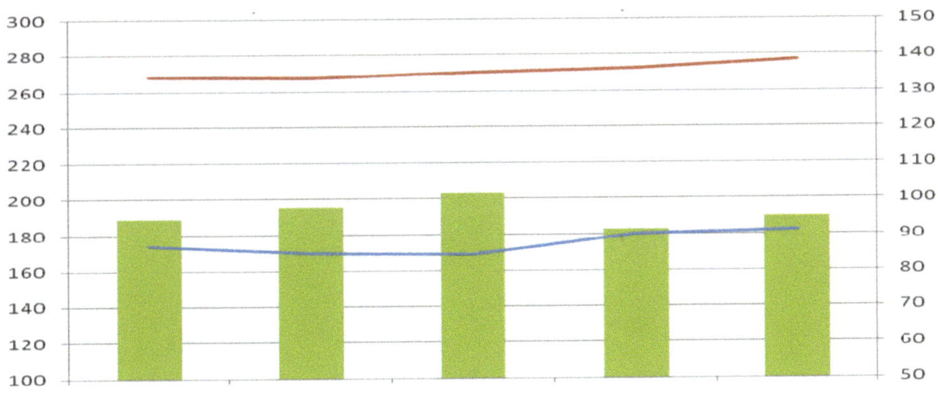

Figure 14.10: Trends of seasonal parameters from 1981 to 2012.

The preliminary analysis of the dry spells during 2003–2012 in the cropping season (Figure 14.11) shows that the intensity (duration) of dry spells increased gradually from the first decade after sowing until the third decade when it reaches a first peak. Then it fades in mid-season to increase again later in the season, and it reaches a second peak between 9 and 10 decades after sowing (at the harvest time). This attenuation in mid-season is due to the intensification of rainfall in the period from July to August. This trend is similar for the majority of the stations in the region.

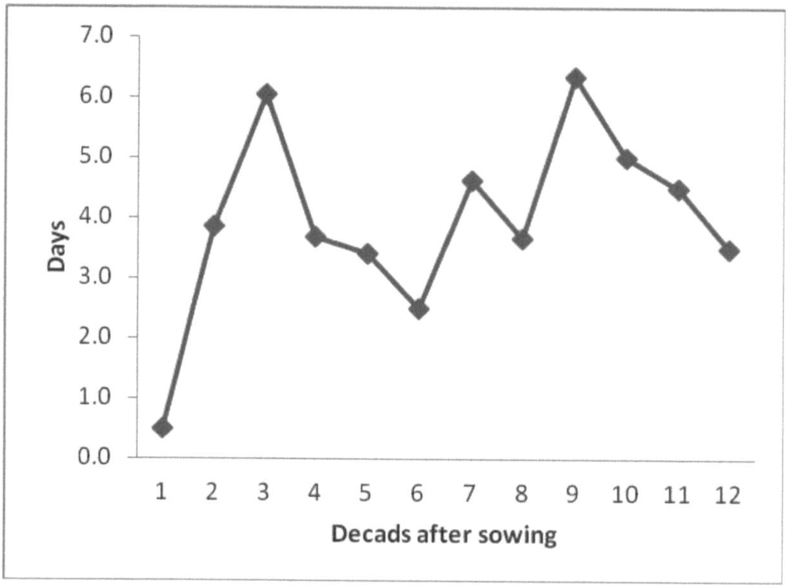

Figure 14.11: Average length of dry spells during the season (average 2003–2012).

Figure 14.12 shows the distribution of the 5 classes of dry spells during the season for the whole Region. Short spells (< 5 days and 5–10 days) are the most frequent, while longer spells have lower frequency (0 to 15 %).

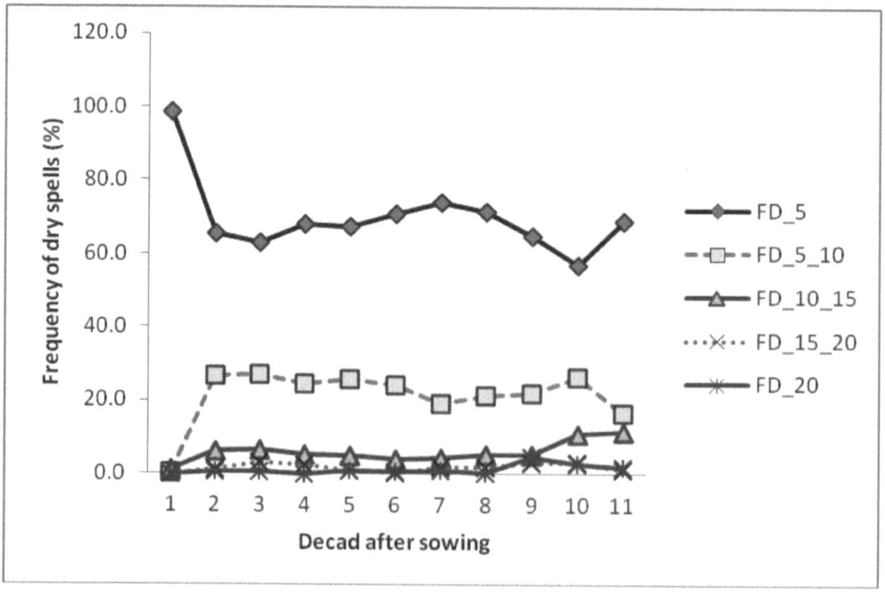

Figure 14.12: Distribution of dry spells during the season (average 2003–2012).

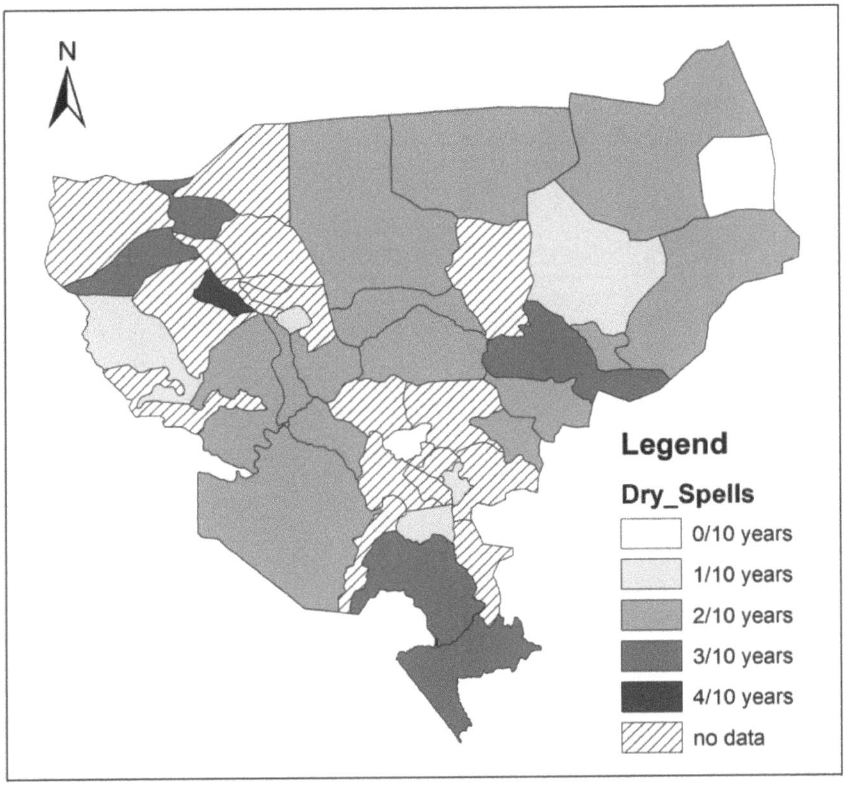

Figure 14.13: Frequency of dry spells longer than 10 days.

Concerning dry spells longer than 10 days, those having stronger impact on crops, Figure 14.13 shows that they are almost infrequent.

From this analysis, we can state that the dry spells of short duration (<5 days or between 5 and 10 days) are the most frequent, while longer dry spells (> 10 days) are less frequent for the majority of the municipalities in the region.

14.5 Drought Risk Indicators

The impact of dry spells on crops' phenology has been analyzed using DHC. Dry spells longer than 10 days have a strong impact on water balance, as shown in Table 14.3, resulting in a significant reduction of decadal water balance. However, in none of the cases studied, did the stress significantly affect the cumulative index.

ZAR outputs confirm DHC ones, indicating that during the growing season, in the last ten years (2003–2012) the Tillabery region has not experienced water stress conditions which significantly affected crop development.

Table 14.3: Number of dry spells > 10 days per municipality over 10 years (2003–2012)

Municipality	Dry_Spells	Municipality	Dry_Spells	Municipality	Dry_Spells
Abala	2	Hamdallaye	nd	Ouro Gueladio	nd
Anzourou	nd	Imanan	2	Parc W	3
Ayerou	3	Inates	nd	Sakoira	nd
Bani Bangou	2	Karma	nd	Sanam	0
Bankillare	3	Kirtachi	nd	Say	1
Bibiyergou	nd	Kokorou	nd	Simiri	2
Bitinkodji	nd	Kollo	1	Sinder	nd
Dargol	2	Koure	nd	Tagazar	2
Dessa	nd	Kourfey centre	2	Tamou	3
Diagorou	nd	Kourtey	2	Tera	1
Diantchandou	2	Libore	nd	Tillabery	1
Dingazi Banda	nd	Mehana	4	Tondikandia	3
Filingue	1	Namaro	2	Tondikwindi	2
Gorouol	nd	N'Dounga	nd	Torodi	2
Gotheye	2	Ouallam	2	Youri	nd

Table 14.4: Water balance index calculated with DHC for millet experiencing prolonged dry spells.

Station	Year	Decade	Dry spell	Idc	idc_dec+1	Ic_dec+1
Dargol	2009	3	15.5	47	6	56
Simiri	2009	3	16.5	21	9	49
Tillabery	2009	3	18.5	59	1	53
Banibangou	2008	4	15.5	98	76	98
Ouallam	2009	5	16.5	19	28	43
Sanam	2009	5	18.5	44	3	52

Decade: number of decades after sowing
Dry spell: length of dry spell in days
Idc: decadal water balance index
Ic: cumulated water balance index
Idc_dec+1: decadal water balance index at following decade
Ic_dec+1: cumulated water balance index at following decade

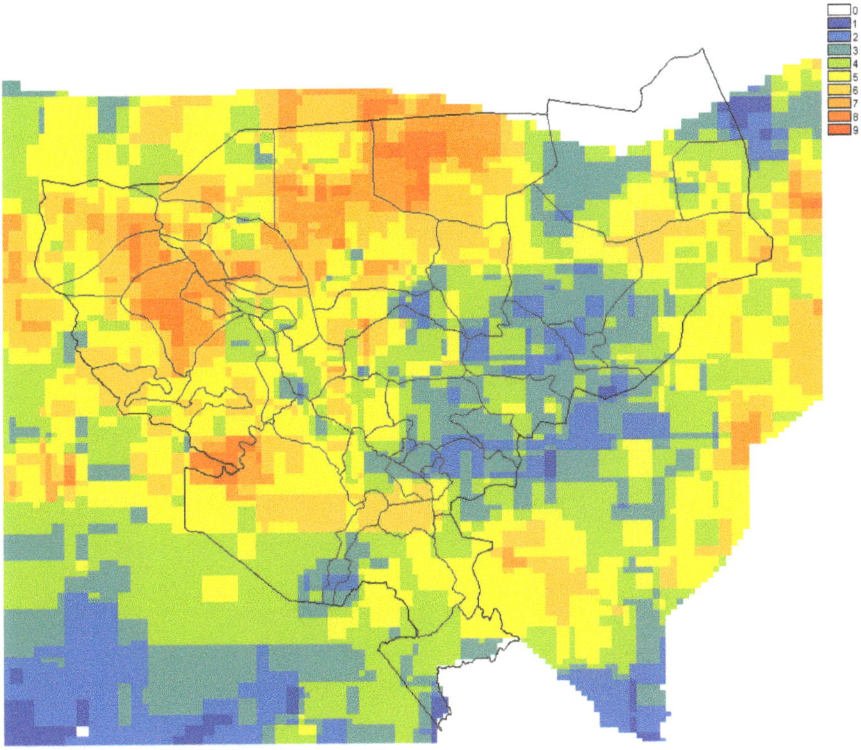

Figure 14.14: Frequency of sowing failures during 2003–2012.

By contrast, ZAR indicated that many areas have been affected by sowing failures. Figure 14.14 shows the frequency of sowing failures for the 2003–12 period. The north-western part of the Region is particularly affected by this phenomenon (more than one year over two), while the eastern part seems to be less affected.

The frequency averaged per municipality (Table 14.5) shows that the municipality of Kokorou in the north-west has been affected by 7 sowing failures during last 10 years, while the municipality of Abala in north-east only 2 times.

Table 14.5: Average frequency of sowing failures per municipality during 2003–2012.

Municipality	Failures over 10 years	Municipality	Failures over 10 years	Municipality	Failures over 10 years
Abala	2.4	Hamdallaye	3.3	Ouro Gueladio	4.3
Anzourou	5.8	Imanan	2.8	Parc W	3.5
Ayerou	5.7	Inates	5.2	Sakoira	5.1
Bani Bangou	6.2	Karma	4.9	Sanam	4.5

continued **Table 14.5:** Average frequency of sowing failures per municipality during 2003–2012.

Municipality	Failures over 10 years	Municipality	Failures over 10 years	Municipality	Failures over 10 years
Bankillare	6.3	Kirtachi	4.6	Say	5.8
Bibiyergou	6.2	Kokorou	7.0	Simiri	4.3
Bitinkodji	4.4	Kollo	3.8	Sinder	5.3
Dargol	5.4	Koure	3.0	Tagazar	3.7
Dessa	6.5	Kourfey centre	4.5	Tamou	4.2
Diagorou	5.5	Kourtey	4.6	Tera	5.2
Diantchandou	3.0	Libore	2.8	Tillabery	4.4
Dingazi Banda	4.5	Mehana	6.7	Tondikandia	2.8
Filingue	3.7	Namaro	4.7	Tondikwindi	5.9
Gorouol	5.6	N'Dounga	2.8	Torodi	5.1
Gotheye	4.5	Ouallam	4.6	Youri	4.4

Table 14.6: Delay of flowering: average frequency per municipality during 2003–2012 period.

Municipality	Delay > 30 days over 10 years	Municipality	Delay > 30 days over 10 years	Municipality	Delay > 30 days over 10 years
Abala	5	Hamdallaye	2	Ouro Gueladio	0
Anzourou	2	Imanan	2	Parc W	0
Ayerou	3	Inates	4	Sakoira	2
Bani Bangou	5	Karma	1	Sanam	5
Bankillare	2	Kirtachi	0	Say	0
Bibiyergou	2	Kokorou	2	Simiri	3
Bitinkodji	1	Kollo	1	Sinder	2
Dargol	2	Koure	0	Tagazar	1
Dessa	2	Kourfey centre	3	Tamou	0
Diagorou	2	Kourtey	2	Tera	2
Diantchandou	0	Libore	1	Tillabery	2
Dingazi Banda	3	Mehana	2	Tondikandia	2
Filingue	3	Namaro	1	Tondikwindi	4
Gorouol	3	N'Dounga	1	Torodi	1
Gotheye	2	Ouallam	2	Youri	2

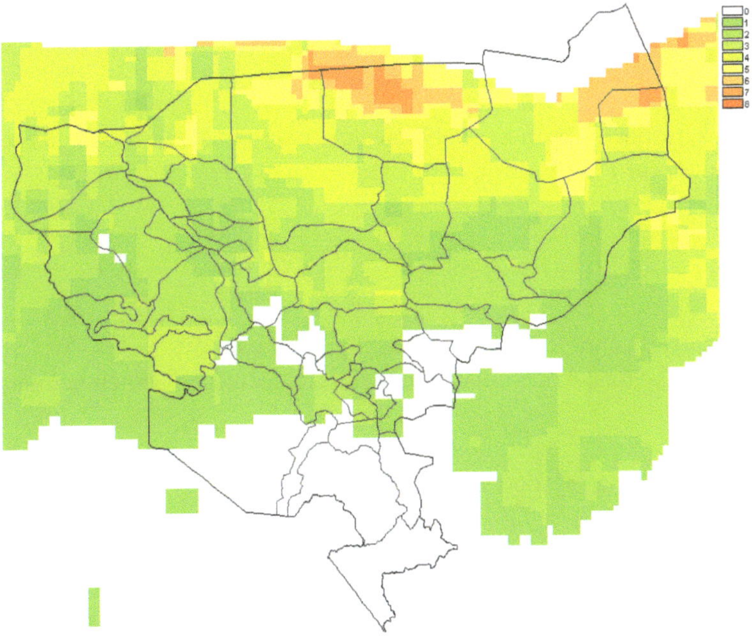

Figure 14.15: Number of years with millet flowering at the 3rd decade of September during the period 2003–2012.

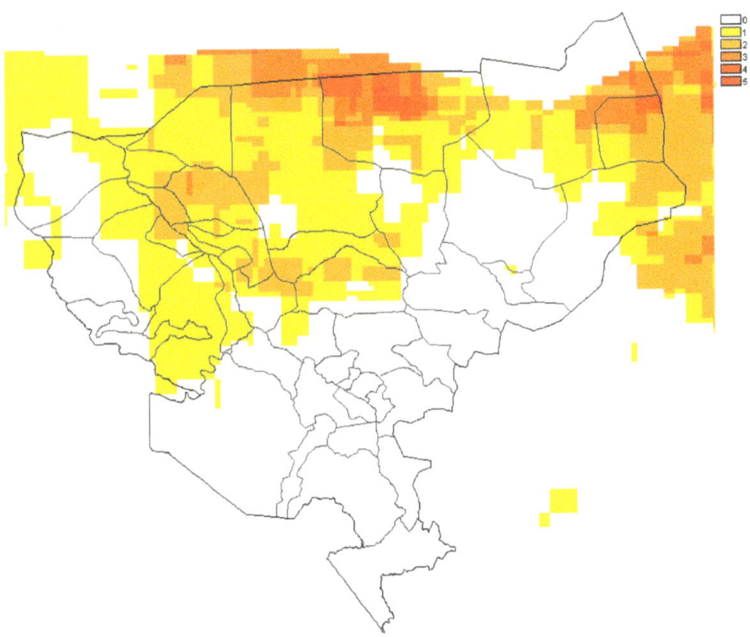

Figure 14.16: Number of years with millet bolting at the 3rd decade of September during the period 2003–2012.

The analysis of phenological stages with ZAR shows that in the northern part of the Region, it is more likely that risks related to length of the season occur. Indeed, areas where millet, at the third dekad of September, is still flowering or bolting are at medium and high risk, respectively, to not complete their phenological cycle, as in the case of the municipalities of Tondikwindi, Bani-Bangou, Abala, and Sanam.

14.6 Conclusions

The agro-climatic analysis showed that in the region of Tillabery the start of the growing season tends to be late, while the end of the season comes in September and is stable.

The frequency analysis of dry spells shows that in Tillabery protracted dry spells are less frequent, while those of short durations are observed almost every year. However, these short dry spells have little negative impact on crops.

The evaluation of a dry spells' impact on crops through water satisfaction indices simulated by the DHC model shows a sensible impact on the decadal index early in the season, but without affecting the cumulative index, and thus grain production.

This is also confirmed by the results obtained with the ZAR model during the last ten years: there have been no cases of major water stress in crops' sensitive phases. We can therefore exclude, a priori, that in the region there are areas that are more at risk of dry spells during the course of the rainy season than others. By contrast, sowing failures and delays in the start of the season for certain municipalities are a frequent threat to crop production. Sowing failures are particularly frequent, in fact 18 municipalities over 45 have a frequency higher than 0.5 (one year over two).

In practice, based on the results of ZAR on phenological phases, we can conclude that in the northern part of the Tillabery region, the drought risk is associated with delaying the length of the season; in fact; if a crop is still in bolting or flowering phases at the end of September, there is a concrete risk that the cycle will not be completed. The southern part of the Region has lower risk related to the length of the season, but it still has the risk of sowing failure due to a false start of the rainy season.

Crop production and yield data disaggregated to the municipality could have confirmed such results, but unfortunately the data doesn't exist.

This chapter has provided a more nuanced understanding of how different areas of an administrative Region are exposed and vulnerable to drought risk. Findings in this paper will help to guide a more general discussion on food production systems' adaptive capacity to observe, along with future climate changes. The implication of the results is that policy makers need to formulate a given area's specific and targeted climate adaptation policies and programmes that foster asset building to increase the capacity of vulnerable households to adopt agricultural practices that are less likely to be adversely impacted by climate variability and change. In order to support this process, national and sub-national development strategies require strong risk

Table 14.7: Drought risk indices per municipality.

Municipality	%_Dry_spells > 10 days	%_Sowing Failures	%_Delay > 30 days	Cumulative index*
Abala	0.2	0.24	0.5	0.94
Anzourou	nd	0.58	0.2	0.98
Ayerou	0.3	0.57	0.3	1.17
Bani Bangou	0.2	0.62	0.5	1.32
Bankillare	0.3	0.63	0.2	1.13
Bibiyergou	nd	0.62	0.2	1.02
Bitinkodji	nd	0.44	0.1	0.74
Dargol	0.2	0.54	0.2	0.94
Dessa	nd	0.65	0.2	1.05
Diagorou	nd	0.55	0.2	0.95
Diantchandou	0.2	0.3	0	0.5
Dingazi Banda	nd	0.45	0.3	0.95
Filingue	0.1	0.37	0.3	0.77
Gorouol	nd	0.56	0.3	1.06
Gotheye	0.2	0.45	0.2	0.85
Hamdallaye	nd	0.33	0.2	0.73
Imanan	0.2	0.28	0.2	0.68
Inates	nd	0.52	0.4	1.12
Karma	nd	0.49	0.1	0.79
Kirtachi	nd	0.46	0	0.66
Kokorou	nd	0.7	0.2	1.1
Kollo	0.1	0.38	0.1	0.58
Koure	nd	0.3	0	0.5
Kourfey centre	0.2	0.45	0.3	0.95
Kourtey	0.2	0.46	0.2	0.86
Libore	nd	0.28	0.1	0.58
Mehana	0.4	0.67	0.2	1.27
Namaro	0.2	0.47	0.1	0.77
N'Dounga	nd	0.28	0.1	0.58
Ouallam	0.2	0.46	0.2	0.86
Ouro Gueladio	nd	0.43	0	0.63
Parc W	0.3	0.35	0	0.65

continued **Table 14.7:** Drought risk indices per municipality.

Municipality	%_Dry_spells > 10 days	%_Sowing Failures	%_Delay > 30 days	Cumulative index*
Sakoira	nd	0.51	0.2	0.91
Sanam	0	0.45	0.5	0.95
Say	0.1	0.58	0	0.68
Simiri	0.2	0.43	0.3	0.93
Sinder	nd	0.53	0.2	0.93
Tagazar	0.2	0.37	0.1	0.67
Tamou	0.3	0.42	0	0.72
Tera	0.1	0.52	0.2	0.82
Tillabery	0.1	0.44	0.2	0.74
Tondikandia	0.3	0.28	0.2	0.78
Tondikwindi	0.2	0.59	0.4	1.19
Torodi	0.2	0.51	0.1	0.81
Youri	nd	0.44	0.2	0.84

* The Cumulative index was calculated using the average value of "%_Dry_spells > 10 days" when the data was not available (nd).

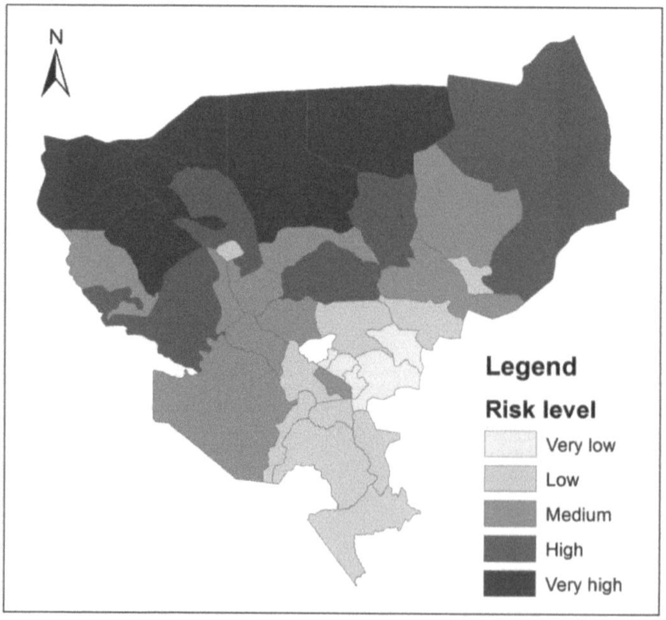

Figure 14.17: Map of drought risk cumulative index.

information. Considering that current risk information on drought is fragmented and not sufficiently utilized in decision making, a better understanding of drought likelihood, food production losses, and risks to crop production systems would encourage informed investments by the agricultural public and private sector, and greatly assist agricultural planning by governments.

References

Amogu, O., M. Esteves, L. Descroix, K. Souley Yéro, J.-L. Rajot, M. Malam Abdou, S. Boubkraoui, J.M. Lapetite, N. Dessay, I. Zin, J.-P. Vandervaere, A. Bachir, I. Bouzou Moussa, E. Gautier, E. Le Breton, and I. Mamadou. 2014. Runoff evolution according to land use change in a small Sahelian catchment. *Hydrology and Earth System Sciences* 8, 1569–1607. doi:10.1080/02626667.2014.885654.

Bacci, M., V. Tarchiani, P. Vignaroli, L. Genesio, and A. Di Vecchia. 2009. Identification et suivi des zones à risque agro-météorologique au Sahel. Rome: Aracne Editrice.

Beguería, S., S.M. Vicente-Serrano, J.I. López-Moreno, and J.M. García-Ruiz. 2009. Annual and seasonal mapping of peak intensity, magnitude and duration of extreme precipitation events across a climatic gradient, North-east Iberian Peninsula. *International Journal of Climatology* 29: 1759–1779.

Bruce, J.P. 1994. Natural Disaster Reduction and Global Change. *Bull. Amer. Meteor. Soc.* 75, 1831–1835.

Desinventar, Corporacion OSSO. http://www.desinventar.org/en/database

Diallo, M.A. 2001. Mise en place de procédures d'analyses statistiques agroclimatiques et la banque de données sur les paramètres de la saison agricole pluviale. Niamey: Centre Régional Agrhymet.

EM-DAT, The OFDA/CRED international disaster database. Bruxelles: Université Catholique de Louvain. www.emdat.be.

Garvelink, W.J., F. Tahir. 2012. Misunderstood: Getting the right response to food shortages in the Sahel. Washington: Center for strategic and international studies.

Gubbels, P. 2012. Breaking the hunger cycle: Pathways to resilience in the Sahel. World Vision and Save the Children.

Mahé, G., G. Lienou, L. Descroix, F. Bamba, J.E. Paturel, A. Laraque, M. Meddi, N. Moukolo, H. Hbaieb, O. Adeaga, C. Dieulin, F. Kotti, and K. Khomsi. 2013. The rivers of Africa: witness of climate change and human impact on the environnement. *Hydrological Processes* 27 (15): 2105–2114.

McKee, T.B., N.J. Doesken, and J. Kleist. 1993. The relationship of drought frequency and duration of time scales. Eighth Conference on Applied Climatology, American Meteorological Society, Jan 17–23, 1993, Anaheim CA, 179–186.

Mousseau, F., and A. Mittal. 2006. Sahel: a prisoner of starvation? A Case study of the 2005 food crisis in Niger. Oakland: The Oakland Institute.

Munich Re, NatCatSERVICE. 2011. Natural catastrophe know-how for risk management and research.

Nicholson, S.E., C.J. Tucker, M.B. Ba. 1998. Desertification, drought, and surface vegetation: an example from the West African Sahel. *Bulletin of the American Meteorological Society* 79: 815–829.

Pini, G., V. Tarchiani. 2007. Les systèmes de production agro-sylvo-pastoraux du Niger : la caractérisation agro-écologique. Working Paper n. 21 CCTM-Politecnico di Torino, Turin.

Redmond, K.T. 2002. The depiction of drought: A commentary. *Bull. Amer. Meteor. Soc.* 83: 1143–1147.
RN-République du Niger, INS. 2012. RGPH-Recensement General de la Population et de l'Habitat, République du Niger.
Rinaudo, T., S. Yaou. 2009. Agricultural task force report: World Vision Niger agricultural development. Friedrichsdorf: World Vision.
Samba, A. 1998. Les logiciels Dhc de diagnostic hydrique des cultures. Prévision des rendements du mil en zones soudano-sahéliennes de l'Afrique de l'Ouest. *Sécheresse* 9: 281–288.
Tarchiani, V., A. Di Vecchia, L. Genesio, and V. Sorbi. 2006. From food security to development in the Sahel region. In *Natural resource management in Sahel - Lessons learnt*, ed. M. Møllegaard. Copenhagen: Serein.
UNISDR. 2007. Hyogo framework for action 2005–2015: Building the resilience of nations and communities to disasters. Geneva: UNISDR.
UNISDR. 2009. Terminology on disaster risk reduction. Geneva: UNISDR.
Vicente-Serrano, S.M., J.I. López-Moreno, and B. Van der Hurk. 2005. Hydrological response to different time scales of climatological drought: an evaluation of the Standardized precipitation index in a mountainous mediterranean basin. *Hydrology & Earth System Sciences* 9(5): 523–533.
Wilhite, D.A. ed. 1993. *Drought assessment, management, and planning: Theory and case studies.* Springer. Isbn: 978-1-4615-3224-8.
WMO-World Meteorological Organization. 2011. Climate knowledge for action: a global framework for climate services. Geneva: WMO.

Silvia Macchi[39] and Liana Ricci[40]

15 Climate Change Adaptation Through Urban Planning: a Proposed Approach for Dar Es Salaam, Tanzania

Abstract: The need for climate change adaptation is increasingly influencing the discourse about spatial development strategies throughout the world. Nevertheless, several gaps still exist in our understanding of the spatial dimensions of climate change vulnerability and how to incorporate them into planning practices. Firstly, attention has been mostly focused on how to adjust physical assets to climate change, while the question of how to strengthen local adaptive capacity remains rather neglected. Secondly, while many cities have institutionalized climate change, integration of adaptation considerations into existing urban planning and governance systems is still lacking. As a result, not only do the plans and programs in place for urban development and environmental management often fail to address adaptation needs, they may even jeopardize current adaptive capacity. The latter has particularly serious consequences for Sub-Saharan cities, where people's capacity for autonomous adaptation is a crucial resource, given the limited capacity of local government institutions to fulfill their responsibilities. This chapter proposes a methodology for mainstreaming adaptation into existing planning documents, developed specifically for the city of Dar es Salaam, Tanzania. After providing a brief review of approaches and challenges in adaptation mainstreaming, the main features of the proposed methodology and preliminary results of its application are presented. Lessons learned from the experience are examined in the conclusions.

Keywords: Adaptation, Mainstreaming, Sub-Saharan Africa

15.1 Introduction

The need for climate change adaptation is increasingly influencing the discourse about spatial development strategies throughout the world (Davoudi *et al.* 2009). Nevertheless, several gaps still exist in our understanding of the spatial dimensions of vulnerability to climate change induced impacts and how to incorporate them into planning practices.

Firstly, attention has been mostly focused on how to adjust physical assets to climate change, encompassing issues related to climate proofing, robustness, and

39 DICEA-Sapienza University of Rome, silvia.macchi@uniroma1.it
40 DICEA-Sapienza University of Rome, liana.ricci@gmail.com

resilience of the built environment and infrastructure. Meanwhile, the question of how to strengthen the adaptive capacity of local communities and authorities remains rather neglected by urban planners and policy makers. It should be noted that the latter requires a change in the way vulnerability is interpreted: from a linear result of climate change impacts on an exposure unit (outcome vulnerability), to a dynamic state resulting from the interaction between climate change and contextual conditions associated with an exposure unit (contextual vulnerability) (O'Brien *et al.* 2007). Such a change broadens the spectrum of adaptation action to include the structural inequalities of the context in order to change vulnerability circumstances, thus paving the way for "transformational adaptation" (IPCC 2013a:1).

Secondly, while many cities have institutionalized climate change strategies by establishing a dedicated climate unit either within an existing department or as a separate cross-cutting office (Anguelovski and Carmin 2011) and preparing a local adaptation plan for action (Carmin *et al.* 2012a, 2012b), integration of adaptation considerations into existing urban planning and governance systems is still lacking or immature. In fact, most adaptation mainstreaming research and practices have focused on development policy at the national level (Klein 2002; Huq *et al.* 2003; Agrawala 2005; Persson and Klein 2008). Although this provides a valuable theoretical and practical basis for advancing at the sectoral and local level, a considerable amount of work is still needed to operationalize the concept of adaptation in urban planning, thus enabling the identification of context specific adaptation options.

As a result, not only do the plans and programs in place for urban development and environmental management often fail to address adaptation needs, they may even jeopardize current adaptive capacity. The latter is especially threatening in the Sub-Saharan context, where people's capacity to adapt to change in their living environment is often a necessary substitute for insufficient institutional capacity to provide adequate infrastructure and services to a rapidly growing population (Ricci 2011, 2014).

This chapter proposes a methodology for mainstreaming adaptation into existing urban development and environmental management plans of cities in Sub-Saharan Africa.

According to Friedmann (2005), urban planning in Africa faces four major challenges: an average urban population growth of at least 5% annually; implosion of the informal economy; local government's financial inability to adequately service the population; and allocation of the majority of the land without regard for planning regulations. Climate change is expected to exacerbate existing problems, as it threatens the natural resources upon which livelihoods of the majority depend, and is likely to give further impetus to the vicious circle linking environmental degradation to urban sprawl (Macchi *et al.* 2013). In light of recent climate change predictions, there is an increasingly urgent need to improve sustainable human settlement and infrastructure development in the least developed countries (Satterthwaite *et al.* 2007), while reframing urban policy and governance in an adaptation perspective.

However, the question arises as to whether mainstreaming adaptation into existing plans can contribute to this improvement, when such plans are usually in default of implementation due to an unsuitable "culture of planning" (Friedmann 2005). The present work adopts an incremental perspective, emphasizing the importance of context in identifying viable ways to change the culture of planning in Sub-Saharan Africa. This perspective associates the need for adaptation to climate change with a shift in how the Sub-Saharan city is conceptualized. Adaptation mainstreaming is seen as a means to improve the effectiveness of existing plans and programs while bringing the autonomous adaptive capacity of people to the center stage of urban planning in Sub-Saharan Africa (Macchi 2014).

The city of Dar es Salaam was chosen as a case study for developing mainstreaming methodology[41]. It is the most populous city in the country (4.4 million inhabitants in 2012) and the main engine of the national economy: 16.9% of national GDP in 2001–2012; +20.9% of regional GDP in 2011-2012 against a national increase of +17.9% (URT 2013). In the last decade it has expanded far beyond any planning projections: +75% of population in 2002–2012; +76% of continuously built-up areas, and +192% of discontinuously built-up areas in 2002-2011 (Congedo and Munafò 2014), and today a new policy strategy for urban development and environment management is under consultation (Halloran and Magid 2013; Dodi Moss *et al.* 2013). Due to over-pumping of groundwater in the coastal plain and subsequent intrusion of seawater in the shallow aquifer, people living in that area are already experiencing limited access to fresh water, and the entire coastal socio-ecological system is at risk (Faldi and Rossi 2014; see Chap. Sappa). Although the lack of long term meteorological data and the inherent complexity of climate dynamics prevent accurate downscaling of climate change projections to tropical East Africa, available observations show an increase in temperature and a decrease in rainfall (Rugai and Kassenga 2014), which will amplify the need for water resource conservation and promotion of alternatives to the coastal aquifer.

A brief review of approaches and challenges in adaptation mainstreaming are provided below. The main features of the proposed methodology and a few preliminary results from its application are then presented, with a focus on the planning provisions for peri-urban areas in the Dar es Salaam 2012-2032 Draft Master Plan (Dodi Moss *et al.* 2013). Lessons learned from the experience are examined in the conclusions.

41 The study has been conducted within the framework of the EU funded project "Adaptation to Climate Change in Coastal Dar es Salaam", which aims to improve the effectiveness of local authorities in supporting the autonomous efforts of coastal peri-urban populations to adapt to climate change. For further details see www.plannning4adaptation.eu

15.2 Understanding Adaptation Mainstreaming

In policy literature, the term "mainstreaming" is often used interchangeably with "integration" and "incorporation" to designate a strategy for dealing with cross-cutting issues, like adaptation to climate change. According to Working Group II of the Intergovernmental Panel on Climate Change (IPCC WG2), which focuses specifically on climate change impacts, adaptation, and vulnerability, there is robust evidence that "integration of adaptation into planning and decision-making can promote synergies with development and reduce the possibility of maladaptive actions" (IPCC 2013b: 31). In the context of this chapter, adaptation mainstreaming is understood as the process of systematically integrating adaptation needs into existing urban planning documents, while avoiding maladaptation.

Before examining the arguments for and against the adoption of a mainstreaming approach to climate change adaptation, it is worth mentioning three key notions as defined by the WP2 in the IPCC Fifth Assessment Report: adaptation, maladaptation, and adaptive capacity. Adaptation is "the process of adjustment to actual or expected climate and its effects. In human systems, adaptation seeks to moderate harm or exploit beneficial opportunities. In natural systems, human intervention may facilitate adjustment to expected climate and its effects" (IPCC 2013b: 1). Maladaptive actions (or maladaptation) are "actions that may lead to increased risk of adverse climate-related outcomes, increased vulnerability to climate change, or diminished welfare, now or in the future" (ibid.: 18). Adaptive capacity is "the ability of systems, institutions, humans, and other organisms to adjust to potential damage, to take advantage of opportunities, or to respond to consequences" (ibid.: 2).

15.2.1 Mainstreaming Versus Stand-alone Adaptation

Although "there are few assessments of adaptation delivery and effectiveness" (IPCC 2013c: 22), cross-sectoral integrated approaches are viewed as more effective than stand-alone efforts. This is mainly due to the multi-dimensionality and multiscalarity of the notion of vulnerability when applied to socio-ecological systems, as well as to the uncertainty of local vulnerability trajectories, which are highly influenced by the combined effects of climate change and socio-economic scenarios. Such considerations are particularly relevant to the research presented here, which seeks to reduce vulnerability by developing the autonomous adaptive capacity of private individuals, as opposed to improving the climate proofing of public decision-making and spending.

Adaptation as a self-standing action usually consists of measures that address a single policy sector in a mono-disciplinary way. This involves a limited number of actors in a linear decisional process where inputs and outputs are clearly defined ex-ante, and favors solutions that can be repeated in other locations, are achievable in the short term, and use standardized design and plan-driven (conformative) implementation methods. By contrast, adaptation mainstreaming has the potential for multi-sectoral action, and requires the involvement of all levels of governance as well as a broader range of stakeholders in decision-making. It mobilizes different knowledge and competences in an iterative and open-ended process, favors context specific solutions that are achievable in the medium-long term, and uses experimental design and target-driven (performative) implementation methods (Macchi and Ricci 2014).

Given its characteristics, the stand-alone strategy is undoubtedly easier to apply than the mainstreaming strategy, particularly in terms of planning, acquisition of financial resources and other necessary means, decision-making, implementation, and evaluation of results. In addition, it provides the institutions involved with shared, short-term goals that foster motivation to effect change, something that is missing in the mainstreaming approach and may end up compromising its main value-added, i.e. its transformative potential.

However, the efficacy of a stand-alone and sector-specific action in reducing vulnerability in the medium-long term is far from guaranteed in rapidly changing contexts, like those in Sub-Saharan cities, or where highly uncertain risks are involved. Conversely, by ensuring coherence and seeking synergies across policy domains and institutional levels, the mainstreaming strategy allows the plurality of interconnected, multisectoral factors that shape vulnerability (Adger *et al.* 2007) to be addressed in an integrated way and with a long-term perspective.

Finally, according to the literature, integrating adaptation into development and sectoral decision-making in sectors affected by climate risks also facilitates the leverage of the much larger financial flows than would be available to finance adaptation separately (Agrawala 2005). In general, integrating adaptation allows for more sustainable, efficient, and effective use of resources (Persson and Klein 2008).

The mainstreaming methodology and exercise reported in this chapter are intended as a first step towards the full development of an adaptation mainstreaming strategy involving all sectors and levels of government in Dar es Salaam. Rather than beginning with a detailed reconstruction of decision-making among local authorities, a few provisions of existing urban development and environmental management plans and programs are used as entry points for the mainstreaming process. The reasons for this are twofold. Firstly, the research team sought to address issues of direct concern to local officers involved in the project in order to stimulate their interest and increase the likelihood that change would be implemented. Secondly, regarding requests for funding, the officers involved expected help in identifying which measures already in place would qualify as adaptation with little or no change. The funding issue is a bottleneck for Dar es Salaam's local authorities when implementing planning decisions, as their budget depends on national transfers which are limited and uncertain (Tanzania is classified as a least developed country, with a per capita GDP of USD 652 in 2012). As a result, a hybrid approach was adopted where each sectorial planning provision was considered separately, but with a view to advancing adaptation mainstreaming.

15.2.2 Focusing on Adaptive Capacity

This work assumes that the purpose of adaptation is to strengthen people's capacity to autonomously adapt to a changing environment rather than to secure the physical environment through improved infrastructure and measures for impact mitigation[42]. The specific focus is on populations living in peri-urban areas within the coastal plain and their capacity to adjust livelihood strategies and practices in response to actual or expected variations in living conditions. The "adaptive capacity" concept is central to this approach and deserves some further clarification.

Firstly, the relation of adaptive capacity to vulnerability should be examined. Depending on the author, adaptive capacity may be seen as being one component of overall vulnerability (Füssel and Klein 2006) or as a separate, though related, factor, as is usually the case in disaster risk management studies. The recent IPCC WP2 contribution to the Fifth Assessment Report provides useful insight into this subject by distinguishing two types of vulnerability: contextual or starting-point vulnerability and outcome or end-point vulnerability. The former is defined as "a present inability to cope with external pressures or changes, such as changing climate conditions" and "a characteristic of social and ecological systems generated by multiple factors and processes" (IPCC 2013a: 8). The latter describes vulnerability levels after adaptation has taken place (ibid.: 19).

Given the scope of this work, the notion of contextual vulnerability has been adopted. Considerable time and energy was spent gathering people's views on environmental changes of concern and current strategies adopted to cope or adapt to them, as well as to outline the main factors and conditions influencing households' in the peri-urban (Ricci 2011, 2014).

Drawing on Amartya Sen's capability approach, adaptive capacity can be defined as the sum of all the strengths, attributes, and resources available to a society (institutions, local groups, individuals, etc.) that could play a positive role in facilitating adjustment to expected climate chance and its effects. Building this kind of capacity is acknowledged as "a means, among other objectives, to shift the analytical balance from the negative aspects of vulnerability to the positive actions by people" and "fundamental to imagining and designing a conceptual shift favoring disaster risk reduction and adaptation to climate change" (IPCC 2012: 33). In other words, it encourages adoption of a planning attitude, as capacity building requires "a clear image of the

42 A full justification of this choice is given in Macchi (2014), including scarce financial resources, a rapidly changing environment, the high uncertainty of climate change impacts, and recognition of the peri-urban as the main modality of urban development in "the century of urbanization". In this context, the concept of adaptive capacity is seen as a heuristic tool for analysis of the Sub-Saharan city (Ricci 2011, 2014).

future with clearly established goals" (ibid.), and forces planners to focus on social systems and their relation to resource scarcity.

Notwithstanding Sen's teaching that human well-being depends on actual access to resources rather than the mere availability of such resources, in most of the literature the availability of human, social, technological, manufactured, and financial capital is considered a proxy for adaptive capacity (Nelson *et al.* 2007; Preston *et al.* 2008; Dunford *et al.* 2013). Only a few authors (Adger *et al.* 2007) challenge this view, drawing on the idea that vulnerability is the result of a process in which the system of social interactions and power relations influences people's access to resources, and therefore contributes in a determinative manner to defining the kind of vulnerability of a given social group in a given time and place (O'Brien *et al.* 2007; Simon 2010). These different perspectives bring into discussion the role of institutions in building up local adaptive capacity, which is key to defining the specific objectives to be pursued through adaptation mainstreaming. While the former perspective influences the deployment of resources through planning and management by existing governance systems, the latter emphasizes the role that institutions play in determining the conditions of resource access.

This work adopts the second perspective. Moreover, a special activity was designed and implemented to allow the community to play a role in the adaptation mainstreaming exercise. The community was involved in the development of vulnerability scenarios through implementation of participatory backcasting workshops (Faldi 2013 and 2014). The backcasting scenario approach was chosen as an antidote to the risk of placing too great a weight on people's current adaptation strategies while overlooking their aspirations for change. In other words, vulnerability is not only changing with the context, but is also shaped by the community's aspirations. The core idea of participatory backcasting is to start by creating a shared, desired vision of the future and to then look backwards to the present to determine what challenges will need to be faced and to assess the potential for change. Through this process the community was able to provide indications of what kinds of action should be taken to change current environmental management and decision-making in order to achieve the desired future.

15.2.3 Operationalizing Adaptation Mainstreaming at the Local Level

At the operational level, a mainstreaming strategy involves four types of change in policy making: procedural, organizational, normative, and policy reframing (Persson and Klein 2008). The first consists of introducing new or modifying existing decision-making procedures while providing targeted information to those tasked with policy development and implementation. The second involves amendments to formal responsibilities and mandates, creating new or merging existing institutions, networking among diverse departments, and structural changes of budgets. The third entails the formalization of the issue to be mainstreamed in existing strategies and policy frameworks as well as the allocation of additional targeted resources. The forth aims at reshaping the policy frame of traditional sectors to embed the issue at stake into the thinking of relevant stakeholders.

Given the scope of this study, only changes applicable at the local government level have been considered (i.e. the identification of initiatives for changing existing procedures and organizational structures). Indeed, in the Tanzanian context, any change in policy frameworks requires the involvement of the national government, an action that goes beyond the reach of the ACCDar project. This represents a major limitation of the study. However, although contact with national authorities has been quite limited, the mainstreaming exercise that was carried out to test the proposed methodology also tackled the analysis of planning documents that fall within the competence of state ministries (i.e. master plans for urban development and water supply at the city level) while also being of great concern to municipal councils. The purpose of such analysis was to better understand the developed methodology's potential to enable local authorities to advocate change with state ministries and agencies.

As highlighted in the literature, "climate adaptation is context dependent and it is uniquely linked to location, making it predominantly a local government and community level of action" (IPCC 2013d: 6). However, adaptation efforts undertaken by local governments may be hampered by a variety of factors, including institutional, social, informational, financial, and cognitive barriers (Adger and Barnett 2009). Here the focus is on barriers in the existing institutional framework, since mainstreaming requires institutional changes of various kinds and at various levels. In particular, the relationships between local authorities and the national government are considered.

One barrier to effective mainstreaming of adaptation at the local level is the inadequate capacity of governments to effectively coordinate the range of adaptation initiatives being implemented in their territory, which is especially true in Africa (IPCC 2013e). This is due mainly to unclear and often overlapping roles and responsibilities between levels and actors, which inhibits knowledge production and circulation, and is aggravated by obstacles in national, subnational, and local governmental approaches to addressing complex and multidimensional problems, such as climate change adaptation. Besides lack of capacity, there may also be a lack of political will to coordinate. In this respect, it bears mentioning that in Sub-Saharan Africa the tension between national interests in the city as a motor of economic growth and the interests of the majority of the urban population, for whom the city is a resource for achieving individual plans, is increasingly palpable. The existence of competing values and policy priorities among local authorities and the national government may seriously hamper any decision by local authorities to build, support, and reinforce the autonomous adaptive capacity of their citizens.

Secondly, the decentralization and devolution of power and functions from central to local authorities also plays a role in enabling or undermining mainstreaming efforts. In Tanzania, despite a number of reforms currently being deliberated by the parliament, decentralization of decision-making and administration remains extremely slow (Lerise 2000). Local government authorities were established following *Local Government Act 7* and *12* of 1982, but the first Local Government Reform Programme only began in 1998, with a second one initiated in 2011. As a result, the decentralization process has not yet been completed. Key decision making on urban development and major infrastructure remains firmly in the hands of the state. Local urban governments have sole responsibility for providing most services, while a few (electricity, hospitals, secondary education, police, economy, and tourism) are shared with other levels of government (Smit and Pieterse 2014).

This situation leads to a high degree of compartmentalization between sectors, actors, and policies operating at similar administrative levels. In Tanzanian local administrations, although the Municipal Council is attempting to prepare a three-year Strategic Plan and a related Medium Term Expenditure Framework following the national devolution policy, single departments depend more on guidance from their respective ministry than from the local authority. Obviously, the limited financial power of local authorities together with the inadequacy and uncertainty of funding from the national government play a role. However, mainstreaming across different sectors seems more practical at the local scale since conflicts between competing policy goals are more evident and precise than at the national level, as people living in the same place have a shared interest in avoiding local socio-ecological crises that could threaten local commons.

Lastly, power relationships within formal institutions are to be included among the factors that may hinder or enable effective mainstreaming. Local government officers may oppose changes that threaten their interests, including power hierarchies, or are inconsistent with their values and beliefs. The adoption of the mainstreaming approach may also be a cause for concern because it often implies a reduction in the amount of funding dedicated to climate change adaptation. Moreover, the request for greater coordination between different levels and sectors of government may be associated with the threat of increased control of local authorities by national government and donors.

Nevertheless, local authorities represent a formidable resource in terms of guaranteeing that mainstreaming occurs from the bottom-up. Moreover, the need for adaptation can be an impetus for institutional change as well as an opportunity to strengthen the capacities of local governments, both in addressing climate challenges and in advocating for adaptation with higher-level authorities. The present research takes on those challenges and proposes a mainstreaming methodology that has been developed by involving local authorities at all stages while providing their officers with opportunities to share concerns and ideas regarding different policy sectors (including health, waste management, water and sanitation, fire and rescue, disaster management, transportation, urban planning, agriculture and livestock, natural resources, civil and environmental engineering) (see Chap. Shemdoe *et al.*).

15.3 Developing a Mainstreaming Methodology

A mainstreaming methodology has been developed and tested on four planning documents. This methodology represents only one stage of a lengthier process that involves: (i) assessing the need to build local authorities' adaptation capacity, (ii) exploring the adaptive capacity of households and identifying their main concerns in a changing environment (i.e. a decrease in freshwater availability in the coastal aquifer), (iii) assessing the coastal aquifer's sensitivity to seawater intrusion caused by climate change and urban sprawl, (iv) defining adaptation objectives in cooperation with the community, and (v) developing a strategy for adaptation mainstreaming with a selected group of municipal officers. Following the mainstreaming exercise, key initiatives will be identified and implemented by local authorities to improve their capacity to support peri-urban households' efforts to adapt to a changing environment.

15.3.1 Methodology Outline

A mixed team of urban planners, environmental engineers and a hydrogeologist from Sapienza (Italy) and Ardhi (Tanzania) universities[43] developed the methodology and carried out the mainstreaming exercise. Although the whole activity took six months to complete, the mainstreaming exercise is designed to last for a period of two months.

The methodology is composed of three phases, preceded by the initial selection of the planning provisions to be analyzed. Phase 1 consists of assessing the selected planning provisions to identify adaptation needs. In phase 2, a set of amendment options is identified for each planning provision under review. In phase 3, the most feasible or suitable amendment options are chosen and recommendations are formulated for their implementation (Figure 15.1).

Assessment criteria for phase 1 include: (i) two adaptation concerns (ACs), both related to water resources because access to fresh water was a top priority due to the increasing salinization in the study area; (ii) three possibilities for autonomous adaptation (PAAs), identified on the basis of a series of household interviews and participatory workshops; and (iii) several criteria related to mitigation, included in order to integrate this goal into the process. Planning provisions were assessed against each of these criteria to identify negative and positive impacts. More details are provided in Table 15.1.

43 The team included Liana Ricci, Carlo Norero, and Giuseppe Sappa from Sapienza University of Rome, and Riziki Shemdoe and Gabriel Kassenga from Ardhi University.

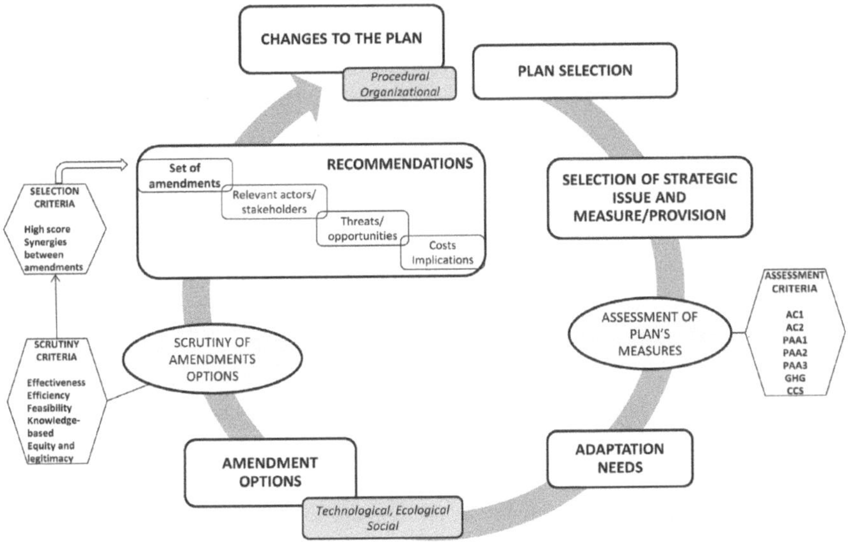

Figure 15.1: Methodology Diagram.

Table 15.1: Phase 1 criteria for assessment of planning provisions.

ADAPTATION CONCERNS		
AC1	water resource conservation	to assess whether a planning provision interferes positively or negatively with the recharge rate of coastal aquifers, prevents or increases the risk of groundwater pollution, and causes a decrease or increase in groundwater extraction
AC2	access to fresh water	to assess whether a planning provision implies a drop or a rise in household costs to access water
POSSIBILITIES FOR AUTONOMOUS ADAPTATION		
PAA1	water source diversification	to assess whether a planning provision increases or reduces the variety of water sources upon which residents can rely
PAA2	changes in income generating activities	to assess whether a planning provision facilitates or impedes residents in adapting their economic activities to cope with environmental changes
PAA3	changes in settlement patterns & relocation	to assess whether a planning provision supports or hinders household capacity to make structural changes in their living place or relocate
MITIGATION CONCERN		
GHG	greenhouse gas emissions	to assess whether a planning provision implies a reduction or an increase in greenhouse gas emissions
CCS	carbon capture and sequestration	to assess whether a planning provision implies a reduction or an increase in local capacity to capture or sequester carbon dioxide

Next, a list of Adaptation Needs (AN) was drawn up on the basis of their potential impacts. According to the nature of the impacts that a given planning provision is expected to induce, three types of Adaptation Needs were proposed:

- need to completely revise – the provision only has negative implications (AN1);
- need to strengthen or adjust the provision to better address the threats detected (AN2);
- no need to change – the provision has no negative impact on any issues considered (AN3).

Planning provisions were selected for analysis at the request of local officers, with a focus on provisions that, with minor changes, could qualify as adaptation measures for funding purposes. As such, only cases of AN2 were likely to be analyzed. The following methodological phases were developed, with a special focus on that kind of adaptation need.

In phase 2, a set of amendment options was designed for each of the AN2 identified in the previous stage. From the literature (IPCC 2012), three different approaches for tackling adaptation can be discerned: (i) applied technological and infrastructure-based approaches (e.g. provide new water infrastructure); (ii) investing in natural capital and ecosystem-based adaptation (e.g. preserve, maintain and expand natural habitat); and (iii) human development and vulnerability reduction (e.g. improve regulation of access to water). Technological, social, and ecological options can be combined since they are often interdependent and synergetic. The zero-option (no change) should also be considered.

In the third stage of the process, each set of amendment options is scrutinized in order to select those that are most feasible and suitable. To that end, the following criteria were considered:

- Effectiveness: sustainability and flexibility
- Efficiency: costs and benefits, low-regret, no regret, and win-win-win sub-criteria
- Feasibility: technical, social, and institutional barriers to implementation
- Knowledge base: knowledge gaps limiting amendment implementation, and potential of the amendment to bridge the gap between knowledge and action
- Equity and legitimacy

Through a scorecard, the highest scoring amendment options in each set were selected. Potential synergies were also considered.

Finally, instructions were provided on how to implement the selected amendments, including how to identify the actors and stakeholders to be involved, the opportunities and threats (e.g. technical, social, institutional, etc.) associated with the option, and the cost implications to be considered.

15.3.2 Data Collection

In keeping with scope of the ACCDar project, which focuses on under-serviced peri-urban neighborhoods in the Dar es Salaam coastal plain, the planning documents used in the mainstreaming exercise were selected with a focus on peri-urban areas facing the problem of groundwater salinization due to seawater intrusion. Accordingly, a pilot mainstreaming exercise was conducted with Temeke Municipal Council, since most of Temeke lies within the coastal plain and consists of peri-urban and rural areas, with the latter expected to convert into peri-urban in the coming years. The Strategic Plan and Medium Term Expenditure Framework documents for years 2010/2011–2012/2013 were provided by the Temeke Municipal Council.

Hydrogeological surveys indicate that seawater intrusion already represents a major vulnerability concern in neighborhoods where boreholes are the main source of water, and intrusion is expected to progress even faster under future conditions of climate change and urban sprawl. It should be highlighted that the actual rate of water extraction from the coastal aquifer already exceeds the recharge rate (see Chap. Sappa). In the future, the combined effects of climate change and urban development are expected to further reduce the natural recharge capacity of the coastal aquifer, while groundwater withdrawal is likely to increase due to the growth of both domestic and productive demand. In light of these considerations, researchers focused on existing planning documents relevant for future urban development and water supply in the Temeke territory. Draft versions of two additional documents were obtained: the "Dar es Salaam Master Plan 2012–2032"; and the "Strategic Water Supply and Sanitation Plan for Dar es Salaam".

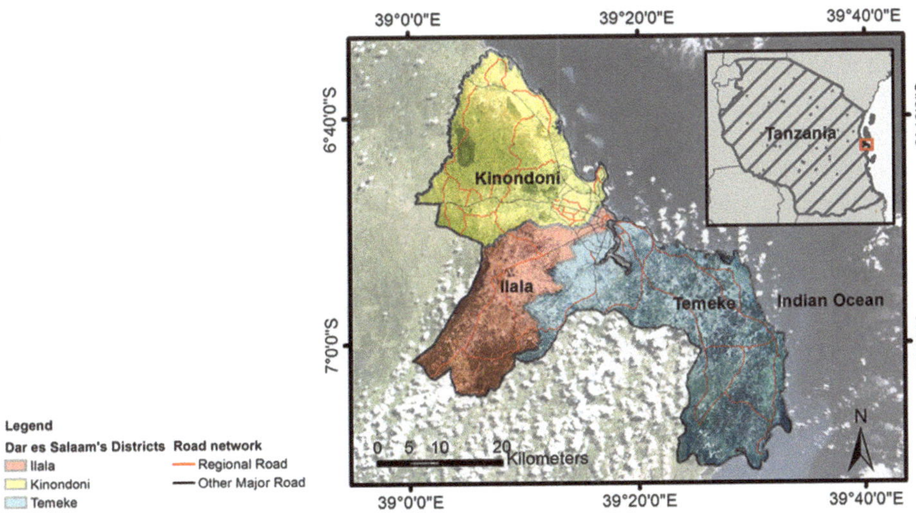

Figure 15.2: Location of Dar es Salaam and its three municipalities.

All four Temeke plans were reviewed. For each plan, one key issue and two related provisions were selected for further assessment.

From the two municipal planning documents, the objectives "management of natural resources" and "environmental improvement" were selected as the key issues to focus on, and the specific provisions chosen for assessment were protection of green areas, forest conservation, tree plantation, and demonstration toilet construction.

From the Water Supply and Sanitation Plan, the objective "Develop a long term water supply strategy improvement plan (25 year horizon) for improving/expanding water supply services for Dar es Salaam" was selected as a key issue, with a focus on the provisions: (i) improving surface water sources from 276,000 m^3/d to 576,000 m^3/d ultimate capacity by 2032; and (ii) installation of 20 deep wells with a minimum depth of 600 m in Kimbiji and Mpera.

Under the Dar es Salaam Master Plan 2012–2032, the key issue selected to focus on was "Design Guidelines" and the building provisions selected for assessment related to: (i) consolidation process zone; and (ii) peri-urban areas and urban agriculture zone.

Results from the analysis of the latter provision are presented below as an example of the methodology outputs, while the whole process, from impact detection to recommendation of a single adaptation need, is summarized in Table 15.2.

15.4 Preliminary Results: Amending Building Provisions for Peri-Urban Areas

The Dar es Salaam Master Plan 2012–2032, which still was under consultation at the time of writing, sets out the objectives and policies aimed at achieving a shared vision of the metropolitan city for the next twenty years. It defines the direction of territorial development and provides for a system of rules and procedures for its implementation.

The proposed mainstreaming methodology has been tested with two planning provisions under Section 2 "Proposed Land Use Zones" and the related parts of Section 3 "Town Planning and Building Standards" of the Design Guidelines. Special attention has been paid to the consequences of those provisions for the future development of peri-urban areas in the southern part of the Dar es Salaam region, which falls under the authority of the Temeke Municipal Council. Recommendations for adaptation mainstreaming are likely to be more productive in areas that are still mostly undeveloped and therefore provide a more favorable context for transition to sustainable settlement patterns.

The results reported below relate to Art.18 of the Design Guidelines, as formulated in the Draft Final Report of the Master Plan (kindly provided by the planning team).

Table 15.2: From impact assessment to recommendation. An example from the mainstreaming exercise for the measure "Protection of environment and reserved areas in 4 wards enhanced by 2013", Temeke Municipal Council Strategic Plan for years 2010/2011–2012/2013.

ASSESSMENT CRITERIA	EXPECTED IMPACTS	ADAPTATION NEEDS	AMENDMENT OPTIONS		RECOMMENDATIONS following evaluation of amendment options
AC1	...	...	...		...
AC2	NEGATIVE IMPACT: When natural water sources in the reserved area currently provide free water to residents, the provision might increase water access costs	AN2: Need to provide options for ensuring no additional freshwater access costs	No change option		When natural water sources within the reserved areas currently provide free water to residents, the measure shall be amended to include the set-up of a local water committee to guarantee equitable access to and distribution of natural water sources (including participatory monitoring of groundwater and surface water bodies) (Option 4).
			TECHNOLOGICAL OPTIONS	1. Review reserved area boundaries in a way that will not prevent residents from accessing existing natural water sources 2. Provide new infrastructure for pumping freshwater from within the reserved area to a free water point located outside	
			ECOLOGICAL OPTIONS	3. Set up a water body monitoring system to inform decisions on the quantity of water that residents can extract while respecting conservation goals (e.g. minimum water table level or minimum river flow)	
			SOCIAL OPTIONS	4. Set up a local water committee to guarantee equitable access to and distribution of natural water sources (including participatory monitoring of water bodies) 5. Identify alternative free water sources outside the reserved areas 6. Provide an adequate amount of free freshwater to poor households (change in water service tariff)	While there is a need to limit water extraction from the shallow aquifer in general, the committee shall be provided with clear direction on the quantity of water that residents can withdraw from surface water sources without affecting the ecosystem functioning (Option 3).
PAA1	...	...	...		...
PAA2	...	...	...		...
PAA3	...	...	...		...
GHG	...	...	...		...
CCS	...	...	...		...

Article 18 – Peri-urban areas / urban agriculture

18.1 – These are the parts of the territory outside the urban perimeter, characterized by a strong prevalence of agricultural or potentially agricultural areas and low residential density.

18.2 – In these areas, all possible transformations of an agricultural nature are allowed, including the construction of residential and / or service buildings, related to the agricultural activity.

In the case of dispersed settlements, the new residential buildings may not exceed the density of one new dwelling per hectare.

18.3 – The Municipalities may decide to establish a perimeter around existing settlements at the date of approval of the Plan, to which the prescriptions of Article 7[44] *of the present Rules will apply.* (Dodi Moss *et al.* 2013: 300).

This set of planning provisions is of particular importance with regard to mainstreaming CC adaptation into the Master Plan, as it will impact large areas in Dar es Salaam region where people's livelihoods are expected to remain highly dependent on natural resources. Climate change will particularly affect these areas, and special efforts are therefore needed to maintain and develop their adaptive capacity while preventing mal-adaptation. To do so, multiple amendments are required.

15.4.1 AC1: Water Resource Conservation

This article focuses specifically on areas where the predominant land use is agriculture. This may put groundwater resources at risk for several reasons. Firstly, increased use of fertilizers and pesticides will result in water contamination and soil pollution. Secondly, increased demand for water for farming uses (i.e. irrigation and livestock breeding) will increase the rate of water withdrawal from the shallow aquifer. To avoid these impacts, sustainable cultivation techniques should be promoted and the use of chemical fertilizers discouraged or prevented. We therefore propose amending the measure to include the development of pilot projects on sustainable cultivation techniques, including organic (chemical free) farming and water saving techniques (i.e. micro-irrigation and net-houses). To complement this action, an additional amend-

44 *Article 7 – Urban redevelopment areas*
7.1 – These are the parts of the city, mainly residential, characterized by low quality of settlements, low building density and the lack of any urban structure.
7.2 – In these parts of the city the replacement of existing buildings with more appropriate ones, that do not exceed the height of three storeys, is to be carried out. All developments should meet the planning and building standards as defined in Section 3.
7.3 – In these parts of the city all measures are planned to provide them with adequate roads, the necessary infrastructure networks (water, sewerage system, electricity), adequate space for urban facilities and green areas. (Dodi Moss *et al.* 2013: 296–297)

ment could be introduced to provide for the development of initiatives to facilitate learning and sharing of experiences from pilot projects, thereby raising awareness of the benefits of sustainable cultivation techniques among peri-urban communities.

The expansion of built-up areas in the peri-urban zone may aggravate the aforementioned impacts on water source conservation. This is especially true where the redevelopment of existing settlements will occur without simultaneous provision of adequate water supply, sanitation systems, and waste management. To prevent increased water source contamination, soil pollution and groundwater overexploitation, the article should be as amended to require the existence of adequate water supply and sanitation infrastructure and solid waste management as a condition for issuing any new building permit in existing settlements. Meanwhile, at least two additional amendments should be considered to lay the foundation for the design of locally tailored, sustainable infrastructure. First, a monitoring system for underground water levels and quality must be created. Second, a local committee should be set up to ensure community participation in the design, construction and stewardship of new infrastructure.

15.4.2 AC2: Improve Access to Fresh Water

Proper provision of water supply, although highly desirable for water conservation, may entail an increase in household water costs. The same may also occur in areas of increasing competition for water due to the combination of inadequate water service and growing water demand. It is therefore necessary to ensure that no additional freshwater access costs are charged to residents as a consequence of water supply upgrading and population growth. In order to keep freshwater affordable for all residents after redevelopment, it is crucial that the measure be amended to provide protection for the cheapest source of potable water (i.e. community water storage facilities) against contamination and vandalism. The measure should also be amended to include the set-up of local committees in charge of guaranteeing equitable and affordable access to fresh water for residents. Such committees may also initiate steps towards establishing economic agreements with high water consuming companies (e.g. intensive stock-breeders) to keep domestic water bill low.

15.4.3 PAA1: Possibility to Diversify Water Sources

In peri-urban areas, competition for water between domestic and agricultural uses is likely to intensify, and may decrease the variety of water sources available to households for domestic uses. To combat the risk of reducing the diversity of water sources for households, conflict-resolution institutions and tools are needed. We highly recommend complementing the set of provisions under Article 18 with the set-up of local committees to manage conflicts between households and farmers over access to freshwater. Such committees could also represent peri-urban communities in negotiations with high water consuming companies, where compensation for ecological damage could be established and contribution to the development of new sources of water through run-off harvesting and water reuse could be requested.

15.4.4 PAA2: Possibility to Change Income Generating Activities

Regarding the negative impacts on income generating activities, it should be noted that implementation of the planning provisions under consideration may affect the agricultural practices of residents, both in existing settlements and in rural areas throughout the region.

The redevelopment of existing settlements may lead to the exclusion of agricultural uses from residential areas and, in general, a disconnect between agricultural and urban activities. The need therefore arises to preserve agricultural uses within urban areas while ensuring connections between agriculture production and food markets. To that end, we suggest amending the article to require the preparation of a special plan for the protection and development of agricultural and agriculture-related uses near and within urban boundaries. Such a plan should consider water availability as a limiting factor and should secure adequate space for future provision of market facilities. In addition, as an incentive towards more sound agricultural practices, an amendment may be introduced to issue land titles to residents who adopt sustainable farming and water management techniques.

In the case of inadequate management of wastewater and solid waste within dispersed settlements, food-producing farmers may experience an income reduction due to the decreased quality of their products. To prevent crop contamination, we highly recommend providing for awareness raising initiatives on the health and economic risks associated with uncontrolled discharge or improper reuse of wastewater and solid waste in agricultural areas. In addition, the set-up of a local committee to control and promote the quality and safety of food production may be introduced.

15.4.5 PAA3: Possibility for Relocation or Changes in Current Settlement Patterns

A further problem is that the regulations for new settlements defined under the measure may be rejected by residents, and would therefore be completely ineffective and/or may cause residents to migrate elsewhere. It should also be noted that these regulations pay little attention to the environmental impacts of new settlements. Therefore, there is a need to ensure residents' involvement in decision-making that impacts their settlement needs, while enhancing the environmental performance of decisions made according to the measure. We suggest providing for the set-up of a local committee responsible for managing potential conflicts that may arise during implementation. In addition, the regulations provided for new settlements should be amended to include the preservation of natural areas with high ecological value (e.g. wood- and wetlands), and to protect highly productive farmland from residential encroachment.

15.4.6 Contribution to Greenhouse Gas Emissions (GHG) and Carbon Capture/Sequestration (CCS)

Lastly, the article does not consider that future growth in farming activities and the settled population within peri-urban areas will lead to an increased mobility and energy demands, thus causing a negative impact on the environment in terms of GHG emissions. The GHG emissions associated with these new demands must be contained through the promotion of low carbon and energy efficient techniques and systems in the sectors of transportation, agriculture, and energy production. To meet this need, several amendments should be introduced. Firstly, increased emissions could be offset by innovative farming techniques to minimize release of soil carbon, such as organic agriculture and minimum tillage techniques. Secondly, raising awareness initiatives of the environmental impacts of private car transport and fossil based energy production is highly recommended to create a more favorable context for the diffusion of low carbon transport (including public transport service, non-motorized mobility and low carbon vehicles) and energy production (e.g. renewable energy), as well as more energy efficient engines.

15.5 Conclusions

Although results from the analysis of the four selected planning documents still need further examination, the proposed methodology for mainstreaming adaptation into existing urban development and environmental management plans and programs at the local level is definitely valid. In-depth examination of a selection of planning provisions enabled the research group to identify a few key mainstreaming initiatives whose reach goes far beyond the improvement of a single planning document. In other words, what emerges from the analysis of a specific planning provision has the potential to be generalized, and provides clear directions as to how to proceed in order to mainstream adaptation into the whole planning system, which is the ultimate scope of this work.

Those directions include: (i) the development of pilot projects to encourage locally the adoption of best available technologies in a range of fields such as agriculture, forestry, construction, transport, energy, water supply, and waste treatment; (ii) the creation of locally based, participatory monitoring systems to allow for adaptive management of natural resources; (iii) the set-up of local committees with the twofold role of guaranteeing wise and equitable use of resources within the community while also representing the community at local meetings; and (iv) facilitating increased use of ecosystem services payment schemes, such as Equitable Payment for Watershed Services (EPWS), as a way of financing local development while preventing irreversible environmental damages.

It is beyond the scope of this chapter to evaluate the acceptability and potential effectiveness of the mainstreaming initiatives that arose from this exercise. In order to do so, a systematic assessment of the barriers and opportunities that may arise would be necessary (Moser and Ekstrom 2010), to be carried out with direct involvement of all government levels and stakeholders. Considering the present state of the research, the principal conclusions that may be drawn from this preliminary application of the proposed methodology mainly concern possible improvements of the methodology itself. Indeed, there is high agreement among the research group that some revisions would be necessary to simplify the process, particularly concerning the second phase, and to ensure more robust results.

Firstly, the number of criteria for evaluating the amendment options could be reduced, and unnecessary repetition and redundancy eliminated. The use of those criteria require too much time and effort. Also, a preparatory stage should be introduced to ensure that criteria are agreed upon and understood by all participants. Although a number of revisions were made to criteria definitions during the course of their application, differing interpretations can still occur.

Secondly, the formulation of amendment options is largely based on the experience and intuition of the people involved. The results would be more balanced if developed through a focus group of stakeholders, experts from different disciplines, and policy makers. However, while the identification of adaptation needs was quite easy thanks to the mass of in-depth knowledge available, the design of amendment options has been affected by the lack of a clear understanding of the existing institutional framework, both formal and informal. We intend to fill this gap in the coming months as a basic step towards more effective assessment of the barriers and opportunities in implementing the mainstreaming initiatives identified within this study.

Finally, the nature of the planning documents selected for analysis seems to make a difference in terms of ease of application. It appears that the more executive the planning documents under consideration, such as the Medium term expenditure framework and the Strategic plan of Temeke Municipal Council, the more targeted and viable the indications for amendment. This may depend on the structure of the proposed methodology. However, it offers an argument in favor of the importance of "the local" in determining the efficacy of adaptation mainstreaming.

References

Adger, WN., S. Agrawala, M.M.Q. Mirza, C. Conde, K. O'Brien, J. Pulhin, R.S. Pulwarty, B. Smit, and K. Takahashi. 2007. Assessment of adaptation practices, options, constraints and capacity. In *Climate change 2007: Impacts, adaptation and vulnerability. Contribution of working group II to the fourth assessment report of the IPCC*, ed. M.L. Parry, O.F. Canziani, J.P. Palutikof, P.J. van der Linden and C.E. Hanson. Cambridge: Cambridge University Press.

Adger, W. N., and J. Barnett. 2009. Four reasons for concern about adaptation to climate change. *Environment and Planning A* 41(12): 2800–2805.

Agrawala., S. ed. 2005. *Bridge over troubled waters: linking climate change and development*. Paris: OECD publishing.

Anguelovski, I., and J. Carmin. 2011. Something borrowed, everything new: Innovation and institutionalization in urban climate governance. *Curr. Opin. Environ. Sustainability* 3(3): 169–175.

Carmin, J., I. Anguelovski, and D. Roberts. 2012a. Urban climate adaptation in the global South: Planning in an emerging policy domain. *Journal of Planning Education and Research* 32(1): 18–32.

Carmin, J., N. Nadkarni, and C. Rhie. 2012b. Progress and challenges in urban climate adaptation planning: Results of a global survey. Cambridge, MA: MIT.

Congedo, L., and M. Munafò. 2014. Urban sprawl as a factor of vulnerability to climate change: Monitoring land cover change in Dar es Salaam. In *Climate change vulnerability in southern african cities: Building knowledge for adaptation*, ed. S. Macchi, and M. Tiepolo, 73–88. Springer. doi: 10.1007/978-3-31900672-7.

Davoudi, S., J. Crawford, and A. Mehmood. 2009. Climate change and spatial planning responses. In *Planning for climate change. Strategies for mitigation and adaptation for spatial planners*, ed. S. Davoudi, J. Crawford, and A. Mehmood, 7–19. London- Sterling (VA): Earthscan.

Dodi Moss, Buro Happold, Afri Arch, Q-Consult. 2013. Dar es Salaam Masterplan 2012–2032. Main Report. Final Draft, Feb.

Dunford, R., P.A. Harrison, J. Jäger, M.D.A. Rounsevell, and R. Tinch. 2013. Report on assessment of vulnerability across Europe and the identification of vulnerability hotspots. The CLIMSAVE project climate change integrated assessment methodology for cross sectoral adaptation and vulnerability in Europe. http://www.climsave.eu/. Accessed 15 May 2014.

Faldi, G. 2013. The use of backcasting scenario for planning adaptation to climate change in Sub-Saharan urban areas. Paper presented at the AESOP Conference, 15–19 July 2013, UCD – Dublin. http://aesop-acspdublin2013.com/uploads/files/AESOP-ACSP%20Congress%20 Book%20of%20Abstracts.pdf. Accessed 15 May 2014.

Faldi, G. 2014. A Backcasting scenario methodology for planning adaptation to climate change in coastal Dar es Salaam. Paper presented at the AESOP Conference, 9–12 July 2014, Utrecht. http://www.congrexprojects.com/docs/default-source/aesopdocs/aesop-abstract-book.pdf?sfvrsn=0. Accessed 28 July 2014.

Faldi, G., and M. Rossi. 2014. Climate change effects on seawater intrusion in coastal Dar es Salaam: Developing exposure scenarios for vulnerability assessment. In *Climate change vulnerability in southern african cities: Building knowledge for adaptation*, ed. S. Macchi, and M. Tiepolo, 57–72. Springer. doi: 10.1007/978-3-31900672-7.

Friedmann, J. 2005. Globalization and the emerging culture of planning. *Prog Plann* 64(3): 183–234

Füssel, HM., and R.J.T. Klein. 2006. Climate change vulnerability assessments: An evolution of conceptual thinking. *Climatic Change* 75: 301–329.

Halloran, A., and J. Magid. 2013. Planning the unplanned: incorporating agriculture as an urban land use into the Dar es Salaam master plan and beyond. *Environment and Urbanization* 25: 541–558.

Huq, S., A. Rahman, M. Konate, Y. Sokona, and H. Reid. 2003. Mainstreaming adaptation to climate change in least developed countries (LDCs). London: IIED.
IPCC. 2012. Managing the risks of extreme events and disasters to advance climate change adaptation. A special report of working groups I and II of the Intergovernmental Panel on Climate Change. Cambridge-New York: Cambridge University Press.
IPCC. 2013a. Glossary. In *Final draft of the working group II contribution to the IPCC fifth assessment report climate change 2014: Impacts, adaptation, and vulnerability*. http://ipcc-wg2.gov/AR5/report/final-drafts/. Accessed 15 May 2014.
IPCC. 2013b. Technical summary. In *Final draft of the working group II contribution to the IPCC fifth assessment report climate change 2014: Impacts, adaptation, and vulnerability*. http://ipcc-wg2.gov/AR5/report/final-drafts/. Accessed 15 May 2014.
IPCC. 2013c. Foundations for decision-making. In *Final draft of the working group II contribution to the IPCC fifth assessment report climate change 2014: Impacts, adaptation, and vulnerability*. http://ipcc-wg2.gov/AR5/report/final-drafts/. Accessed 15 May 2014.
IPCC. 2013d. Adaptation planning and implementation. In *Final draft of the working group II contribution to the IPCC fifth assessment report climate change 2014: Impacts, adaptation, and vulnerability*. http://ipcc-wg2.gov/AR5/report/final-drafts/. Accessed 15 May 2014.
IPCC. 2013e. Africa. In *Final draft of the working group II contribution to the IPCC fifth assessment report climate change 2014: Impacts, adaptation, and vulnerability*. http://ipcc-wg2.gov/AR5/report/final-drafts/. Accessed 15 May 2014.
Klein, RJT. 2002. Climate change, adaptive capacity and sustainable development. Paper presented at the expert meeting on adaptation to climate change and sustainable development, OECD, Paris, 13–14 March 2002.
Lerise, F. 2000. Urban governance and urban planning in Tanzania. In *Urbanising Tanzania: issues, initiatives, and priorities*, ed. S. Ngware, and J.M.L. Kironde. Dar es Salaam: University of Dar es Salaam Press.
Macchi, S., L. Ricci, L. Congedo, and G. Faldi. 2013. Adapting to climate change in coastal Dar es Salaam. Proceedings of the AESOP-ACSP joint congress. 15–19 July 2013. Dublin. http://www.planning4adaptation.eu/Docs/newsInfoMaterial/01-2014/Paper-AESOP-Track3-MACCHI.pdf. Accessed 15 May 2014.
Macchi S. 2014. Adaptation to incremental climate stress in urban regions: Tailoring an approach to large cities in Sub-Saharan Africa. In *Climate change vulnerability in Southern African cities: Building knowledge for adaptation*, ed. S. Macchi, and M. Tiepolo, 3–18. Springer. doi: 10.1007/978-3-31900672-7.
Macchi S., and L. Ricci. 2014. Mainstreaming adaptation into urban development and environmental management planning: a literature review and lessons from Tanzania. In *Climate change vulnerability in southern african cities: Building knowledge for adaptation*, ed. S. Macchi, and M. Tiepolo, 109- 124. Springer. doi: 10.1007/978-3-31900672-7.
Moser, S.C., and J.A. Ekstrom. 2010. A framework to diagnose barriers to climate change adaptation. *Proceedings of the National Academy of Science* 107: 22026–22031.
Nelson, R., P.R. Brown. T. Darbas, P. Kokic, and K. Cody. 2007. The potential to map the adaptive capacity of australian land managers for NRM policy using ABS data. CSIRO, Australian Bureau of agricultural and resource economics.
O'Brien, K.L., S. Eriksen, L. Nygaard, and A. Schjolden. 2007. Why different interpretations of vulnerability matter in climate change discourses. *Climate Policy* 7: 73–88.
Persson, Å., and R.J.T. Klein. 2008. Mainstreaming adaptation to climate change into official development assistance: integration of long-term climate concerns and short-term development needs. In Proceedings of the Berlin conference on the human dimensions of global environmental change, Berlin, 22–23 Feb. 2008.

Preston, B.L., T. Smith, C. Brooke, R. Gorddard., T. Measham, G. Withycombe, K. McInnes, D. Abbs, B. Beveridge, and C. Morrison. 2008. Mapping climate change vulnerability in the Sydney coastal councils group. Prepared for the Sydney coastal councils group and the Australian greenhouse office. Melbourne.

Ricci, L. 2011. Reinterpreting Sub-Saharan cities through the concept of "adaptive capacity". An analysis of "autonomous" adaptation practices to environmental changes in peri-urban areas. Rome: Sapienza University of Rome (Italian), PhD thesis. http://padis.uniroma1.it/handle/10805/1375. Accessed 15 May 2014.

Ricci, L. 2014. Linking adaptive capacity and peri-urban features: The findings of a household survey in Dar es Salaam. In *Climate change vulnerability in southern african cities: Building knowledge for adaptation*, ed. S. Macchi, and M. Tiepolo, 89–107. Springer. doi: 10.1007/978-3-31900672-7.

Rugai, D., and G.R. Kassenga. 2014. Climate change impacts and institutional response capacity in Dar es Salaam, Tanzania. In *Climate change vulnerability in southern african cities: Building knowledge for adaptation*, ed. S. Macchi, and M. Tiepolo, 39–56. Springer. doi: 10.1007/978-3-31900672-7.

Satterthwaite, D., S. Huq, M. Pelling, H. Reid, and P. Romero-Lankao. 2007. Adapting to climate change in urban areas: the possibilities and constraints in low- and middle-income nations. Human Settlements Discussion Paper Series, 1. London: IIED.

Simon, D., 2010. The challenges of global environmental change for urban Africa. *Urban Forum* 21(3): 235–248.

Smit, W., and E. Pieterse. 2014. Decentralisation and institutional reconfiguration in urban Africa. In *Africa's urban revolution: Policy pressures*, ed. S. Parnell, and E. Pieterse, 148–186. London: Zed Books.

United Republic of Tanzania. 2013. National account of Tanzania mainland 2001–2012. Dar-es-Salaam: National Bureau of Statistics.

Riziki S. Shemdoe[45], Dionis Rugai[46] and Laura Fantini[47]

16 Climate Change Adaptation in Dar es Salaam: Local Government Opinions and Proposed Interventions

Abstract: This chapter presents climate change adaptation (CCA) interventions proposed for implementation in the three municipalities of Dar es Salaam, namely Kinondoni, Ilala and Temeke. The proposed measures were developed through participatory approaches involving consultative stakeholder workshops. The research presented below is one of the outputs of the Adapting to Climate Change in Coastal Dar es Salaam (ACC Dar) project. The aim of the ACC Dar project is to improve the effectiveness of municipal council initiatives in Dar es Salaam in supporting the efforts of coastal peri-urban dwellers, partially or totally dependent on natural resources, to adapt to climate change (CC) impacts. The interventions proposed by the municipalities focus on reducing saltwater intrusion into the shallow aquifer, since this is one of the main factors of vulnerability. Indeed, there is evidence that this environmental phenomenon is already contributing to the degradation of the ground water upon which a large portion of peri-urban inhabitants relies for access to water. Specifically, those proposals are: (i) to conserve water resources along the coastal belt; (ii) to harvest rainwater as a CCA strategy; and (iii) to build capacity and community awareness of CC impacts on saltwater intrusion. The chapter also highlights the important role played by local government authorities (LGAs) in designing and implementing CCA and how their proposals fit into national strategies linked to CC.
Keywords: Climate change, National policies, Rainwater harvesting, Dar es Salaam

16.1 Introduction

The present research has been developed as a part of the Adapting to Climate Change in Coastal Dar es Salaam (ACC Dar) project. That project aims to support LGAs seeking to improve the effectiveness of CCA initiatives in Dar es Salaam, with a focus on coastal peri-urban areas where dwellers are partially or totally dependent on natural resources. More specifically, this entails enhancing the capacities of Dar es Salaams' municipalities by increasing their understanding of adaptation practices and by providing them with methodologies for mainstreaming adaptation into their

45 Riziki S. Shemdoe is at Ardhi University, Dar es Salaam, Tanzania, shemdoes@gmail.com
46 Dionis Rugai is at Ardhi University, Dar es Salaam, Tanzania, dionisr@gmail.com
47 Laura Fantini is at DICEA-Sapienza University of Rome, laura.fantini76@gmail.com

Urban Development and Environment Management (UDEM) strategies and plans. The achievement of these objectives is expected to contribute to the overall goal of improving implementation of the National Adaptation Programme of Action (NAPA) of the United Republic of Tanzania (URT). In this respect, capacity building and municipal CCA proposal development are essential.

In Africa, the need to focus on and develop strategies for dealing with the immediate to long-term consequences of CC is a priority. Challenges that communities face as a result of CC, such as more frequent and intense floods, water scarcity, declining agricultural productivity, and poor health pose serious challenges for coastal communities whose livelihoods are closely related to the environment. In order to face such challenges, both communities and local government must be involved in addressing CC induced impacts. To that end, the Intergovernmental Panel on Climate Change (IPCC 2014) has prescribed an approach to integrating disaster risk reduction (DRR) and CCA in order to support communities facing CC impacts.

Water resources are highly affected by CC. Potential effects include flooding, where water related infrastructure is put at risk, as well as droughts resulting in low water availability to downstream users. Access to safe and clean water is a basic human right, as declared by the United Nations in 2011. Tanzania has signed Agenda 21, an outcome of the UN environment meeting in Rio de Janeiro, which stated that all nations must protect natural resources, including water, and foster conservation of the ecosystems.

Although Tanzania is blessed with a variety of surface water resources (7% of the land area is covered by lakes and rivers draining into four major river basins), surface water is limited throughout the country for most of the year due to uneven rainfall distribution, a prolonged dry season and arid or semi-arid conditions in most of the country. As a result, groundwater plays a major role in meeting demand, especially in rural and peri-urban areas. In Tanzania, water is used mainly for domestic purposes, watering livestock, power production, irrigation, and industry. In many areas of the country, ground water resources are becoming limited due to the increasing demands of a growing population, intensifying farming and livestock, and industrialization. The negative impacts of CC, such as floods and droughts, further detract from the welfare of communities.

Dar es Salaam is among the coastal Sub-Saharan cities experiencing rapid urbanization, and it has an annual population growth rate of 5.6%. Below, the key role to be played by LGAs in addressing CCA under such circumstances is explored in the context of interventions to be implemented by three municipalities in Dar es Salaam. The proposed interventions were developed via participatory approaches through the ACC Dar Project.

A capacity-building strategy was developed involving the four local authorities governing the city of Dar es Salaam. A consultative survey was conducted in 2012 in Dar's LGAs to investigate existing ways of addressing CC related issues at the local level (Kassenga and Mbuligwe 2012).

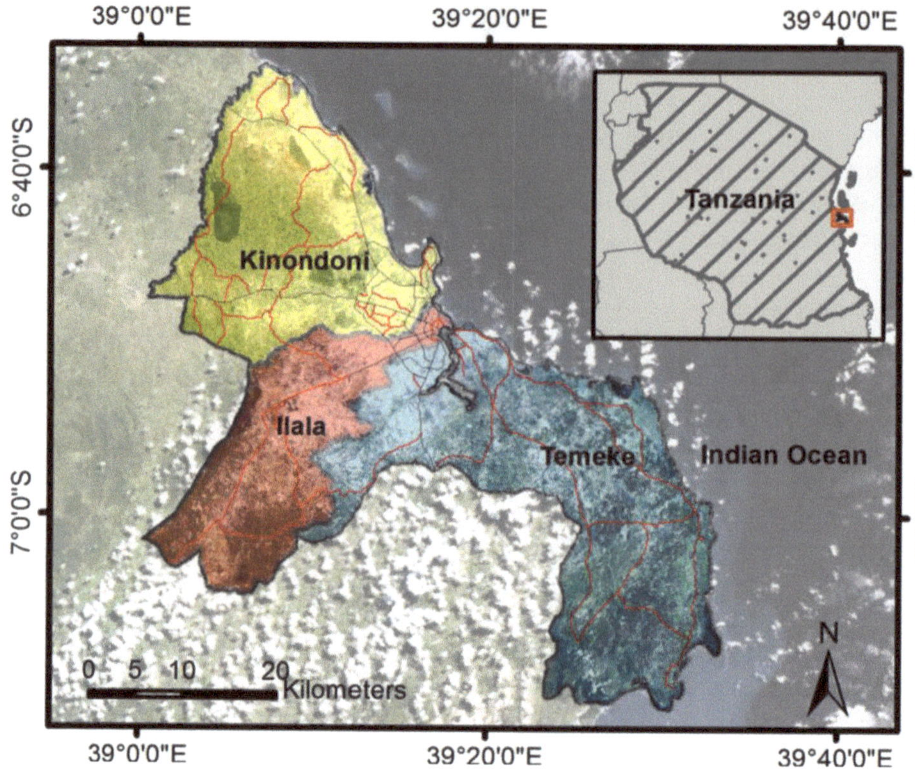

Figure 16.1: Dar es Salaam City and its three Municipalities (Rugai & Kassenga, 2012).

Subsequently in 2013, 30 technical cadres from the three municipalities were involved in two residential trainings. The departments involved deal with sectors pertinent to CC issues, namely agriculture and livestock development, health, urban planning, natural resources, waste management, works, and water. The objectives of the trainings were threefold: i) to improve understanding of CC impacts; ii) to enhance methodologies for assessing communities' adaptive capacities, monitoring changes in peri-urban areas, and exploring CC vulnerability scenarios; and iii) to foster changes within existing sector plans to mainstream adaptation initiatives.

As a result of this process, the municipalities designed three CCA measures to minimize saltwater intrusion into ground water, to be implemented in key areas. The measures designed are grass roots oriented adaptation innovations.

16.2 Climate Change Governance in Tanzania

As highlighted in Shemdoe (2013), Tanzania does not yet have a CC policy in place. However, in response to climate variability and CC impacts on natural and social systems, several national programs and strategies have been devised to address climate change issues both directly and indirectly. These plans and programs are in line with international agreements, such as the United Nations Framework Convention on Climate Change (UNFCCC) and its Kyoto Protocol. Below are some of the relevant legal frameworks, plans and strategies.

In Tanzania, CC and environmental issues are regarded as cross-cutting issues, which are under the Division of Environment (DoE) of the Vice President's Office (VPO). Although various CC sector related policies address CC indirectly, policies and programs for addressing current vulnerability and promoting CCA at the city level are limited. In addition, the mainstreaming of CCA concerns in development plans, strategies, programs, projects, and routine activities has yet to be done (Rugai & Kassenga, 2014). Planning and implementing CCA strategies, plans, and operational measures at the local level offers a range of advantages and opportunities not available at the national level. Some adaptation interventions are much easier to implement at the local level. LGAs have a strong relationship with other local stakeholders, such as community groups, NGOs, and other grass roots organizations, which can communicate and mediate with the community and engage civil society (Rugai, Fantini and Shemdoe 2014).

The city is governed by four LGAs: the Dar City Council (DCC) and three municipal councils, namely Kinondoni, Ilala and Temeke. Each municipality is organized in sector departments, which are in turn composed of service units. Their functions are carried out through further decentralization in *wards*, in turn divided in *sub-wards*. The institutional framework in Dar es Salaam also has several levels of powers and functions, which in some cases can lead to coordination difficulties. The surveys conducted within Dar's LGAs highlighted many constraints in addressing CCA related issues, particularly: i) lack of knowledge and awareness of issues related to climate change among local officers, ii) limited capacity to effectively assess the long-term sectoral impacts of CC and to develop viable adaptation measures, iii) lack of coordination among sector departments and units within municipalities, and iv) low capacity for investing in adaptation activities (Kassenga and Mbuligwe 2012; Rugai and Kassenga 2014). Capacity building among LGAs to support their ability to develop interventions regarding CC impacts was therefore needed.

16.3 Planning Process in Tanzania

Planning practice in Tanzania follows the opportunities and obstacles to development (O&OD) approach, which is essentially a bottom-up approach. The O&OD provides the general framework within which issues and areas of actions are identified in the development planning and budgeting process. Issues are identified at the subward level (the lowest governance unit in the country) and are then forwarded to the ward level where they are consolidated into a prioritized Ward Development Plan. Plans are then submitted to LGAs at the district level (in the context of this chapter, this is the municipal level), where they are again consolidated into the Municipality Development Plan. The consolidated plan and budget are then submitted to the parliament where the national budget is discussed. Work at the municipal level is normally coordinated by the Planning and Coordination Unit under the Director of the municipality. There is currently a gap in the inclusion of CCA interventions in planning and budgeting systems at the LGA level. This is due to the fact that awareness of CC impacts and the need for adaptation is low. This hinders the inclusion of CCA in the O&OD as well as in the whole planning and budgeting system of local government.

Based on the capacity enhanced through the ACC Dar project, local officers were able to re-conceptualize past and existing municipal plans as CCA interventions. The following sections describe the CCA measures proposed by the three municipalities of Dar es Salaam city.

16.4 Climate Change Adaptation Measures Proposed by Municipalities in Dar es Salaam

16.4.1 Kinondoni Municipality

Of the three municipalities in Dar es Salaam, Kinondoni is most affected by poor access to water, despite the fact that it has the broadest water infrastructure network. Water availability problems aggravated by CC impacts are prompting an increasing number of people to invest in groundwater pumps. One of the areas in this municipality affected by a lack of piped water is Goba ward, located outside the city center and characterized by high levels of urban sprawl. For almost ten years, the area has received no water through the existing pipes. Residents in this ward must therefore extract ground water for consumption through boreholes. Meanwhile urban sprawl in the area has led to increased water requirements. This situation has resulted in over pumping of ground water, thereby increasing salinity levels.

16.4.2 Opinions and Proposed Interventions for Climate Change Adaptation in Kinondoni

Municipal technical personnel, facilitated by the ACC Dar project, developed a proposal to reduce saltwater intrusion into the shallow aquifer and improve health in the Goba ward community. The proposal also sought to develop capacity by training 20 teachers in four primary schools on rainwater harvesting. The expected results of this adaptation strategy are: (i) reduced time spent fetching water, (ii) improved health of residents, (iii) enhanced economic status of the community (poverty minimization), and (iv) minimized water user conflicts. Expected outputs also include availability of clean and safe water and existence of a trained Community water user committee. Actions to achieve those outputs include: construction of rainwater harvesting infrastructure, provision of training on improved water use techniques, and the establishment of a functional water user committee.

16.4.3 Ilala Municipality

Vulnerability to the negative impacts of CC, such as groundwater depletion and seawater intrusion, is exacerbated in Ilala Municipality due to over pumping and deforestation around catchment areas. This situation leads to reduction of municipal water supply with adverse impacts on health, ecosystems, and consequently on the well-being of the inhabitants. Traditionally, Ilala Municipality experiences two rainy and two dry seasons per year. During the rainy seasons, most of the excess rainwater is lost through surface runoff as there is no rainwater harvesting system. When the dry season sets in, water becomes scarce. To address the problem, the government has invested in construction of reservoirs at the catchment in Kidunda Morogoro to increase supply. However, this has not been able to fulfill the requirements of Ilala's fast growing population. More and diversified interventions are therefore needed to assist communities in the municipality to access water for various domestic, sanitation, and agriculture purposes, the latter of which is becoming an important source of livelihood for community segments in the municipality.

16.4.4 Opinions and Proposed Interventions for Climate Change Adaptation in Ilala

Consultations with the municipal technical staff improved understanding of the options for increasing and diversifying water availability in the municipality. One of their proposals entails looking for financing to address sanitation and water availability issues, with the goal of improving sanitation and water harvesting schemes in 50 primary schools to support adaptation to CC impacts in the municipality.

The specific purpose is to raise awareness within the community of the importance of harvesting rainwater as a CCA intervention. Based on their analysis, the municipal technical teams indicated as main outputs: (i) 50 primary schools equipped with water harvesting schemes, (ii) 1,200 teachers and 15,000 pupils trained to use rainwater harvest technology, (iii) 200 neighborhood communities that replicate the same technology, and (iv) 2,500 communities sensitized to rainwater harvesting practices. The foreseen outcomes of the proposed initiative are the improved availability of water and the reduction of waterborne diseases in the area.

16.4.5 Temeke Municipality

Temeke covers an area of 656 km^2 with a 70 km long coastline. Temeke is located in the south of Dar es Salaam, borders coastal regions in the south, north and west, and is bounded to the east by the Indian Ocean. Most of the area is covered by sandy soils. The main natural vegetation is coastal shrubs, Miombo woodland, coastal swamps and mangrove trees. Temeke lies in the tropical coastal belt of Dar es Salaam. It is influenced by two major climatic seasons: rainfall and temperate. Its rainfall pattern is of bimodal type with erratic conversional rains. Whereas Monsoon rains occur between December and February, long heavy rains occur in the period from March to June. The amount of rainfall received ranges from 800–1,200 mm per annum. High temperatures prevail throughout the year, ranging from 25°C during the period of June to August up to 35°C in the period from January to March. According to the URT (2012) population census, there are 1,151,865 inhabitants with an estimated growth rate of 4.6% per year. Over 90% of the population uses water from the boreholes. There are an estimated 1,000 existing boreholes in the municipality, both public and private. This has considerable implications with regard to over pumping of ground water and thus saltwater intrusion into the shallow aquifer, which is the main water source for communities in various areas of the municipality. Based on this, the municipality has developed a proposal to improve adaptive capacity to respond to saltwater intrusion. The proposal is in line with the Government strategy for urgent action to address environmental challenges in marine, lake, river and dam environments (URT 2008). The strategy has identified eight challenges facing the marine environment, eight challenges facing lakes, rivers and dam environments, eight cross-cutting environmental challenges and related strategic actions to address them. Among the identified environmental challenges in protecting the marine environment are: mangrove degradation, dynamite fishing, disappearance of endangered species and pollution. The identified challenges facing lake, river and dam environments include: sedimentation, pollution, use of prohibited fishing nets, and disappearance of satellite lakes. Cross-cutting environmental challenges include: poor public participation in environmental conservation efforts, communicable disease among communities living in the environs, and low levels of public awareness. The Strategy has also set short, medium, and long-term actions and has designated which institutions are responsible for their implementation. This must all be factored into the Temeke Strategic plan 2013–2018, the purpose of which is to provide social-economic services to its population so that they can achieve a good livelihood.

16.4.6 Opinions and Proposed Interventions for Climate Change Adaptation in Temeke

As already stated, as part of the ACC Dar project Temeke Municipality has proposed an intervention to improve adaptive capacity to respond to saltwater intrusion. The main purpose is to minimize saltwater intrusion along the coastal belt of the municipality. Three main outcomes are expected: minimized seawater intrusion, improved coastal environment, and improved human health. Outputs proposed by the municipality's technical teams include: (i) 400 hectares of mangroves planted, (ii) members of 10 Ward Development Committees trained, (iii) members of Water User and Health Committees from 63 sub-wards trained, (iv) 20 schools provided with rainwater harvesting infrastructure, and (v) 10 wards with access to clean and safe water.

16.5 Suitable Rainwater Harvesting Techniques in Each Municipality

The proposed rainwater harvesting may use different types of infrastructure. Depending on the municipality, the most suitable rainwater harvesting technique may include channeling runoff water into a small ground reservoir to be owned by individual households or the community in general. This type of rainwater harvesting can be used for irrigation, especially in light of growing urban agriculture. This would be an opportunity to take advantage of the water that runs from roads, using simple underground tanks with silt traps and trenches to reduce sediment inflow. This kind of design is widely used, for example, in Malawi. The type of structure that has been tested in Malawi follows various designs, depending on the economic status of the community (UNEP n.d.).

Another potentially suitable rainwater harvesting technique is the rooftop system. This is a common practice that is used in Tanzania, including in the three municipalities of Dar es Salaam, and could easily be adopted by the communities for which these interventions have been proposed. It could be useful in a variety of institutions, particularly primary and secondary schools where there is a wider catchment. Experience in Dar es Salaam and elsewhere in Tanzania indicates that rainwater harvesting storage can be done in two ways: (i) above ground, and (ii) sub-surface. The size of storage facilities will depend on the catchment area and the economic capacity of the community. The commonly used tanks are plastic (PVC), ferro-cement, masonry and reinforced concrete.

16.6 Alternatives to Rainwater Harvesting

In Dar es Salaam, the main source of water – apart from the water supplied by the DAWASCO (Dar es Salaam Water and Sewerage Corporation – is ground water. Individual households in most areas of the city have their own deep wells to access water for domestic purposes and gardening. During extended dry seasons, even if rainwater-harvesting technology is available, the alternative source will be ground water. As reported in Shemdoe *et al.* (2014) the Strategic Water Supply Plan for Dar es Salaam aims to ensure that water supply for Dar es Salaam is improved and that community members can access water with limited problems. The goal set by the plan is to increase surface water sources from the current capacity of 276,000 m^3/d to 576,000 m^3/d by 2032. One of the planned measures is the installation of 20 deep wells with a minimum depth of 600 m in Kimbiji and Mpera to produce 260,000 m^3 and 130,000 m^3 per day, respectively. This is likely to have both positive and negative impacts on communities and ground water ecosystems. The identified positive impacts of the measure include improved access and diversified water sources, reduced costs of treating waterborne diseases and improved livelihoods in the areas. This is a clear indication that the intended alternative source during extended dry spells when rainwater cannot be harvested is to use ground water.

16.7 Rainfall and Runoff in Dar es Salaam

Dar es Salaam forms part of the coastal catchment of the Wami/Ruvu basin. According to JICA (2005), the city has four major coastal rivers: Msimbazi, Mzinga, Kizinga, and Mpiji. The Msimbazi River originates from Pugu forest reserves and flows towards the Indian Ocean. It has a total length of about 35 km and a catchment area of about of 289 km^2 with Sinza, Ubungo, and Luhanga as its major tributaries (IUCN 2010).

The second river system in the city consists of the Kizinga and Mzinga rivers, which originate in the Pugu/Kisarawe hills. These rivers have a total length of 17.5 km and 10.4 km with catchment areas of 432 km^2 and 41 km^2, respectively (IUCN 2010). The final river system in the city is Mpiji river. This river forms a border between Dar es Salaam and the coastal regions. It has an approximate length of 12.7 km and catchment area of about 52 km^2.

Increases in the rainfall-runoff ratio are expected due to increasing impervious surface areas in the city. Land cover change analysis indicated substantial increases in built-up areas during 2002–2012 (Figure 16.3). Expected increases in surface runoff makes rainwater harvesting a desirable option, as water that is just being lost could be used both domestically and in crop production.

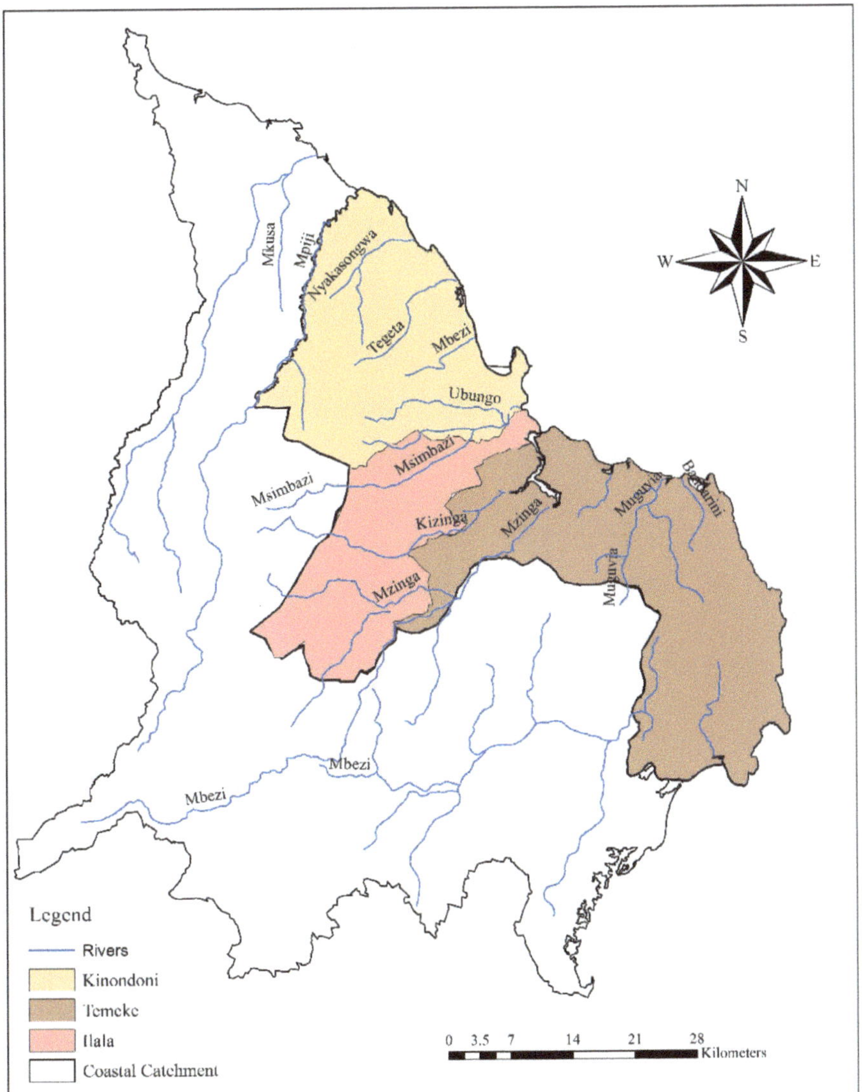

Figure 16.2: Catchment and Major Rivers in Dar es Salaam.

16.8 Rainfall seasons in Dar es Salaam

Dar es Salaam has a bimodal rainfall distribution with two main rainy seasons. The long rains (*Masika*) occur from mid-March to the end of May, and the short rains (*Vuli*) from mid-October to late December. Although June to September is typically a dry season for most parts of the country, coastal areas tend to receive a small amount of rainfall in this period.

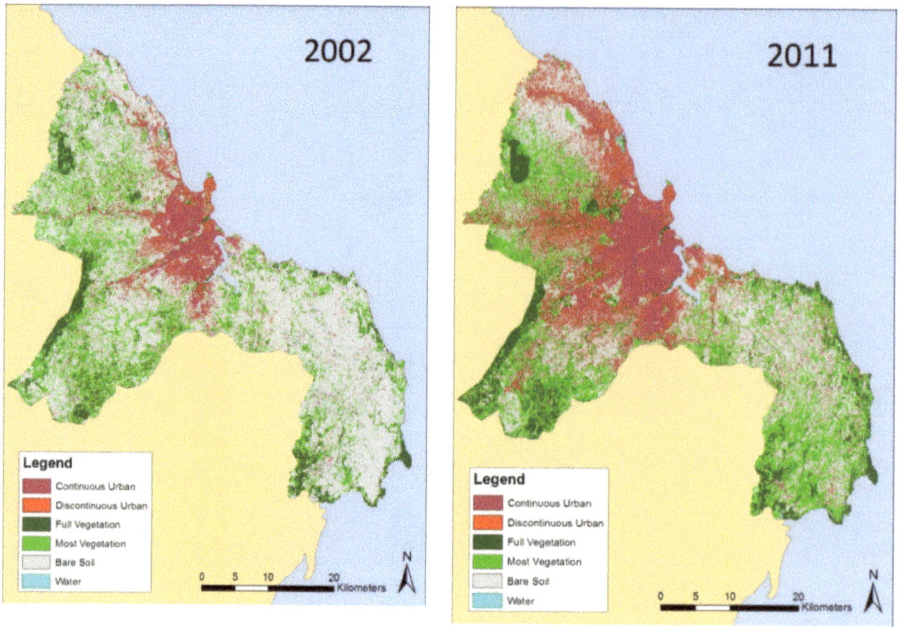

Figure 16.3: Land cover change, Dar es Salaam (Source: Congedo 2012 using Landsat Imagery).

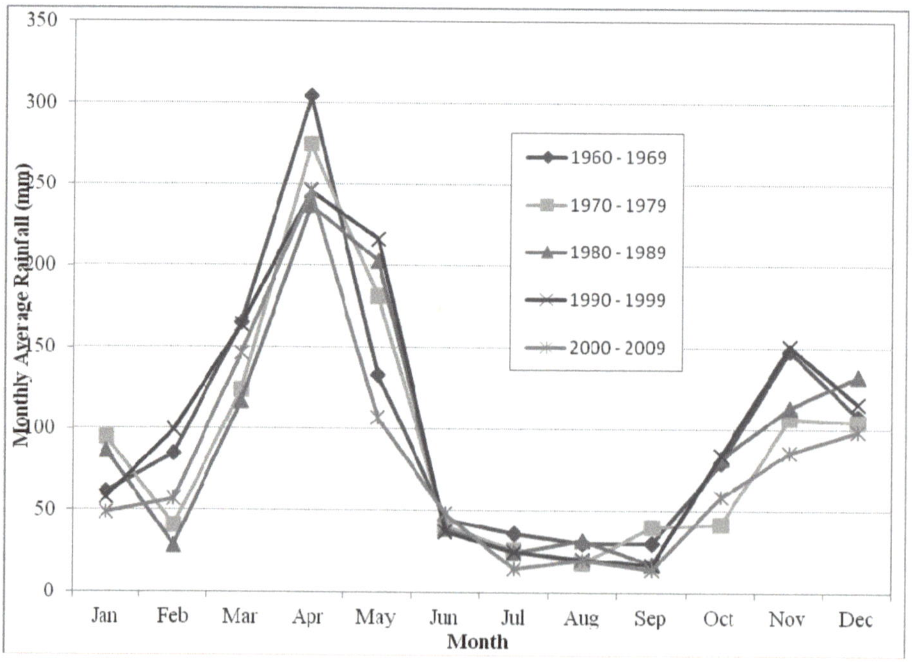

Figure 16.4: Average monthly Rainfall for Dar es Salaam.

16.9 Implementation Procedures for the Proposed Interventions

Implementation procedures for the interventions proposed by the three municipalities depend on the availability of resources, especially financial ones. Municipal governments intend to include some of these interventions in their plans and budgets funded by both internal and central government sources. Although finances from these sources are limited, there is a willingness among municipal and city councils to invest in some of these interventions. At the same time, other efforts are being made to secure more funds from other sources, such as development partners and other funding organizations.

16.10 Proposal Implications for the Policy and Legal Framework

The interventions outlined above are in line with various existing policies and programs in Tanzania, including:

16.10.1 National Strategy for Growth and Reduction of Poverty (NSGRP) (2010)

The proposed projects intend to implement the operational goals of the NSGRP. In the short term, the NSGRP commits Tanzania to achieving the Millennium Development Goals for sustainable environment. In collaboration with other service providers, the municipalities are entrusted to promote and implement projects that will improve the wellbeing of their respective communities.

16.10.2 National Water Policy, 2002

The proposed interventions are also consistent with the National Water Policy (NAWAPO) of 2002 and the water sector objectives. These projects seek to improve health and reduce poverty among the population living in peri-urban areas through improved and sustained access to reliable, adequate, safe and affordable water and sanitation services through decentralized district-based implementation and management.

16.10.3 National Forest Policy, 1998

The overall goal of the National Forest Policy (1998) is to enhance the contribution of the forest sector to the sustainable development of Tanzania and the conservation and management of its natural resources for the benefit of present and future generations. As such, the mangrove conservation proposed by Temeke Municipality could help to implement some of the objectives of the national forest policy.

16.10.4 The Tanzania Development Vision 2025

The Tanzania Development Vision 2025 foresees the alleviation of widespread poverty through improved socio-economic opportunities, good governance, transparency and improved public sector performance. These objectives not only deal with economic issues, but also include social challenges such as education, health, environment, and increased involvement of people in their own development. The proposed projects would contribute to some of these goals.

16.10.5 National Adaptation Programme of Action – NAPA, 2007

The overall vision of Tanzania's NAPA is to identify immediate and urgent climate change adaptation actions that are robust enough to lead to long-term sustainable development in a changing climate, as well as to identify climate change adaptation activities that most effectively reduce the risks that a changing climate poses to sustainable development (URT 2007). Rainwater harvesting is one of the adaptation strategies that could help in addressing climate change impacts to the community in the respective areas for which these projects are proposed.

16.11 Conclusions and Recommendations

In developing countries, the need to focus on and develop strategies for dealing with the immediate to long-term consequences of CC is inevitable. Challenges that communities face as a result of CC, such as more frequent and intense floods, water scarcity, declining agricultural productivity, and poor health, also pose serious challenges at the national level. Without proper integration of adaptation innovations into local and national government plans and budgets, implementation of research results and proven best practices will not be sustained. This chapter has highlighted local government proposals for CCA interventions in Dar es Salaam to reduce saltwater intrusion into ground water. The three interventions proposed here need facilitation in order to be implemented.

We recommend piloting the interventions by including adaptation components in the Medium term expenditure framework (MTEF) guidelines, which municipalities use as a budgeting tool, and by soliciting financial support from various sources, including internally generated local government revenues and development partners.

References

Congedo, L. 2012, Dar es Salaam land cover classification maps. http://www.planning4adaptation.eu/042_Maps.aspx

IPCC. 2007. *Climate change 2001: Impacts, adaptation, and vulnerability, Contribution of working group II to the fourth assessment report of the IPCC*. Cambridge: Cambridge University Press.

Kassenga, G., and S. Mbuligwe. 2012, Investigation of Dar es Salaam's institutional activities related to climate change. Working paper ACCDAR Project, 1.

Rugai, D., and G. Kassenga. 2014. Climate change impacts and institutional response capacity in Dar es Salaam, Tanzania. In *Climate change vulnerability in southern african cities. Building knowledge for adaptation*, ed. S. Macchi, and M. Tiepolo, 39–56. Springer. doi: 10.1007/978-3-31900672-7.

Rugai, D., L. Fantini, and R. Shemdoe. 2014. Capacity building for climate change adaptation at local government level in Dar es Salaam. Working paper ACCDAR project, 3.

Shemdoe, R., G. Kassenga, S. Macchi, C. Norero, L. Ricci, and G. Sappa. 2014. Mainstreaming adaptation into existing urban development and environmental management plans. Working paper ACCDAR Project, 3.

Shemdoe, R. 2013. Climate governance and development case study of Tanzania. Heinrich Boll Stiftung. http://za.boell.org/sites/default/files/downloads/Tanzania_long_doc_low_res.pdf.

URT-United Republic of Tanzania. 2002. NAWAPO-National water policy. Ministry of Water and livestock development.

URT. 2006. National livestock policy. Ministry of livestock development. Dar es Salaam.

URT. 2007. NAPA-National adaptation programme of action. Vice President's office, Division of environment.

URT. 2008a. National forest policy. Ministry of natural resources and tourism. Dar es Salaam.

URT. 2008b. Strategy for urgent action on the conservation of marine, lakes, rivers, and dams environment.

IUCN-Eastern and Southern Africa programme. 2010. The Ruvu basin: A situation analysis.

UNEP.-. Rain water harvesting case studies. http://www.unep.org/pdf/RWH/casestudies.pdf. Accessed 10 Oct. 2014.

Madhav Giri[48]

17 Household Level Vulnerability to Climate Change in Nepal – A Comparison of a Semi-urban and a Rural Village Development Committee

Abstract: Climate change-related events have emerged in recent years in Nepal. The effects of climate change are particularly adverse for sectors like agriculture and water management, which are dependent on climatic variables. Subsistence farmers in developing countries like Nepal, where agriculture is mainly rain-fed, and who have very few resources, and have weak adaptive capacity, and may be unable to cope with changing climatic conditions. These factors increase farmers' vulnerability to climate change. Local level vulnerability assessment is very important to formulate suitable policy measures to address their livelihood. Household level vulnerability to climate change depends on different factors, so there is still uncertainty in methodology to measure vulnerability. However, this research has adopted the concept of integrated vulnerability assessment and the indicator method to analyze the vulnerability of farmers of semi-urban areas-Pragatinagar Village Development Committee (VDC) of Nawalparasi, and rural areas – Kagbeni VDC of Mustang districts of Nepal utilizing the data collected from 155 households and VDC profiles from these 2 VDCs. Different socioeconomic and biophysical factors were collected and classified into three classes (exposure, sensitivity and adaptive capacity). Principal component analysis (PCA) was used to prioritize the indicators. Household analysis of vulnerability indicated that poor households are vulnerable anywhere due to low adaptive capacity, regardless of where they are located. Policy measures should focus on improving the adaptive capacity of rural households.

Keywords: Climate change, Vulnerability, Adaptive capacity, Exposure, Sensitivity, Principal Component Analysis (PCA)

17.1 Introduction

The impact of climate change affects developing countries more severely than developed countries because of their generally low adaptive capacities (IPCC, 2007). Rural communities in developing countries like Nepal are more vulnerable to the climate change and its impacts due to their limited capacity to cope with hazards associated with changes in climate (UNFCCC 2009).

---

48 Madhav Giri is a PhD student in Environment and territory at DIST-Politecnico and University of Turin, Italy, madhav.giri@polito.it

Vulnerability has been considered as a function of adaptive capacity, sensitivity and exposure (Fussel and Klein 2006; Adger 2006; IPCC 2001; Kelly and Adger 2000). Adaptive capacity is the ability of people to cope with or adjust to the changing context. It is explained mainly by socioeconomic indicators. Sensitivity is the degree to which a system is affected adversely or beneficially – by climate stimuli, and exposure is the nature and degree to which a system is exposed to significant climatic variation (IPCC 2001).

Vulnerability literature has mainly been concentrated on contributing to theoretical measurement at a regional or national scale, with selected indicators for each region or state, and identifying adaptation strategies that might have implications for national or regional adaptation planning (Brooks *et al.* 2005; Paavola 2008; Thomas 2008; Salami *et al.* 2009). However, assessing vulnerability at household and community level is urgent for local-level planning and prioritization of adaptation strategies.

Nepal is geographically fragile due to rugged terrain. It is situated in between India and China with an area of 147,181 km^2. Though its average width is only about 150 km, altitude varies from 60 meters above sea level (masl) to 8,848 masl (CBS 2004). The southern part of Nepal is low land and covers 23% of the area of Nepal. Around 40% of the land is under cultivation. The middle region of Nepal consists of 42% of the land area, where only 10% of the land is suitable for agriculture. The northern part covers one third of total area, where only 2% land is cultivable (Maharjan 2003). A large portion of Nepalese populations depend on natural resource based livelihoods, and they have low levels of adaptive capacity because of higher incidence of poverty. So, Nepal is placed among the most vulnerable countries to climate change (Oxfam 2009). It is vulnerable to natural disasters, mainly drought, floods and landslides. If poverty exists, the adverse impacts of climate change and extreme events will certainly increase the vulnerability.

This chapter focuses on the community of semi-urban and rural areas of Nepal. Vulnerability to climate change depends upon adaptive capacity of a wide range of attributes, and adaptive capacity is explained by socio-economic indicators. The importance of indicators varies from place to place. The study intends to compare the vulnerability and importance of adaptive capacity of semi-urban and rural areas of Nepal. So, Pragatinagar VDC of Nawalparasi was chosen as a semi-urban area and Kagbeni VDC of Mustang district as a rural area of Nepal. Both of these study areas lie in the same political region of Nepal but are in different geographic regions. Pragatinagar lies in the southern region of Nepal where population density is increasing due to suitable weather for settlement, fertile land and access to services and facilities. Instead, Kagbeni lies in mountain district-Mustang, northern part of Nepal where the temperature is cold, and it is a semi-arid region. Their adaptive capacity and adaptation practices are different than Pragatinagar. This paper will conduct in-depth analysis of the ward level vulnerabilities of both areas by using quantitative analysis with qualitative information obtained from primary field surveys and secondary data.

17.2 Methodology

17.2.1 Study Area and Data Source

This study covers 2 Village Development Committees (VDC) – Pragatinagar VDC of Nawalparasi district and Kagbeni VDC of Mustang district. VDCs are the lowest administrative tiers in Nepal which are composed of nine wards. These two VDCs were selected on the basis of their location, settlement of the people and access to services and facility. The southern part of Nepal has plain areas, and has access to services and facilities. So, the population density is also higher in Pragatinagar compared to Kagbeni.

Pragatinagar VDC has a total area of 15.77 square kilometers and the total population is 15,494, with a large favorable agricultural land (12.3 km^2). Climatically it lies in a sub-tropical zone. On average, 2,300 ml rain occurs annually. People have migrated from different places to Pragatinagar. So, it has a heterogeneous ethnic composition, but Kagbeni has a homogenous and indigenous Gurung community. Kagbeni VDC is situated in the Lower Mustang, with an area of 285.45 km^2 and the total population of only 1,140. The cultivated irrigated land constitutes 2.98 square kilometers. Climatically, Kagbeni falls in alpine climatic zone (CBS 2011).

This study is based on the primary data collected by household surveys conducted in two phases, and secondary data collected from respective VDC offices. Eighty-four

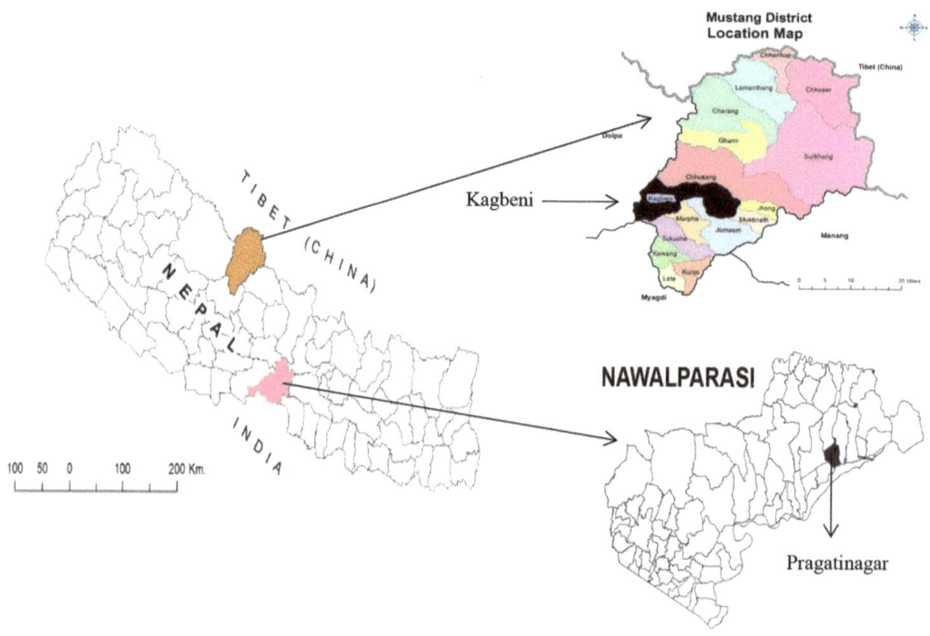

Figure 17.1: Map of study districts with VDCs (Pragatinagar and Kagbeni).

households from Pragatinagar and 64 households from Kagbeni VDC were selected using the following formula given by Arkin and Colton (1963):

$$\text{Sample size } (n) = \frac{N * Z^2 * P(1-P)}{N * d^2 + z^2 * P(1-P)}$$

N = Total number of households; z = Value of standard variation at 95% confidence level (1.96); P = Estimated population proportion (0.05); d = Error limit of 10% (0.1)

All the respondents were over the age of 30. Since wards are considered as the unit of analysis, information was collected about every housess of each ward from the VDC office. To analyze exposure of the study area, meteorological parameters data were obtained from Department of Hydrology and Meteorology (DHM) in Kathmandu, Nepal for the time period of 30 years, from 1981–2010. The nearest meteorological stations-Dumkauli for Pragatinagar (500meters east from study area) and Jomsom for Kagbeni (four km south from study area) were taken to analyze the climatic variables. Missing data were less than one percent which was interpolated using ordinary kriging method in ArcGIS10.

17.2.2 Selection of Vulnerability Indicators

Vulnerability to climate change was analyzed at the ward level by a complex interrelationship of multiple factors: exposure, sensitivity and adaptive capacity. Many variables of these components cannot be quantified directly. In this study, selection of indicators for adaptive capacity was based on the CARE climatic vulnerability and capacity analysis methods and DFID sustainable livelihoods framework, where adaptive capacity is taken as a function of asset possession by the households (Nelson *et al.* 2010b).

17.2.2.1 Exposure

Exposure is the nature and degree to which a system is exposed to significant climatic variation (IPCC 2001). Historical changes in climate variables (maximum and minimum temperature, rainfall amount, rainfall days and 24-hour extreme rainfall) and occurrence of extreme climatic events were taken as indicators of exposure for this study. The coefficient of the trends of climate variables was calculated. This chapter has considered floods and drought as climate change events for this study.

17.2.2.2 Sensitivity

Sensitivity is the degree to which a system will respond to a change in climatic conditions (IPCC 2001; Paavola 2008). Marshall *et al.* (2009) and Daze, Ambrose, and

Ehrhart (2009) have taken sensitivity indicators based on the livelihood impacts of climate change related disasters. Fatalities due to natural disasters, property damage (land, houses and crop) due to natural disasters and share of non-agricultural income of a family are considered as sensitivity of a household.

17.2.2.3 Adaptive Capacity

Adaptive capacity is the ability of people to cope with or adjust to the changing context and is explained by socioeconomic indicators. These indicators have been divided into five types of livelihood assets- human, social, physical, natural and financial. All of these indicators are not directly related to climate shocks only, but support to combat climatic shocks through different ways.

Table 17.1: Indicators for adaptive capacity.

S.No.	Assets	Indicators
1	Human	Knowledge of climate risk, Agriculture and vocational training, education level of family members
2	Social	Women's savings and loans groups, membership in CBOs
3	Physical	Access to services and facilities (road, market, school and medical centre), Irrigated Land, House type, device to access information
4	Natural	Reliable water resources, settlement in un-inhabitant areas
5	Financial	Livelihood Diversification Index, Diversified Income Sources

Financial assets are important. If all income is derived from agriculture only, then such income suffers during the bad weather. Instead, if the income is derived from other sources as well, the risk is reduced because risk is distributed among the sources. The Herfindahl index of diversification is used to calculate Livelihood Diversification Index (LDI) using the formula:

$$D_k = \sum_{i=1}^{N} (S_{i,k})^2$$

where D_k is the diversification index, i is the specific livelihood activity, N is the total number of activities being considered, k is the particular ward, and $S_{i,k}$ is the share of ith activity to the total income for particular ward.

17.2.3 Calculation of the Vulnerability Index

Data in this study were at different scales, from zero to hundred. So, Normalization was done by using the formula:

$$\textit{Normalized Value} = \frac{\textit{Actual value} - \textit{Mean}}{\textit{Standard deviation}}$$

Then, weights should be allocated to these indicators. The literature indicates that there are three methods used to assign weights to indicators: (1) arbitrary choice of equal weight (Lucas and Hilderink 2004), (2) expert judgment (Brooks *et al.* 2005); and (3) statistical methods such as principal component analysis or factor analysis (Cutter *et al.* 2003). Assigning equal weights is too subjective. The literature also shows that all indicators do not affect vulnerability equally (Hebb and Mortsch 2007). The assigning of weights through expert judgment is often criticized because of the availability of experts in the particular field (Lowry *et al.* 1995). Therefore assigning weight by Principal Component Analysis (PCA) to generate weights was thus chosen over compared to the previous two methods (Nelson *et al.* 2010b; Cutter *et al.* 2003). PCA is a technique used to extract few orthogonal linear combinations of variables which most successfully capture information from a set of variables (Gbetibouo and Ringler 2009). The purpose of using PCA use is to reduce the data dimensions and prioritize the indicators.

PCA was run to prioritize the important indicators for the selected indicators of exposure, sensitivity and adaptive capacity separately using Statistical Package for Social Science (SPSS22). As the cumulative percentage of the Eigen value of first component was more than 65%, the loadings from the first PCA component used as the weights for the indicators. A two-step PCA was run separately. The first step PCA was run for each asset group separately to calculate the weight of each asset type. The second step PCA was run using the index values for the five asset groups to analyze the total adaptive capacity. This formula was used to construct all three indices (exposure, sensitivity and adaptive capacity):

$$I_j = \sum_{i=1}^{k} b1\left(\frac{a_{j,i} - x_i}{s_i}\right)$$

where, *I* is the index value, *b* is the loadings from first component from PCA for respective indicators, *a* is the indicator value, *x* is the mean indicator value, and *s* is the standard deviation of the indicators. Finally, the vulnerability index was calculated as: $V = E + S - AC$, where, *V* is the vulnerability index, *E* is the exposure index, *S* is the sensitivity index and *AC* is the adaptive capacity index for respective wards. A higher value of the vulnerability index indicates higher vulnerability.

17.3 Results and Discussion

17.3.1 Exposure

Table two shows the weights and coefficient trend of the exposure indicators across the two study sites. Weights represent the importance and relation of indicators to exposure. For instance: when the weight of rainfall amount, rainfall days and extreme rainfall is high, it means they impact more on exposure of these areas. The coefficient of both maximum and minimum temperature parameters show increasing trends in both areas but Kagbeni shows a slow decreasing trend for annual maximum temperature. The minimum temperature shows a rapid increasing trend relative to the maximum temperature in Pragatinagar, which is supported by the household questionnaire survey where 88% of the respondents feel that temperatures have risen in last decade. The summer maximum temperature is increasing in Pragatinagar, but minimum temperature is decreasing, meaning the gap between maximum and minimum temperature is increasing which is not good for summer crops as well. In contrast, just opposite situation is found in Kagbeni, Mustang. It is very important to notice that both maximum and minimum temperatures in winter are increasing in both areas, except the winter maximum temperature shows a slightly negative trend in Pragatinagar.

Table 17.2: Weights and coefficients for indicators of exposure.

Indicators	Pragatinagar	Kagbeni
Annual maximum temperature	−0.663 (0.006)	−0.559 (−0.001)
Annual minimum temperature	0.026 (0.016)	−0.079 (0.07)
Rainfall amount	0.797 (7.119)	0.920 (1.508)
Rainfall days	0.546 (−1)	0.846 (−0.05)
Extreme rainfall	0.730 (0.233)	0.554 (0.1)
Natural disasters	0.27 (1.75)	0.41 (2.1)

Source: Department of Hydrology and Meteorology, Nepal, Field survey 2012/13
Note: Figures in parenthesis indicate the coefficient trend of indicators.

There is a clear indication of temperature change in Kagbeni. It shows that the summer maximum temperature is in a decreasing trend for the last three decades. In contrast, the summer minimum temperature increased dramatically. After 1993, the minimum temperature is above average minimum temperature except in 2008 and 2009. Winter temperatures are drastically changed in Kagbeni VDC. Since 1994, the winter minimum temperature was always above the average temperature for 16 years. The winter temperature in 1981 was –3.01°C but it increased to −0.33°C in 2010. Overall, the minimum temperature in last three decades was –1°C, but the average minimum temperatures for 1981–1990, 1991–2000 and 2001–2010 were –2.41°C, −0.55°C and −0.05°C respectively. It clearly shows an increasing trend in the minimum temperature for Kagbeni VDC.

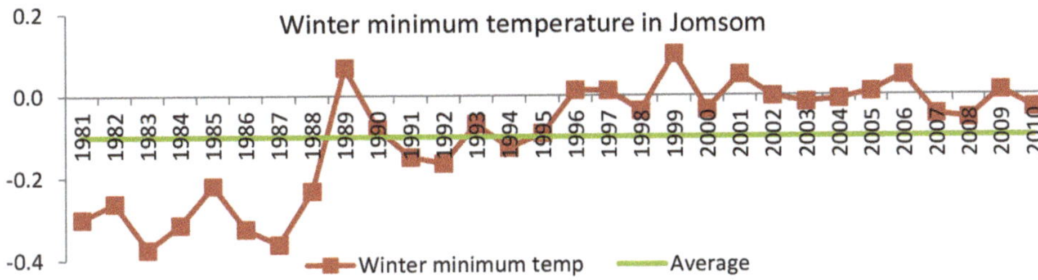

Figure 17.2: Winter minimum temperature in Kagbeni (1981–2010).

Rainfall is one of the main meteorological indicators which affects people and crops directly. The annual rainfall in Pragatinagar is 2379 mm. The rainfall coefficient shows rainfall with an increasing trend amount of 7.12 mm per year; however the data show uneven rainfall in those years. The summer rainfall data also shows an increasing trend with the coefficient of three mm, but winter rainfall show a slow decreasing trend. However, the rainfall days show an opposite trend to the rainfall amount, which is a big problem farmers are currently facing. The rainfall trend was almost constant for the last three decades, but the rainfall days decreased extremely in the last 30 years. The rainfall day's trend coefficient shows that rainfall days is decreasing by one day every year in Pragatinagar and 0.05 days in Kagbeni.

Though average annual rainfall days in Nawalparasi from 1981–2010 was 113.4 days, it was only 94.4 days in last decade (2001–2010). The data also show that rainfall days decreased immensely in last 10 years. The total rainfall days in 1981 was 115 days, but it decreased to 64 days in 2010. Rainfall days were always less than 100 days in last six years. The average for summer rainfall days was 80 days. Surprisingly, the summer average rainfall was only 67.2 days in last decade.

Mountainous VDC-Kagbeni shows almost desert like characteristics for rainfall. Average annual rainfall was only 267 mm during last 30 years. Annual, summer and winter rainfall shows an increasing trend but rainfall days are decreasing remarkably constant. It could lead to landslides and flooding in the study area. Annual rainfall in 1981 was 83 days but it was only 46 in 2010. Similarly, 48 days rainfall occurred in summer, 1981, but decreased to 24 days in 2010.

17.3.2 Sensitivity

Table 17.3: Weights and VDC mean values for indicators of sensitivity.

Indicators	Pragatinagar	Kagbeni
Fatalities	0.910 (1)	0.95 (1.18)
Land damage	0.968 (48.33)	0.93 (27.09)
Household damage	0.998 (14.44)	0.93 (2.1)
Crop damage	0.996 (466.67)	0.99 (230.5)
Non-agriculture based income	−0.231 (56)	−0.491 (75.76)

Table three shows that indicators of sensitivity have contributed to the sensitivity index as per the assumption. Although the number of natural disasters and damage looks high in Pragatinagar, all the indicators show high values in Kagbeni if we compare it by percentage. Household damage, land and crop damage due to flooding and drought in Pragatinagar and Kagbeni are the major problems related to sensitivity. Both VDCs are facing drought frequently in recent years. Over the last decades, they tackled drought 8 times during the summer season in Pragatinagar. More than two-third of the populations of both the VDCs are engaged in agriculture. Agriculture based income is around 38% in Pragatinagar. It may be due to fertile land and favorable weather condition for rice and other crops. The weather is very cold and snow falls heavily in winter in Kagbeni. Agricultural based income is only 24.24%. The weight of non-agriculture based income is negative in both VDCs. It indicates that if non-agriculture income is high, vulnerability decreases due to a high diversified source of income. The main sources of income in Kagbeni are remittances from abroad. Only 14.54% of the total population works abroad but it covers 66.57% of the total income of the family in Kagbeni. In contrast, 17.39% of total populations of Pragatinagar work abroad, mainly in the gulf countries, and their remittance covers around 25% of the total income. During survey, 67% respondents of Pragatinagar and 75% of the respondents of Kagbeni answered that their family members went abroad to work because of decreasing agriculture production and change in climate.

17.3.3 Adaptive Capacity

Figure three and four describes the separate index score for every asset. First step PCA defines the impact of individual indicators within each asset group, while the second step PCA shows the relative importance of the five types of assets which define the overall adaptive capacity.

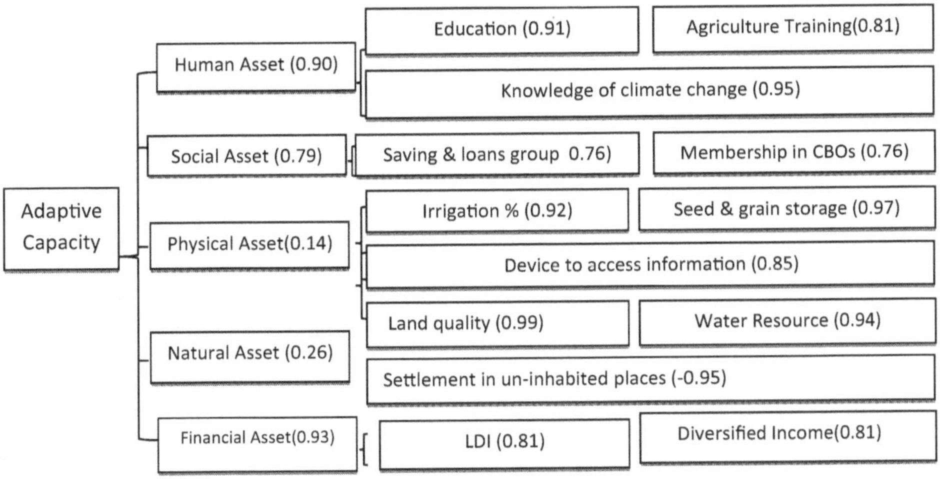

Figure 17.3: Weight of indicators and assets to structure the aggregate adaptive capacity index in Pragatinagar VDC.

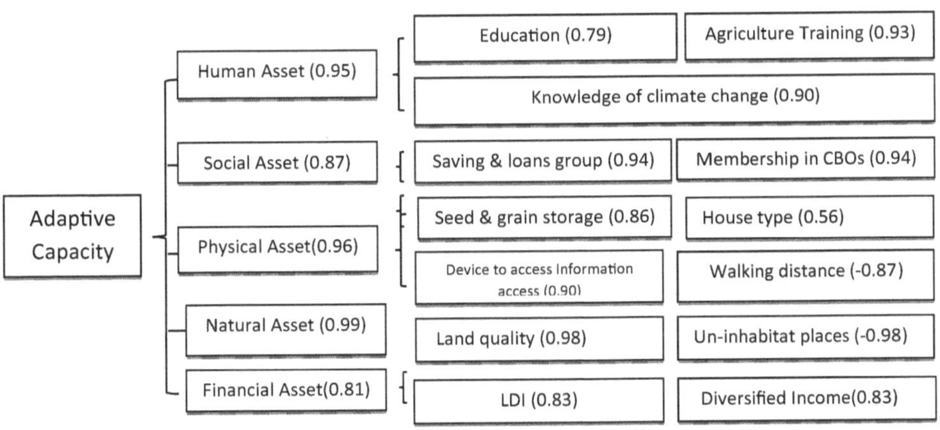

Figure 17.4: Weight of indicators and assets to structure the aggregate adaptive capacity index in Kagbeni VDC.

The higher the score, the higher the impact of indicators in each asset. For instance – for the human assets, knowledge on climate change and highest qualification of the family members received higher weights, which means they are more important indicators in human assets followed by agriculture and vocational training taken by any family members. All three indicators increase the adaptive capacity as shown by the strong positive sign of the weight.

Figure three and four show that financial assets and human assets are the two most important determinants of overall adaptive capacity in Pragatinagar, followed by social assets. Financial assets are the most convenient assets which can be converted into other forms of assets if needed. Indicators of human assets such as – education, knowledge about climate change and skill development trainings are essential to be able to utilize physical and financial assets properly. Besides, Women's savings and loans group and membership in CBOs are also important indicators of social assets. Likewise, as all assets have weights of more than 0.8 in Kagbeni, all assets are important determinants to measure adaptive capacity. However, natural assets, physical assets and human assets are main important indicators in Kagbeni.

17.3.4 Adaptive Capacity Index and Vulnerability Index

17.3.4.1 Pragatinagar

Figure five and six displays the ward-wise index values for adaptive capacity and its five components and vulnerability of Pragatinagar. Most of the assets types are positive in ward number 8, 5 and 6 in Pragatinagar, so the overall adaptive capacity is also high in those wards. As second-step PCA shows that financial assets and human assets are the two most important determinants of adaptive capacity, followed by social assets in Pragatingar,- households of ward number 1, 5 and 8 have positive score on those assets.

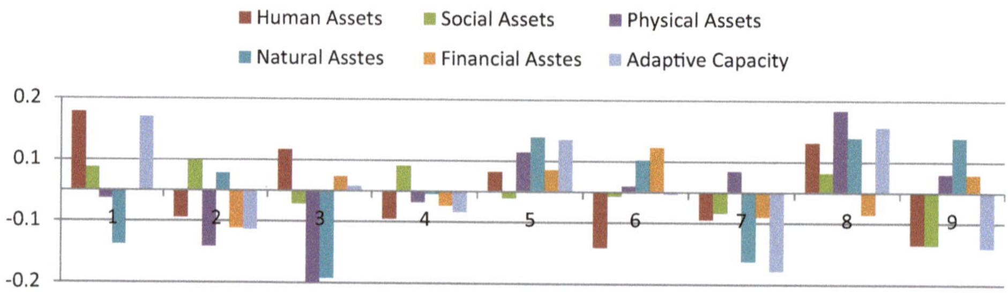

Figure 17.5: Ward-wise index scores for adaptive capacity and it's components in Pragatinagar VDC.

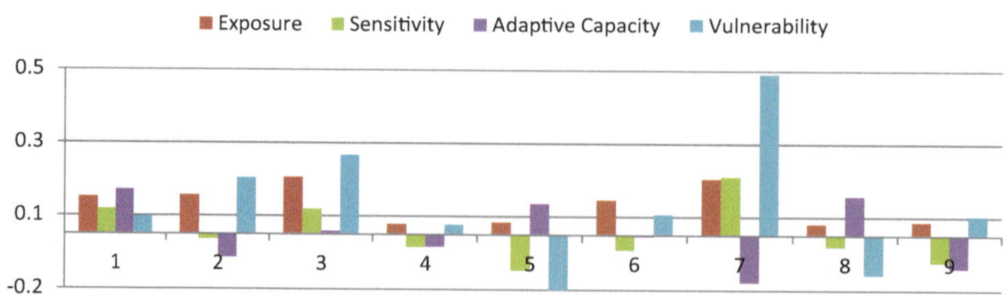

Figure 17.6: Ward-wise index scores for vulnerability and it's components in Pragatinagar VDC.

The higher the value of vulnerability index, the higher the vulnerability of wards. According to the ward-wise vulnerability index of Pragatinagar VDC, ward number 7 and 3 are the most vulnerable wards, while number 5 and 8 are the least vulnerable. Ward numbers 3 and 7 are along the river. The frequency of natural disasters, for instance, as well as house, land and crop damage due to disasters are common in these wards. So, they have high exposure and sensitivity with low adaptive capacity, resulting increased vulnerability.

17.3.4.2 Kagbeni

The mean values of individual indicators in figure seven shows that ward number 7 and 8 rank first and second in terms of possession of all assets, followed by ward number 9 and 6 in Kagbeni VDC. Ward number 1 stands last in terms of all the asset categories (except financial assets) and thus has the least adaptive capacity.

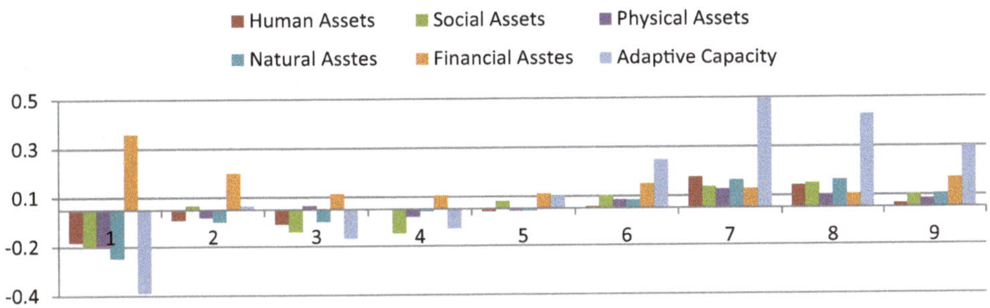

Figure 17.7: Ward-wise index scores for adaptive capacity and it's components in Kagbeni VDC.

All wards of Kagbeni VDC are highly exposed to climate change parameters and sensitive to climate related disasters. Different wards have different adaptive capacities. Based on that, the vulnerability of wards is different to one another. Ward number 1 has highest vulnerability index followed by ward number 3 and 4, as adaptive capacity is low in those wards. In contrast, adaptive capacity is high in ward number 7 and 8, so they are less vulnerable to climate change.

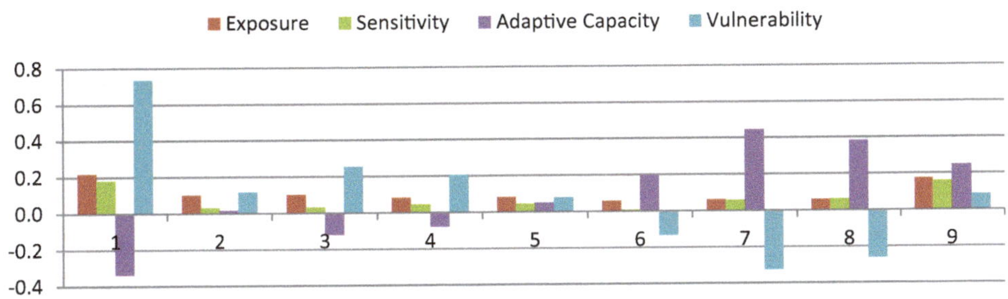

Figure 17.8: Ward-wise index scores for vulnerability and it's components in Kagbeni VDC.

17.4 Household Adaptive Capacity

Financial assets and human assets are the two most important determinants of overall adaptive capacity in Pragatinagar followed by social assets. However, natural assets, physical assets and human assets are the main important indicators in Kagbeni. This indicates that the priority of semi urban-Pragatinagar and rural area-Kagbeni have different priorities of adaptive capacity. Pragatinagar needs to focus mainly on improving financial assets. Agriculture based income is the main source of income (38%) in Pragatinagar. It may be due to fertile land and favorable weather condition for rice and other crops. Weather is very cold and snow falls heavily in winter in Kagbeni. So, agricultural based income is only 24.24%. The main source of income in Kagbeni is remittance from abroad. Only 14.54% of total population work abroad but it covers 66.57% of the total income of the family in Kagbeni. In contrast, 17.39% of the total populations of Pragatinagar work abroad mainly in the gulf countries and their remittance cover around 25% of the total income. 67% respondents of Pragatinagar and 75% respondents of Kagbeni answered that their family members went abroad to work because of decreasing agriculture production and change in climate. Farmers of Pragatinagar have less diversified income sources compared to Kagbeni. Fruit farming, livestock farming, and tourism are other sources of income in Kagbeni.

Secondly, human assets are equally important in both of the study areas. Only one-third and one-fifth of people in Pragatinagar and Kagbeni respectively, have heard about climate change and its effects. The number of literate people is also much less. Only 21% of adult people in Pragatinagar and 18% in Kagbeni have passed grade 10. Similarly, the number of people taking vocational training is also very few, only 7% in Pragatinagar and 12% in Kagbeni. Due to the lack of skilled and vocational training, 92% of the people who work abroad from Pragatinagar works as an unskilled manpower in gulf countries and their salary is 180 dollars a month in an average where as 83% of the people who work abroad from Kagbeni are literate and skilled manpower. So, they all are working in USA, Europe, Japan or India. No one from Kagbeni works in Gulf country as an unskilled labor. Women's savings and loans networks are strong in Pragatinagar. Every ward has Women's organizations, but Kagbeni has weak women's savings and loan networks. Numbers of community based organizations are increasing in recent years.

Rural VDC-Kagbeni needs to be focused mainly on access to services and facility. Access to transportation, medical and market are the main constraints for adaptive capacity in Kagbeni. Only ward number 7 and 8 have road access. Farmers living in Syangtang (ward number 1) need to walk about 10 hours to get road access, high schools, markets and health facilities. Farmers living in Ward number 2, 3, 4 and 5 also need to walk about one hour for road access. House type is also an important indicator to be focused on in Kagbeni. Eightyeigth % of the houses have mud roofs. There was continuing rainfall for 42 hours in Kagbeni in July, 2012. After that, some of the farmers have converted the roof of house from mud roof to stones. Land quality has a higher impact in determining the adaptive capacity. More than half (54.7%) of the land is barren and only 1.05% of the land is cultivable in Kagbeni, but most of the land (80%) is agricultural land in Pragatinagar.

Figure 17.9: a) Traditional house in Ward number 2, Kagbeni; b) Farmers in the field in Ward number 9, Kagbeni; c) Local road in ward number 7, Kagbeni; d) Modern house in ward number 3, Pragatinagar; e) Farmers in the field ward number 1, Pragatinagar; f) Local road in ward number 8, Pragatinagar.

17.5 Conclusion and Recommendations

The results indicate that exposure of a community to long term changes in climate variables and occurrences of natural disasters are important components to determine the overall vulnerability of the study areas. The past 30 years' meteorological data shows that though the rainfall amount is increasing slightly, rainfall days are decreasing remarkably in both the study areas, mainly in Pragatinagar. Rainfall days are decreasing by one day every year in Pragatinagar and 0.05 days in Kagbeni, which is a big problem for farmers nowadays. Annual rainfall days as well as summer rainfall days decreased immensely in last decades. As such, drought and flood increased in last decades.

Both maximum and minimum temperatures show an increasing trend in both study areas. Minimum temperature is increasing significantly, mainly during the winter in Kagbeni. Since 1993, the minimum temperature is always above than average minimum temperature for 16 years. Overall, minimum temperature in last three decades was –1°C, but the average minimum temperature in 1981–1990, 1991–2000 and 2001–2010 were –2.41°C, –0.55°C and –0.05°C respectively. It clearly shows an increasing trend in the minimum temperature in Kagbeni VDC.

Fatalities, land damage, household damage and affected crops have a very strong relationship with sensitivity, and non-agriculture based income has the opposite influence on sensitivity. Kagbeni is more sensitive than the semi-urban area – Pragatingar. Both VDCs are different in nature. Pragatinagar lies in plain area, so flood and drought are the major disasters related to climate change. In contrast, Kagbeni is the mountainous VDC, where more than half of the land is barren, and snowfall, flood and wind are the main problems. Due to this, flood, drought, landslides and soil erosion are the common disasters. Both VDCs are facing drought frequently in recent years. Over the last decades, Pragatinagar and Kagbeni tackled drought eight and six times, respectively, in the summer season.

However, biophysical elements that determine the exposure for instance, temperature, rainfall, rainfall days, extreme rainfall and natural disasters are the most important components to determine the vulnerability; – they are beyond the immediate influence of the policy makers. Adaptive capacity is the component which is directly focused on for policy implications, out of three components. Adaptive capacity has indirect implications on sensitivity. Improving adaptive capacity helps to improve the sensitivity of the community. For instance, improving knowledge of farmers about climate change and literacy rate (physical assets) may decrease fatality, land and crop damage in the community. Improving the irrigation facilities (physical assets) in the community decreases the sensitivity of crops to droughts.

Financial assets and human assets are the two most important determinants of overall adaptive capacity in Pragatinagar followed by social assets. Indicators of human assets are – education, knowledge about climate change and skill development trainings; it is essential to be able to utilize physical and financial assets properly. Around 1,000 young people go abroad daily from Nepal to work as unskilled manpower. So, vocational training should be given to the people. Plumbers, electricians, basic computer training, cooking, and language training would all have a great impact to reduce vulnerability. Also, public awareness training on climate change and its effects on health as well as family planning, gender equality, leadership programs, safe drinking water and sanitation would also help to increase adaptive capacity. Besides, Women's savings and loans group and membership in CBOs are also important indicators of social assets. However, the case of Kagbeni is different. Natural assets, physical assets and human assets are the main important indicators in Kagbeni. As the ward level is considered as the unit of analysis, results show that wards having high exposure and sensitivity and low adaptive capacity are more vulnerable to climate change (ward number 3 and 7 of Pragatinagar). Thus, vulnerability of the households of these wards can be decreased by improving the adaptive capacity, which will indirectly lower sensitivity. Though financial assets are the most convenient assets, other assets are also equally important. Ward number 1 of Kagbeni has very high financial assets, but it is considered to be the highest vulnerable ward of Kagbeni VDC due to the lower value of other assets and indicators.

Both VDCs are similar in terms of exposure to flooding and drought, but different for sensitivity and adaptive capacity. The semi-urban area- Pragatinagar, is less sensitive and has high adaptive capacity, but the mountainous rural VDC – Kagbeni, is more sensitive to climatic disasters and has low adaptive capacity. So, Kagbeni is more vulnerable than Pragatinagar. Kagbeni is far behind, mainly for transportation facility, access to technology, social networks, hospitals and markets. So, efforts from government or non-governmental organizations in Kagbeni should be geared towards improving road facilities and house types (physical assets). Human capacity should be improved in both the VDCs through education and trainings, which would facilitate adoption of non-agriculture income sources. Policy needs to focus on increasing poor people's access to technologies, skills for sustainable agriculture production. To manage natural resources such as soil, land and water is also another challenge. Agriculture is still the backbone of Pragatinagar VDC, thus irrigation facilities should be established as this would help facilitate higher production and the possibility of growing cash-crops. Also, policy should emphasize the non-agriculture livelihoods option because it helps to improve the cash income of the community and reduce their dependency on natural resources.

References

Baral, N. 2005. Resource use and conservation attitudes of local people in the Western Terai landscape, Nepal. Miami: Florida International University, masters thesis.

Brooks, N. 2003. Vulnerability, risk and adaptation: A conceptual framework. Working Paper 38, Tyndall Centre for Climate Change Research.

Brooks N., W.N. Adger, and P.M. Kelly. 2005. The determinants of vulnerability and adaptive capacity at the national level and the implications for adaptation, *Glob Environ Change A* 15(2): 151–162.

CBS-Central bureau of statistics. 2003. National sample census of agriculture Nepal, 2001/02: Highlights. Kathmandu: National planning commission secreteriat.

CBS. 2004. Statistical pocket book of Nepal. Kathmandu.

Cutter, S.L., B.J. Boruff, and W.L. Shirley. 2003. Social vulnerability to environmental hazards. *Social Sciency Quarterly* 84(2): 242–261.

Daze, A., K. Ambrose, and C. Ehrhart. 2009. Climate vulnerability and capacity analysis (handbook). London: CARE International.

DFID-Department for International Development. 1999. Sustainable livelihoods guidance sheets. London: DFID.

Ellis, F. 2000. Rural livelihoods and diversity in Developing Countries. New York: Oxford University Press.

Füssel, H.M., and R.J.T. Klein. 2006. Climate change vulnerability assessments: An evolution of conceptual thinking. *Climatic Change* 75(3): 301–329.

Fussel, H.M. 2007. Vulnerability: a generally applicable conceptual framework for climate change research. *Glob Environ Change* 17(2): 155–167.

Gallopin, G.C. 2003. A systemic synthesis of the relations between vulnerability, hazard, exposure and impact, aimed at policy indentification. In *Handbook for estimating the socio-economic and environmental effects of disasters*. Mexico: ECLAC-Economic Commission for Latin American and the Caribbean.

Gbetibouo, G.A., and C. Ringler. 2009. Mapping South African farming sector vulnerability to climate change and variability. A subnational assessment. IFPRI Discussion paper 885. Washington: International Food Policy Research Institute.

Hebb, A., and L. Mortsch. 2007. Floods: Mapping vulnerability in the upper Thames watershed under a changing climate. Final report of the Canadian foundation for climate and atmospheric sciences (CFCAS) project: Assessment of water resources risk and vulnerability to changing climatic conditions.

IPCC-Intergovernmental Panel on Climate Change. 2007. *Climate change 2007: Impacts, adaptation and vulnerability. Contribution of working group II to the fourth assessment report of the IPCC*, ed. M.L. Parry, O.F. Canziani et al.

IPCC. 2001. *Climate change: The scientific basis*. https://www.ipcc.ch/ipccreports/tar/wg2/index.php?idp=8

IPCC. 2001. *Climate change 2001: Impacts, adaptation, and vulnerability*, ed. J. McCarthy, O. Canziani, N. Leary, D. Dokken, and K. White. Cambridge: Cambridge University Press.

Kelly, P.M., and W.N. Adger. 2000. Theory and practice in assessing vulnerability to climate change and facilitating adaptation. *Climatic Change* 47, 325–352.

Kimenju, S.C., and D. Tschirley. 2009. Agriculture and livelihood diversification in Kenyan rural households. Nairobi: Tegemeo Institute of agricultural policy and development.

Lowry, J., H. Miller, and G. Hepner. 1995. A GIS-based sensitivity analysis of community vulnerability to hazardous contaminants on the Mexico/U.S. Border. *Photogrammetric Engineering & Remote Sensing* 61(11): 1347–1359.

Lucas, P.L., and H.B.M Hilderink. 2004. The vulnerability concept and its application to food security. Report 550015004/2004. Netherlands: National institute for public health and the environment.

Maharjan, K.L. 2003. Peasantry in Nepal - A study on subsistence farmers and their activities pertaining to food security. Research center for regional geography. Hiroshima University.

Marshall, N.A., P.A. Marshall, J. Tamelander, D. Obura, D. Malleret-King, and J.E. Cinner. 2009. A framework for social adaptation to climate change: Sustaining tropical coastal communities and industries. Gland: IUCN.

Nelson, R., P. Kokic, S. Crimp, H. Meinke, and S.M. Howden. 2010a. The vulnerability of Australian rural communities to climate variability and change: Part I - Conceptualizing and measuring vulnerability. *Environmental Science and Policy* 13, 8–17.

Nelson, R., P. Kokic, S. Crimp, P. Martin, H. Meinke, S.M. Howden, et al. 2010b. The vulnerability of Australian rural communities to climate variability and change: Part II - Integrating Impacts with adaptive capacity. *Envrionmental Science and Policy* 13, 18–27.

Oxfam. 2009. Even the Himalayas have stopped smiling. Climate change, poverty, and adaptation in Nepal. Kathmandu: Oxfam International.

Paavola J. 2008. Livelihoods, vulnerability and adaptation to climate change in Morogoro, Tanzania. *Environ Sci Policy* 11(7): 642–654.

Thomas R.J. 2008. Opportunities to reduce the vulnerability of dryland farmers in Central and West Asia and North Africa to climate change. *Agric Ecosyst Environ.* 126(1–2): 36–45.
UNFCCC. 2009. Climate change: Impacts, vulnerabilities and adaptation in Developing Countries. www.unfccc.int/resource/docs/publications/impacts.pdf.
Vincent, K. 2004. Creating an index of social vulnerability to climate change for Africa. Working Paper 56 Tyndall Center for Climate Change Research.

Marc Prohom and Oriol Puig[49]

18 Weather Observation Network and Climate Change Monitoring in Catalonia, Spain

Abstract: This paper details the weather and climate monitoring of the Meteorological Service of Catalonia. Following a general overview of the institution, the weather observation network is described, as is the importance of monitoring real-time phenomena efficiently. Finally, the procedures, methods and techniques for achieving reliable climate change monitoring are also described, paying particular attention to the importance of data rescue initiatives and the quality control and homogenization of climate data series. To summarize, the main characteristics of climate change experienced in Catalonia are shown, together with the results of climate projections for the near future.

Keywords: Climate projections, Lighting detector, Radar network, Radiosounding station, Weather stations

18.1 Overview of SMC: History, Background and Characteristics

The Meteorological Service of Catalonia (SMC, in its Catalan abbreviation) is a public sector company within the Generalitat (the autonomous government of Catalonia), with its own legal status. The SMC was established in 1921 by the "Mancomunitat", the Catalan governmental body that grouped together the four provinces of the region, with Eduard Fontserè being appointed the first director. Between 1921 and 1939, the main duty of the institution was weather forecasting, in order to provide improved forecasts of extreme weather conditions and information for certain sectors such as air transportation. During its seventeen-year existence, the SMC gave an extraordinary boost to meteorology in Catalonia, with some of its work proving to be of great scientific value and gaining international prestige. Among the projects with widespread impact outside Catalonia were the SMC's participation in producing the *International atlas of clouds and of states of sky*, its contribution to the International Polar Year (1932–33), by creating two mountain-top observatories, and its design of the Jardí rain gauge. However, in 1939 the SMC was closed down and its archives and belongings confiscated due to the Spanish Civil War.

Many years later, with the return to democracy in Spain, the 1979 Statute of Autonomy gave the Catalan government exclusive rights to reestablish the former

49 Meteorological Service of Catalonia, Barcelona, mprohom@meteocat.com

meteorological service, and thanks to the 2001 Meteorological Law, the service once again became a reality. Since then, the main duties of SMC have been:
- Assisting public authorities and institutions in their needs for meteorological and climate information.
- Managing and maintaining the weather observation network.
- Operating the meteorological database of Catalonia.
- Programming, introducing and managing a system of forecasting and monitoring meteorological phenomena and climate change.

This paper shows the main tools and methods that the SMC uses to achieve such goals, i.e. its weather observation network and climate change monitoring system.

18.2 Weather Observation Network

Every day, thousands of atmospheric observations are made around the world, measuring several meteorological parameters in a variety of ways. In Catalonia, the SMC has developed a weather observation network that provides information on the current state of the weather, providing inputs to numerical forecasting models and information on the intensity, frequency and spatial distribution of extreme weather events. This chapter describes measurements made by the SMC at weather stations, both automatic and conventional, in the upper atmosphere, and via weather radar and lightning detection systems.

18.2.1 Automatic and Conventional Land Weather Stations

Surface observations are particularly important because they provide information about weather that is experienced by people and most of the socioeconomic environment. Two different networks are operated by the SMC: Automatic (AWS) and Conventional (or manual) Weather Stations (CWS).

At present (2012), 174 AWS cover the entire Catalan territory. Observations are made primarily for the purpose of providing information on the current state of the atmosphere for weather forecasting purposes. For this reason, they include measurements of a wide range of meteorological elements, have a uniform spatial distribution and cover the variety of climates found in the region, particularly the mountainous areas of the Pyrenees. A description on the technical specifications of the AWS is provided in Table 18.1.

The measurement of all elements is fully automated, with data being logged at the station and transmitted to a central system based at the SMC headquarters in Barcelona. Each station produces observations at one-minute intervals, and thirty-minute observations are then generated for onward real-time transmission to users worldwide via the SMC website.

Table 18.1: Main technical characteristics of the elements of the AWS network (MR, Measurement Range and Acc, Accuracy).

Element	Equipment branch/model	Technical specifications	Height (m)	Num. of elements
Datalogger	Campbell Scientific/CR1000	Input voltage range: 5Vdc Final storage: 4Mb		174
Wind force and direction	RMYoung/05103	MR: 0–100 m/s Sensitivity: 1.0m/s	2.0 6.0 10.0	58 13 81
Global solar radiation	Kipp&Zonen/CMP6	MR: 285–2800 nm Sensitivity: 5–20 μV/W/m²	2.0	43
Temperature and relative humidity	Vaisala/HMP155	MR: −80–60°C MR: 0-100% Acc: −20/+40°C	1.5	174
Barometric pressure	Vaisala/PTB110	MR: 600-1100 hPa Acc: ±0.3 hPa at +20°C		72
Precipitation	Lambrecht tipping bucket Thies tipping bucket	Acc: 2% Resolution: 0.1 mm	1.4	72 88
Accumulated precipitation	Geonor/T200B Vibrating-wire weighing transducer	Capacity: 600 mm Sensitivity: 0.05mm Acc: 0.1% FS	2.0	11
Snow depth	Campbell Scientific/SR50A Acoustic sensor	MR: 0.5 to 10 m Acc: ±1 cm	3.0	18
Soil temperature	Campbell Scientific/107	MR: −35°C to +50°C Acc: ±0.2°C (0° to 50°C)	−0.05 −0.50	38 35

Although most of the observation process today is automated, a network of 135 CWS provide additional data that is not reported by AWS. This network is called XOM (Catalan abbreviation of Meteorological Observers Network) and is formed by trained and enthusiastic meteorological observers who also provide estimates of elements that are hard to automate, such as visibility, cloud cover and meteors, while also reporting any severe weather events in near real time, i.e. heavy snowfalls, hailstorms, tornadoes, etc. Figure 18.1 shows the spatial location of both networks. Many CWS are maintained to meet the requirements for accurate climate averages at a wide variety of locations around Catalonia. A long uninterrupted record is essential for this purpose (at least 30 years and preferably longer) and this is achieved at a large number of stations.

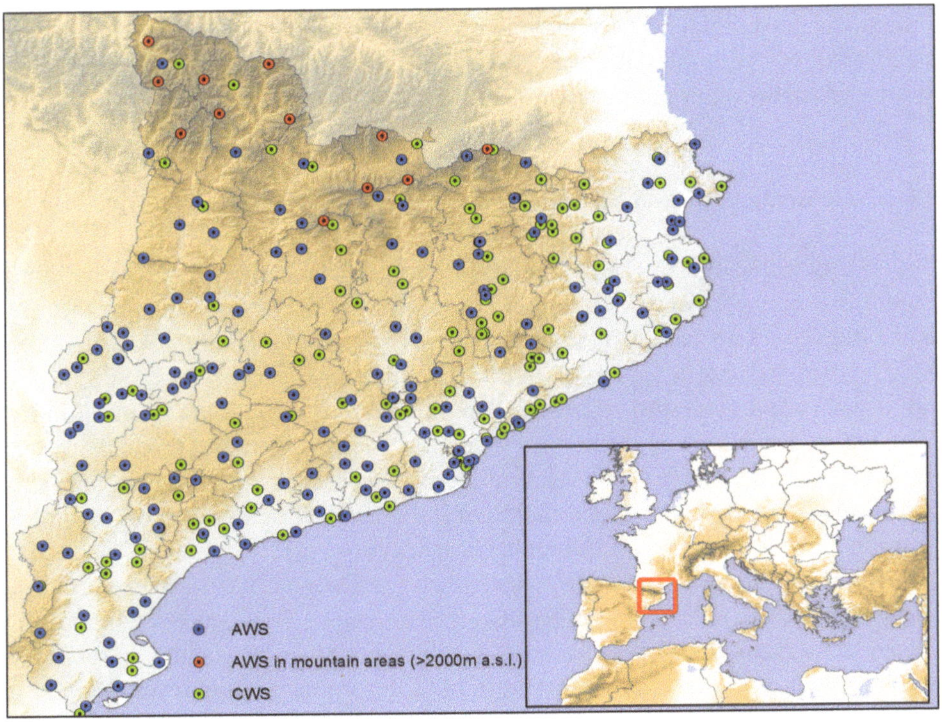

Figure 18.1: Spatial distribution of AWS (in blue) and CWS (in green) operated by SMC, distinguishing high mountain AWS (in red).

18.2.2 Radar Network

Meteorological radars are observation instruments which allow rainfall and its likely intensity to be detected from a great distance. This makes radars an essential tool for monitoring and forecasting extremely heavy rainfall, thus minimizing the effects of downpours on people and property. The radar network of the SMC (XRAD) is composed of four systems, which are Doppler type and operate in C band (5600 to 5650 MHz), with a 240 km observation radius and a 1 km resolution. All the radar systems generate a new volume of data every 6 minutes, and each one is composed of three tasks: the first one, with a range of more or less 250 km and the lowest elevation, is generated for surveillance purposes, while the remaining two (range between 130 and 150 km, and seven elevations to each one) scan practically the whole troposphere with the purpose of collecting data for generating a precise rainfall estimation, and for the characterization of the vertical development of echo structures (Bech *et al.* 2004). Data are collected in polar coordinates, but most of the products are converted into a Cartesian grid. As well as rainfall, mountains also return the energy received from radar systems. As a result, in mountainous areas such as Catalonia, these features have to be taken into consideration when designing the structure of a radar network. Fixed echoes and other associated problems (attenuation, occultation, anomalous propagation, bright band, etc.) require a series of subsequent corrections to be made, especially if a quantitative use of the radar-produced observations is desired.

18.2.3 Lightning Detection Network

The sum of intracloud (IC) and cloud-to-ground (CG), known as the total lightning data, is determined by the Lightning Detection Network (LDN) operated by the SMC since 2003. The LDN covers Catalonia with an efficiency of more than 90%. It is currently composed of four Vaisala LS-8000 total lightning detectors. IC and CG flashes are detected and processed separately. On one hand, IC flashes are detected in the VHF (110 to 118 MHz) and located using interferometry (Lojou and Cummins 2006). On the other hand, CG return strikes, and are mainly observed by a low frequency (LF) sensor and located using a combination of the TOA/MDF (Time-of-Arrival/Magnetic Direction Finding) techniques (Cummins *et al.* 1998).

The SMC LDN is an important step forward in the real-time detection of thunderstorms and the short-term forecasting of their course. This tool gives support for surveillance meteorological tasks and complements other tools such as satellites, weather radar networks and AWS (Figure 18.2). Lightning data adds information to this characterization and can define the most active storms, benefiting a wide range of users such as power companies, telecommunication networks, major infrastructures, outdoor leisure complexes, etc.

18.2.4 Automatic Radiosounding Station

The radiosounding system allows for an understanding of atmospheric conditions by measuring pressure, temperature, relative humidity and wind speed and direction from the ground to heights of more than 15–20 km. This information is of particular importance for diagnosing highly-unstable meteorological conditions which can cause intense precipitation. Since April 1998, radiosounding has been carried out twice a day (at 00.00 GMT and 12.00 GMT) at the University of Barcelona (in collaboration with the Department of Astronomy and Meteorology), and since 2008 this system has been part of the World Meteorological Network (08129 code). Numerical weather prediction models need this information for their initialization as they work with data from the ground, as well as from more than 20 atmospheric layers at altitudes of up to 15,000 m. Data from these soundings is entered daily into mesoscale numerical models operating at the SMC, and provides a substantial improvement in forecasts for Catalonia. Since December 2012, the system has been fully automated, with weather balloons and radiosondes being launched without the presence of a technician (Figure 18.3).

Figure 18.2: Lightning discharges map for a single day, 5th July 2012. Source: Meteorological Service of Catalonia.

Figure 18.3: Automatic Radiosounding Station located on the roof of the Faculty of Physics (University of Barcelona).

18.3 Climate Change Monitoring and Future Projections

The SMC issues its Climate Indicator Bulletin (CIB) every year, which details the latest status of climate change and variability in Catalonia. This report aims to help readers such as policymakers and researchers obtain a better understanding of climate evolution and therefore take measures to prevent global warming and to protect the environment. Where long-term records exist, it is possible to describe how current conditions fit into the long-term context of a variable and changing climate. What is more, the same records can be used to build and test the systems used to make long-term climate predictions. For this reason, in order to obtain a reliable view of the current climate, it is necessary to have long, quality-controlled and homogeneous climate series with sufficient spatial coverage. The steps to achieve these goals are described below.

18.3.1 Data Rescue

Developing data rescue initiatives is of special interest for the SMC. Since 2008, several projects have been undertaken in order to obtain long-term and high-quality instrumental records, focusing on: (1) detecting new long-term mountain climate series (>1,500 m above sea level) in historical archives; (2) improving temporal coverage of currently available records, i.e. filling gaps; and (3) recovering metadata information from different sites (see Aguilar *et al.* 2003 for more information on metadata). More than 1,500 daily climate records have been detected, encompassing the period from 1780 up to the present, measuring mainly temperature and precipitation. Among them, the works carried out to recover the climate records of Barcelona (since 1780), which are the most complete and oldest climate records of the Iberian Peninsula, are particularly relevant (Prohom *et al.* 2012) (Figure 18.4). Data rescue initiatives have also centered on reconstructing extreme events, i.e. heavy snowfalls, thus providing knowledge of the spatial distribution and frequency of such phenomena. Finally, the SMC also participates in several data rescue forums on an international level, such as Mediterranean Climate Data Rescue (MEDARE) and the EUMETNET Data Rescue Program (EUMETNET-DARE).

18.3.2 Quality Control

Data quality control is defined as a system of checks to assess and maintain the quality of a climate series. Quality control is simply a matter of comparing measured and expected values, i.e. what is acceptable, before a value is presumed to be suspicious or an error. Quality control procedures are designed to ensure data consistency, integrity, correctness and completeness, and also to identify and address errors and

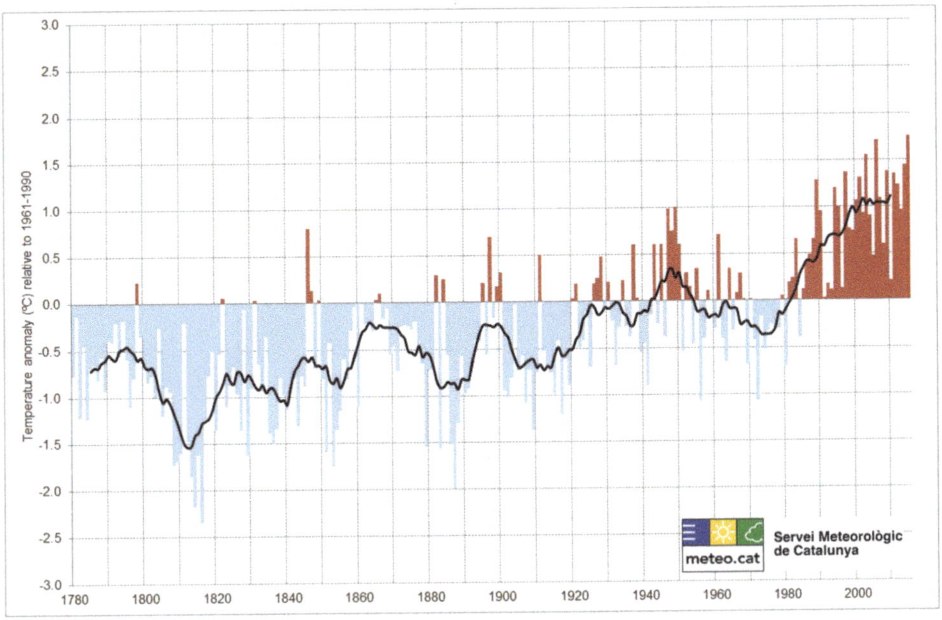

Figure 18.4: Mean yearly temperature in Barcelona (1780–2015) expressed as anomalies taking 1961–1990 as the reference period. The dark curve corresponds to an 11-year moving average.

omissions. This step is essential in detecting likely outliers, and to correct, confirm or remove them from the series.

As an example, on a daily scale for temperature the following checks are monitored: (1) Gross errors (45°C > Temperature < –15°C); (2) Internal coherency (Maximum temperature ≥ Minimum temperature), excessive leaps between two consecutive daily values (> 25°C) and flatline detection (more than 5 consecutive days with the same value); (3) Temporal coherency (more than 4 times the standard deviation and 3 times the interquartile range compared to climatological values); and (4) Spatial coherency (spatial representation of the same element to detect excessive differences in comparison to highly correlated values and nearby stations). Values within these thresholds are flagged as suspicious, checked against the original sources where possible, and finally removed or confirmed. With a visual inspection of the data, gross errors can also be detected.

By means of quality control, false inhomogeneities are eliminated and a homogeneity analysis is made without inferences.

18.3.3 Homogeneity Analysis

As previously said, the accuracy of climate observations is often affected by inhomogeneities due to changes in the technical or environmental conditions of measure-

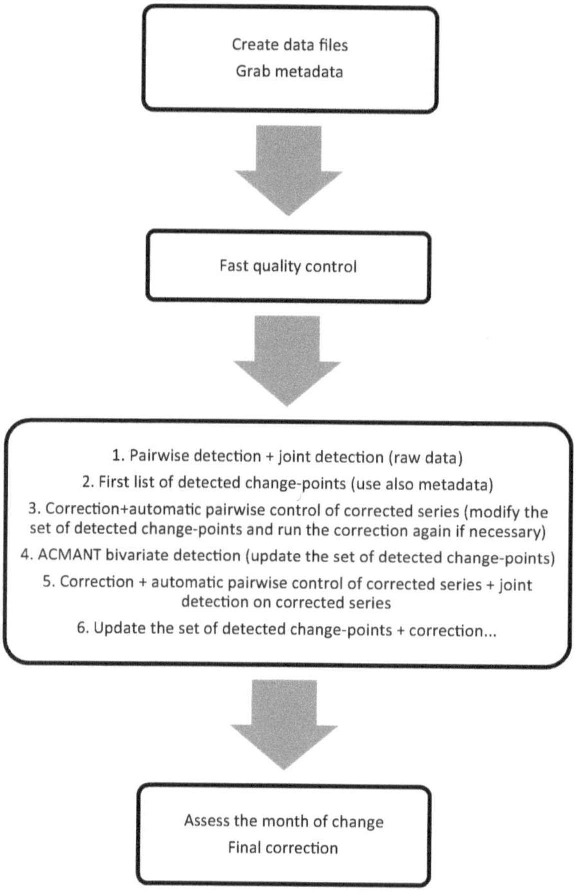

Figure 18.5: Task flow chart from HOMER.

ments (Auer *et al.* 2005). Most of these changes cause sudden shifts (break-points) in the series of local climate data, while others result in gradually increasing biases from real macroclimatic characteristics (urban development). As a result, the correction of inhomogeneities is highly recommended before any climate analysis is performed.

Among the large number of methods developed, the SMC has adopted HOMER as the procedure to homogenize instrumental series, as it includes the best features of other methods, namely PRODIGE, ACMANT and CGHSEG (Mestre *et al.* 2013), which have been ranked among the best methods for homogenizing monthly and annual climate data (Venema *et al.* 2012). HOMER provides a fast quality control of data, including an interactive semiautomatic method that combines three detection algorithms (pairwise-univariate detection, joint detection and ACMANT-bivariate detection), while ANOVA is used for correction (Figure 18.5). At a daily scale, as adjustments present a new level of complexity that are due mainly to the high variability inherent in daily data and the potential need for individual adjustments to each day, SMC has adopted the approach proposed by Vincent *et al.* (2002), whereby monthly adjustments made by HOMER are linearly interpolated to the daily scale, thus preserving trends and variations found in the homogenized monthly series.

These two homogeneity methods provide a reliable view of climate variability and change in Catalonia, both for trends of means (monthly, seasonal and annual scales) and extremes (daily scale).

18.3.4 Recent Climate Trends Detected in Catalonia

The results of how climate is changing in Catalonia are provided by the yearly Climate Indicators Bulletins (CIB). These reports have different chapters depending on the element analyzed: air temperature, precipitation, sea level rise, sea surface temperature, and climate extremes. For most of the elements, the analysis data ranges from 1950 to the present day, but for some series includes a wider time window (from 1905) or shorter one (for sea measurements). Below is a brief description of the CIB results, but for a more detailed, accurate study, the complete report can be downloaded from the SMC website (www.meteo.cat).

For temperature, a significant overall annual increase in mean air temperature in Catalonia has been detected since 1950, at a rate of 0.2°C/decade. Summer is the season recording the highest increases (+0.33°C/decade), while autumn experiences a lower warming rate (+0.15°C/decade). Maximum temperatures have a greater warming trend than minimum temperatures (0.25°C/decade vs. 0.15°C/decade). As for precipitation, no conclusive remarks can be indicated, but as a whole a slight downward trend has been detected in mean annual precipitation since 1950, namely –1.2%/decade, but without reaching the significant level of 95% (Figure 18.6). A significant increase in sea-level temperature has also been noted since the start of measurements in 1974, namely 0.3°C/decade, and the sea level has risen at a rate of 3.6 cm/decade, but over a shorter period, i.e. 1990–2012.

Several climate indices have been calculated using 20 daily homogenized temperature series (1950–2012) and following the criteria of the Expert Team on Climate Change Detection and Indices (ETCCDI) (http://www.clivar.org/organization/etccd). As a whole, all series indicate an increase in indices, thus reflecting a warming trend, i.e. warm nights and days, tropical nights, warm spell duration index and growing season length, while the opposite is experienced for indices indicating colder conditions, i.e. frosty days, icy days, cold nights and cold days. Indices for precipitation are calculated only for two long-term series located in Barcelona and Tortosa (southernmost part of the region), with daily high-quality data available since 1913 and 1905, respectively. No noteworthy results have been found, but a slight increase in days with heavy precipitation (for example, annual account of days when precipitation ≥ 50mm) and in the Simple Daily Intensity Index has been detected.

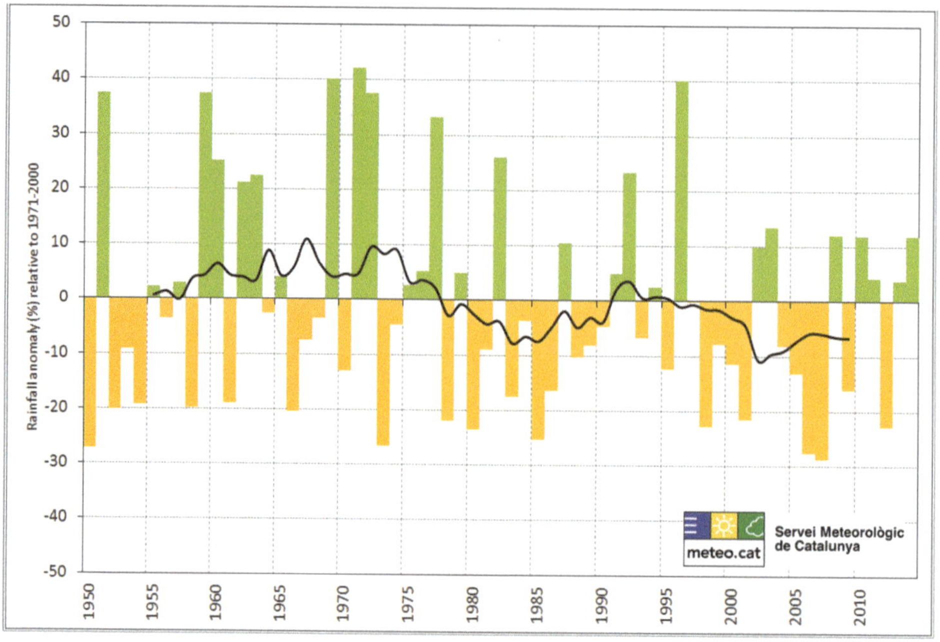

Figure 18.6: Mean annual rainfall in Catalonia obtained as from an average of 68 records over the period 1950–2014. Data is expressed as % anomalies, taking 1971–2000 as the reference period. The dark curve corresponds to an 11-year moving average.

18.3.5 Climate Projections in Catalonia (2001–2050)

Efficient monitoring of recent climate is essential for a better understanding of future climate scenarios. Following this statement, the SMC promotes the ESCAT project (Climate Scenarios in Catalonia, Gonçalvez *et al.* 2014), that derives present time (1971–2000) and future (2001–2050) surface air temperature and precipitation trends for the North Western Mediterranean Basin using dynamical downscaling techniques at high resolution, 10 km. Simulations use the Advanced Research Weather Research and Forecasting (WRF-ARF) model version 3.2.1 (Skamarock and Klemp 2008) and represent the three main families of emission scenarios defined in the IPCC-SRES (Nakićenović *et al.* 2000): A1B (rapid economic and population growth in a globalized world, with introduction of new and more efficient technologies and a balanced use of resources); A2 (divided world characterized by rapid and regionalized economic development, with few technological change and continued population growth); and B1 (a globalized world where environmental policies are promoted, economic development is not a priority, population increases, but not as much as in the A2 scenario). The Spain02 dataset (Herrera *et al.* 2012) is chosen for model assessment, as it provides surface daily maximum, minimum and range temperatures for Iberia and the Balearic Islands, over a regular grid of 0.2º resolution.

Projections show that temperature is expected to rise in all considered scenarios (up to 1.4 K for the annual mean), and particularly during summertime and in high-altitude areas. As for precipitation, annual mean precipitation is likely to decrease (around –5 % to –13 % in the most extreme scenarios). The climate signal for seasonal precipitation is not so clear, as it is highly influenced by the driving GCM (Global Circulation Model) simulation (Figure 18.7). All scenarios suggest statistically significant decreases of rainfall for mountain ranges in winter and autumn.

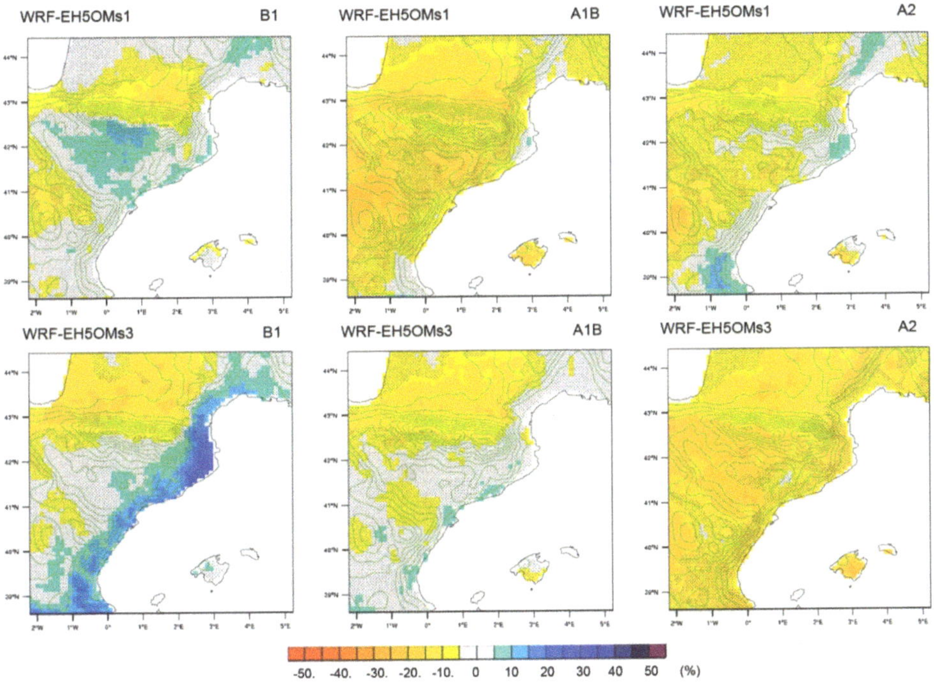

Figure 18.7: Differences in annual precipitation (%) between 2021–2050 and 1971–2000 for Catalonia as projected by WRF-EH5OMs1 (upper panels) and WRF-EH5OMs3 (lower panels) for the B1, A1B and A2 emission scenarios (from left to right).

A representative set of indices for temperature and precipitation extremes are also projected. The model's abilities to reproduce observed extremes are assessed for a control period (1971–2000) using the ERA40 reanalysis to drive the WRF-ARW simulations. The modelling system correctly reproduces the amplitude and frequency of extremes and provides a high degree of detail on variability over neighboring areas. However, it tends to overestimate the persistence of wet events and consequently slightly underestimates the length of dry periods (Barrera-Escoda *et al.* 2014). Drier and hotter conditions are generally projected for Catalonia, with significant increases in the duration of droughts and the occurrence of heavy precipitation events. The projected increase in the number of tropical nights and extreme temperatures could have a negative effect on human health and comfort.

18.4 Conclusions

The Meteorological Service of Catalonia (SMC) has legal status to foster research activities and improve awareness of meteorology and climatology throughout the region. Thanks to the observation network and the efforts made to collect high-quality and homogeneous climate series, as well as the application of robust and reliable analytical methodologies, these objectives are being achieved. Here we have seen an overview of the tools used, proving that in the future more research on climate and weather monitoring should be performed to provide better services and reliable information for citizens and policymakers, in order to better adapt to climate change and severe weather phenomena.

References

Aguilar, E., I. Auer, M. Brunet, T.C. Peterson, and J. Wieringa. 2003. WMO Guidelines on climate metadata and homogenization. WCDMP 53, WMO-TD 1186. Geneva: WMO.

Auer, I., R. Böhm, A. Jurkovic, A. Orlik, R. Potzmann, W. Schöner, M. Ungersböck, M. Brunetti, T. Nanni, M. Maugeri, K. Briffa, P. Jones, D. Efthymiadis, O. Mestre, J.M. Moisselin, M. Begert, R. Brazdil, O. Bochnicek, T. Cegnar, M. Gajic-Capka, K. Zaninivic, Z. Majstorovic, S. Szalai, T. Szentimrey. 2005. A new instrumental precipitation dataset in the greater alpine region for the period 1800–2002. *International Journal of Climatology* 25: 139–166.

Barrera-Escoda, A., M. Gonçalves, D. Guerreiro, J. Cunillera, and J.M. Baldasano. 2014. Projections of temperature and precipitation extremes in the North Western mediterranean basin by dynamical downscaling of climate scenarios at high resolution (1971–2050). *Climatic Change* 122: 567–582.

Bech, J., E. Vilaclara, N. Pineda, T. Rigo, J. López, F. O'Hora, J. Lorente, D. Sempere, and F.X. Fàbregas. 2004. The weather radar network of the Catalan meteorological service: description and applications. ERAD-European conference on radar in meteorology and hydrology- COST 717 Final seminar, ERAD publication, 416–420.

Cummins, K.L., M.J. Murphy, E.A. Bardo, W.L. Hiscox, R.B. Pyle, and A.E. Pifer. 1998. A combined TOA/MDF technology upgrade of the U.S. National lightning detection network. *Journal of Geophysical Research* 103: 9038–9044.

Gonçalves, M., A. Barrera-Escoda, D. Guerreiro, J.M. Baldasano, and J. Cunillera. 2014. Seasonal to yearly assessment of temperature and precipitation trends in the North Western mediterranean basin by dynamical downscaling of climate scenarios at high resolution (1971–2050). *Climatic Change* 122: 243–256.

Herrera, S., J.M. Gutiérrez, R. Ancell et al. 2012. Development and analysis of a 50 year high-resolution daily gridded precipitation dataset over Spain (Spain02). *International Journal of Climatology* 32, 74–85. doi:10.1002/joc.2256.

Lojou, J.Y., and K.L. Cummins. 2006. Total lightning mapping using both VHF interferometry and time-of-arrival techniques. 19th International Lightning Detection Conference, Tucson, USA.

Mestre, O., P. Domokos, F. Picard, I. Auer, S. Robin, E. Lebarbier, R. Böhm, E. Aguilar, J. Guijarro, G. Vertachnik, M. Klancar, B. Gubuisson and P. Stepanek. 2013. HOMER: a homogenization software – methods and applications. *Quarterly Journal of the Hungarian Meteorological Service* 117, 1, 47–67.

Nakićenović, N., J. Alcamo, G. Davis et al. 2000. Emissions scenarios 2000–Special report of the Intergovernmental Panel on Climate Change (SRES). Cambridge University Press, Cambridge. http://www.ipcc.ch/ipccreports/sres/emission/index.php?idp=0.

Prohom, M., M. Barriendos, F. Aguilar, and R. Ripoll. 2012. Recuperación y análisis de la serie de temperatura diaria de Barcelona, 1780–2011. In *Cambio climático. extremos e impactos*, ed. C. Rodríguez-Puebla et al., 207–217. Salamanca: Spanish Association of Climatology. Isbn: 978-84-695-4331-3.

Skamarock, W.C., and J.B. Klemp. 2008. A time-split non hydrostatic atmospheric model for weather research and forecasting applications. *Journal of Computational Physics* 227: 3465–3485. doi: 10.1016/j.jcp.2007.01.037.

Venema, V., O. Mestre, E. Aguilar, I. Auer, J.A. Guijarro, P. Domonkos, G. Vertacnik, P. Stepanek, P. Zahradnicek, J. Viarre, G. Müller-Westermeier, T. Szentimrey, M. Lakatos, C.N. Williams, M. Menne, R, Lindau, D. Rasol, E. Rustemeier, K. Kolokythas, T. Marinova, L. Andresen, F. Acquaotta, S. Fratianni, S. Cheval, M. Klancar, M. Brunetti, C. Gruber, M. Prohom Duran, T. Likso, P. Esteban, and T. Brandsma. 2012. Benchmarking homogenization algorithms for monthly data. *Climate of the Past* 8: 89–115. doi: 10.5194/cp-8-89-2012.

Vincent, L.A., X. Zhang, B.R. Bonsal, and W.D. Hogg. 2002: Homogenization of daily temperatures over Canada. *Journal of Climate* 15: 1322–1334.

Paolo Ronco[50,51], Martina Bullo[50], Valentina Gallina[50], Silvia Torresan[51], Andrea Critto[50,51], Alex Zabeo[50], Elena Semenzin[50], Roland Olschewski[52], Antonio Marcomini[50,51,53]

19 Assessing flood risk in Zurich, Switzerland: the KR-RRA approach

Abstract: Natural disasters due to climate variability and climate change are threatening societies worldwide, posing significant challenges for the scientific community as well as for policy makers and planners in understanding and assessing drivers, dynamics and risks associated to these phenomena.

Floods events, recognized as one of the most threatening water-related disasters, can affect people's lives as well as their properties. In recent years, the number of flood events is growing as a consequence of many factors, both climatic, such as the dramatic intensification of the pattern of precipitations, and non-climatic, such as the change in land use, the increase of population, the presence of relevant human activities in hazard-prone areas and, more general, the uncontrolled urban development. Recent river flooding events in Europe urge the design and implementation of proper flood risk prevention strategies in order to protect society and the environment, and mitigate the impact of these events. The European Flood Directive (FD) 2007/60/EC represents a specific and innovative legislative framework to support the development of these management strategies, in order to reduce the adverse consequences imposed by floods to selected targets. In this context, the recent KULTURisk-FP7 Project developed a state-of-the-art Regional Risk Assessment (RRA) methodology to assess the risk related to flood events. The KULTURisk-RRA methodology, comprehensively presented in Ronco *et al.* (2014) and based on the concept of risk being a function of hazard, exposure and vulnerability, has been applied to the Sihl River valley in Switzerland through a tuning process considering the site-specific context and features (Ronco *et al.* 2015). Flood related risks have been assessed for different targets (namely people, economic activities, natural and semi-natural systems and cultural heritage) lying on the Sihl river valley, including the city of Zurich, which represents a typical case of river flooding in urban areas. The main outputs of the RRA methodology application are GIS-based risk maps that allow for the identification and prioritization of relative hotspots and targets at risk. By means of Multi Criteria Deci-

50 Dept. Environmental Sciences, Informatics and Statistics University Ca' Foscari Venice, Venice, Italy

51 Centro Euro-Mediterraneo su i Cambiamenti Climatici (CMCC), Impacts on Soil and Coast Division, Lecce, Italy

52 Swiss Federal Research Institute WSL, Economics and Social Sciences, Birmensdorf, Switzerland

53 Corresponding author: Antonio Marcomini (marcom@unive.it)

sion Analysis tools, the so-called total risk maps have been produced to visualize the spatial distribution of flood risk within the study area through a tailored participative approach by eliciting feedbacks from expert and local stakeholders. In particular, the total risk map produced for a 300 year return period scenario (selected as reference) demonstrates that the Sihl river valley is generally affected by a relatively low risk, where higher relative risks are concentrated in the deeply urbanized area around the Zurich city centre. Finally, results of the testing of the KR-RRA methodology in a real case study have confirmed its overall consistency, appropriateness and relevance to the specific needs of the region and the objectives posed by the EU Flood directive.
Keywords: Sihl river, Total flood risk index, Zurich

19.1 Introduction

Nowadays, extreme weather and climate dynamics interacting with exposed and vulnerable human and natural systems can lead to severe catastrophes. Only over the last few years research concerning these events, their impacts, and options for dealing with them has become mature enough to support and develop comprehensive assessment strategies (IPCC 2012). In recent years, the frequency as well as magnitude of some types of extreme weather and climate events have increased in parallel with an increased development in natural areas and assets exposed at risk, thereby exacerbating the consequences for water related disasters. The major expected climate change impacts include sea-level rise, coastal erosion, alteration in water quality and increase of flooding (Hirabayashi *et al.* 2013). Recent river flooding events in Europe (i.e. the flood of Elbe river in June 2013) have caused huge damages for people, their properties and the environment, underlining the importance of adequate and sustainable floods management as well as proper prevention measures in order to avoid these dramatic consequences. In Switzerland in particular, severe flood events have occurred in many catchments in the last decade, while periods with frequent floods alternated with quieter periods during the last 150 years (Bründl *et al.* 2009). A study conducted by Hilker *et al.* (2009) in Switzerland estimated a total monetary loss of approximately 8 billion Euros due to floods, debris flows, landslides and rock fall, with 56% of this damage caused by six single flood events from 1978 to 2005.

In this context, the European Flood Directive (FD) 2007/60/EC represents an ad-hoc legislative framework which specifically supports the development of proper flood management strategies, in order to reduce the adverse consequences for human health, the environment, cultural heritage and economic activities.

Actually, innovative and effective research methods and tools are urgently needed to enhance the assessment of exposure, vulnerability and risks associated with flood disasters, in order to improve the development of adequate prevention, mitigation and preparedness measures (Ronco *et al.* 2014). Generally, the risk assessment process encompasses the identification, characterization and evaluation of risks, associated with a specific context and/or system. The outcomes of the risk assessment can be used in a wide variety of decision making processes, such as the realization of new infrastructures (e.g. flood retention basins and/or early warning systems), or determining the acceptability of safety levels and the need for improvement of existing systems (e.g. a flood defence system) (Jonkman 2008). Overall, risk assessment provides a rational basis for flood management decision making at the national scale, as well as at regional and local scales (Hall *et al.* 2003; Apostolakis 2004). Several methodologies have been developed in order to assess flood risk; the choice of one methodology over another largely depends on the objectives of the analysis, the availability of data, the peculiarities of the application context, the level of detail to be achieved, and the dimensions of risk to be addressed (Ronco *et al.* 2014). Recently, Cirella *et al.* (2014) presented a comprehensive review and classification of current approaches and methodologies for the assessment of risks posed by several water-related natural hazards (coastal storms, tsunamis, river floods, avalanches, landslides, etc.). The review underlined that there are very few examples of methodologies that consider the complete suite of elements at risk (targets) through an integrated and multidisciplinary approach, encompassing the entire variety of risk dimensions (i.e. physical/ environmental, social and economic) as requested by the FD. In fact, the elements at risk are mostly buildings, infrastructures and population (e.g. Forte *et al.* 2005; Kubal *et al.* 2009), that are usually analysed separately, in monetary terms and related to damages, only. Moreover, flood risk assessments methodologies were mainly developed for very specific contexts at the local scale, with a high level of complexity and data requirements (e.g. Forte *et al.* 2005; Meyer *et al.* 2009; Kubal *et al.* 2009; Forster *et al.* 2008). Thus, they can hardly be applied to a wider range of case studies.

In this study, the KR-RRA methodology has been applied on a local stakeholder participative manner. It has been developed through the integration of different dimensions of risk and has been performed by successively tailoring the hazard, exposure and susceptibility analysis of flood risk to the specific characteristics of the Sihl river valley, in a local-stakeholder participative manner. Here, the application provided GIS-based relative risk maps and statistics related to different targets to support the (local) decision makers in land planning and decision making processes by providing them with easily understandable hazard, vulnerability, exposure and risk maps.

19.2 Legislative Context: Europe and Switzerland

Urban and rural communities living around floodplains and mountain valleys are increasingly affected by floods through considerable damage to infrastructure, properties and environmental goods (Mazzorana *et al.* 2012). In this context, the European Commission has developed a specific legislative framework for the assessment and management of flood risks based on prevention, protection and preparedness principles.

Traditionally, the European flood control and management practices have been focused on reactive practices based on structural measures, but nowadays it is widely recognized that a paradigm shift is required to move from defensive to proactive action towards a culture of prevention by managing the risk of living with floods (Annamo and Kristiansen 2012).

The European Flood Directive (FD) on the assessment and management of flood risk (EC 2007) came into effect in October 2007 and aims to reduce and manage the flood-related risks to specific targets, namely human health, environment, cultural heritage, economic activities, infrastructures and properties. By clearly distinguishing between hazard and risk concepts, the FD asks for a more sophisticated analysis of natural hazards; in particular, risk maps should allow for visualization of the spatial distribution of (flood) risks in the specific (flood) scenario by considering the risk as the combination of hazard, exposure and vulnerability.

Switzerland is not supposed to adopt European legislation. However, some aspects of the Flood Directive have been transposed into the federal risk prevention strategy and the subsidy policy of the Federal Office for Environment (FOEN), demonstrating a general agreement on the European framework and legislation guidelines, and thus approving the integrated flood management approach.

The federal laws for water protection (Key legislation: 1955, 1971, 1991; SR 814.20) and water management (Key legislation: 1991; SR 721.100) both came into effect in 1991, and represent the main framework for flood management in Switzerland by defining the priority of river protection upkeep and spatial planning measures for flood prevention (Buchecker *et al.* 2011).

These Federal Laws defined the responsibility of Cantonal authorities. They must ensure flood protection and define the space needed for flood water to guarantee its natural functions and use, based on the decisions on risk and hazard given by the confederation strategy (Santato *et al.* 2013).

19.3 The Sihl River Valley: Description and Hydrological Regime

The Sihl is a 68 km long alpine river located in the foothills of the Alps of Switzerland. Its sources (total basin coverage: 336 km^2) are located at Drusberg in the Canton of Schwyz (SZ). Downstream, it flows through the artificial Sihl lake regulated by a concrete dam (upstream basin: 156 km^2) entering the Canton of Zurich (ZH) through the Sihl valley and flowing parallel to Zurich lake. At the end, the Sihl river joins the Limmat river at Platzspitz in the Zurich city centre (downstream basin: 180 km$^{2.}$).

The Sihl river valley is extensively wooded; the forest lying in the hills is classified as coniferous and mixed forest (CORINE Land Cover classification, 2006) and the valley is also cultivated, mainly used as arable land and pastures. The analysed area (77.97 km^2) refers only to the lower part of the valley and in particular to the city of Zurich with its 21 districts and 5 municipalities (Figure 19.1).

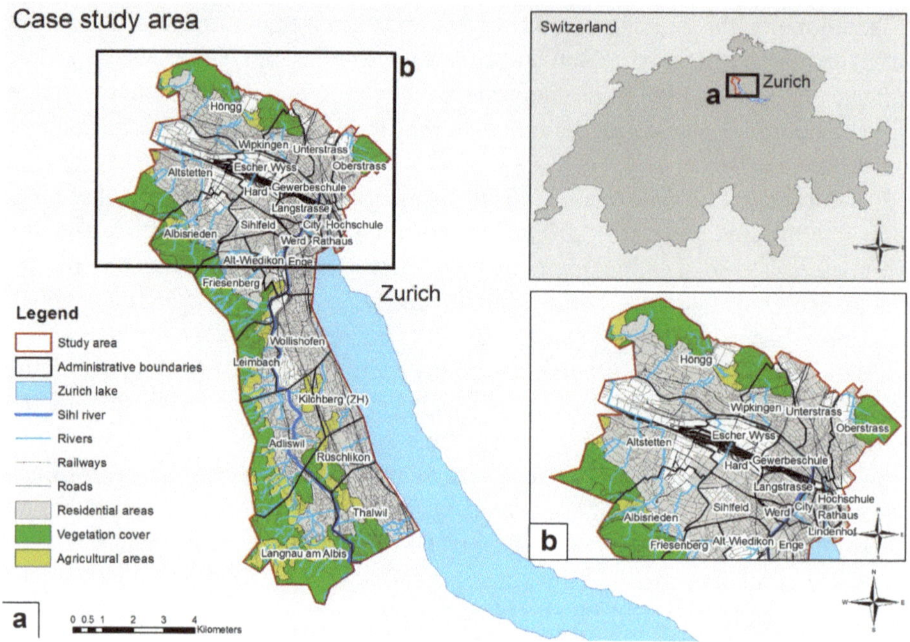

Figure 19.1: The case study area: a) its location in Switzerland and b) focus on the city centre (from: Ronco *et al.*, 2015).

The area is densely populated, especially close to the city of Zurich. According to CORINE Land Cover classification (EEA, 2007), the residential area covers 41.28 km^2 (more than half of the case study area) and the total population is 289,029 (Statistical Office of Canton of Zurich, 2011); 20.19 km^2 are covered by forest and 7.67 km^2 are devoted to agriculture. Several cultural heritage hotspots (e.g. the Swiss National Museum, the Kaspar Escher House, the Fraumünster, the Church of Bühl, etc.) are present in the valley and especially in Zurich city centre. The valley is also characterized by a complex network of infrastructures including railways, roads and pathways, highly concentrated in the lower part of the valley where the city of Zurich is located.

The Sihl river basin is often prone to flash floods. In the lower part of the basin, the Sihl river flows through Zurich, for which it represents the largest flood threat (Addor *et al.* 2011) because, just before joining the Limmat river, it flows beneath the main railway station of Zurich (Zürich Hauptbahnhof HB).

In the past, several flood events hit this area, as listed below (Pro Sihltal 2008).

- 1732: the Sihl river flooded the valley destroying the bridge across the river in Zurich;
- 1846: extensive damage in the Sihl valley and in Zurich city; as a consequence, the Sihl was diverted between the bridge in Brunau and Platzspitz in the city of Zurich;

- 1852: large flood destroyed part of the road in Leimbach district;
- 1876: flood with very high water in the city but no damages were registered;
- 1910: massive flood with a lot of damage also due to large debris input. The Zurich main train station was 30–40 cm under water, some railway tracks were badly damaged and the service was interrupted, Leimbach and Adliswil districts were under 1 m of water and some cellars of the Swiss National Museum at Platzplitz were completely flooded.

In 1937 the artificial Sihl lake was harnessed for both hydropower purposes and as a retention (over-flow) basin to reduce the frequency of flooding downstream (Schwanbeck *et al.* 2010). Until 1999, no more floods were registered in the area but since then, in 2005 and in 2007, severe inundations have demonstrated that the buffer capacity of Sihl lake retention basin is not large enough to prevent the impacts of extreme flood events during heavy rainfall seasons, and especially in spring. These events confirmed the need for a proper flood risk management strategy that includes, among others, effective structural and non-structural protection measures to reduce the risk of flooding for different critical targets of the area, such as the infrastructures. The complexity of flood issues related to the Sihl river and the necessity of planned strategy of prevention measures demands for a broader integrated approach which allows for assessing the risk of flooding related to multiple targets in order to identify and prioritize areas and targets at risk in the region considered and, therefore, to determine the benefits of prevention scenarios (Ronco *et al.* 2015).

19.4 KR-RRA Methodology: Theory and Application

19.4.1 Framework and Background

Within the KULTURisk Project *"Knowledge-based approach to develop a cULTUre of Risk prevention"* funded by the 7th Framework Program of the European Commission, a water risk-based methodology for the evaluation and accounting of risk prevention measures has been developed. By considering three main tiers of analysis, namely the physical/environmental Regional Risk Assessment (RRA), the social and the economic valuation of potential consequences, a conceptual framework developed by Giupponi *et al.* (2014), has been built upon the consolidated formalization of risk being a function of hazard, exposure, and vulnerability. These elements are combined to calculate risk as the combination of the probability of a certain hazard to occur and of its consequences.

Accordingly, the physical-environmental cluster of the risk assessment methodology, based on the RRA approach (Landis *et al.* 2005), integrates four steps of analysis: hazard assessment, characterizing the flood pattern by means of relevant metrics (e.g. flow velocity, water depth, flood extension) according to different scenarios to be investigated; exposure assessment, identifying the elements at risk that could be adversely affected; susceptibility assessment, evaluating the degree to which the targets could be affected; and risk assessment, combining the information about a certain hazard scenario with the exposure and susceptibility of the examined targets, providing a first evaluation of risks related to each receptor through the computation of the relative risk. After normalising of the receptor-related risk, a total (integrated) risk index is calculated by means of Multi Criteria Decision Analysis (MCDA) function (Ronco *et al.* 2014). The main outputs of the KR-RRA are GIS-based maps of receptor-related risks and of the total risk.

19.4.2 Setting of Scenario and Data Characterization

The proposed methodology requires the preliminary and functional analysis of different flood scenarios (baseline and alternative) considering structural and non-structural solutions which could mitigate the risk in the analysed area, in order to evaluate the benefits of different prevention measures, in terms of a "risk-avoidance" perspective.

As far as the Sihl case study is concerned, the available flood hazard scenario refers to three classes of hazards, where flood-prone areas are classified according to different frequency levels (high, medium and low), and based on the concept of return period of hazardous events (30, 100 and 300 years). The 300 year return period scenario has been considered the most relevant for the purpose of this study. The other two scenarios (30 and 100 years), characterized by a flood extension that only marginally affects the typical Sihl prone area, i.e. areas around the main railway station of Zurich (that is a critical hot spot in case of a flood event), have not been considered. Finally, by assessing the most catastrophic configuration, the selected scenario gives the opportunity to plan the mitigation, adaptation, and preparedness actions in a precautionary framework, without considering the mitigation measures included in the Early Warning System established in 2008. The Canton of Zurich is currently developing further prevention measures which, once established, are expected to reduce the flood risk of the Sihl to an extremely low level, but details on the expected impact in terms of flood reduction under different prevention measures have not been calculated yet. Due to this, alternative scenarios were not analysed in this study.

The (relatively complex) database required to fully apply the KR-RRA methodology includes : i) data to characterize the intensity and the frequency of the flood hazard (e.g. hazard metrics such as flow velocity, water depth, flood extension, return period); ii) data referring to the spatial distribution of the investigated targets (e.g. people, economic activities, natural and semi-natural systems, cultural heritage) to perform the exposure assessment; iii) relevant indicators (e.g. percentage of disabled people, slope, soil type etc.) to evaluate the degree to which the different targets could be affected by a flood hazard (susceptibility assessment). In particular, to characterize the affected people, the residential census data has been used to compute the number of people within residential cells of 25 m^2 (basic spatial unit of the analysis), and the CORINE Land Cover dataset has been used to characterize the spatial pattern of the various targets and indicators.

19.4.3 Relative Risk to Relevant Targets

Following the four main steps identified by the KR-RRA methodology (hazard, exposure, susceptibility and risk assessments), the relative risk related to selected targets has been calculated according to the specific method described in Ronco *et al.* 2014. The following paragraphs describe in summary the method, and, in more detail, the results of its application to the Sihl river valley (Ronco *et al.* 2015).

19.4.3.1 People

The approach proposed for the risk assessment to people is based on the methodology developed by the UK Department for Environment, Food and Rural Affairs (DEFRA, 2006). Here, the (flood) hazard assessment identifies water depth and flow velocity as relevant flood metrics. These are considered in a linear relation with the hazard magnitude (i.e. when water depth and flow velocity increase, the hazard level increases linearly). Moreover, the presence of a floating material debris factor is also considered as it can increase the level of danger in relation to depth values. The respective hazard scores for the Sihl river case study have been calculated and range from 0.9 to 6, where increasing values mean an increasing hazard for people.

The exposure assessment considered the presence of the people potentially affected by the hazard and referring to residential areas, only. Here, it is assumed that all the people are present in their homes at low ground, where they do not have safer areas in which to find refuge.

The susceptibility assessment step has been performed by considering the percentage of elder residents (over 75 years) and the percentage of residents with a disability. These factors could increase the susceptibility of people, since they can be more prone to health and stability problems during flood events.

Hazard, exposure and susceptibility have been aggregated in the risk assessment phase in order to provide the number of people injured (R_1) or killed (R_2) during the flood. The risk to people (injured) with classes are reported in Figure 19.2 and the related Table 19.1.

As for the other targets, the intervals of values provided by chromatic tables are obtained through the equal-interval classification methods. The forecasted number of total injuries is 1,000, while the estimated number of total (potential) fatalities is 29. Among the affected areas, Albisrieden and Altstetten districts (densely populated districts with medium scores for susceptibility) are subject to higher values of casualties, with 223 and 155 injuries, respectively, and 5 fatalities each. It should be underlined that these two districts are normally flooded by other tributary rivers to the Limmat river. The rate of injured people relative to the total population of the study area is 0.35% and the percentage of fatalities is 0.01%. These rates suggest that the risk to people is generally low, despite being non-negligible, if we consider the high density of population actually prevailing in the residential area (Ronco *et al.* 2015).

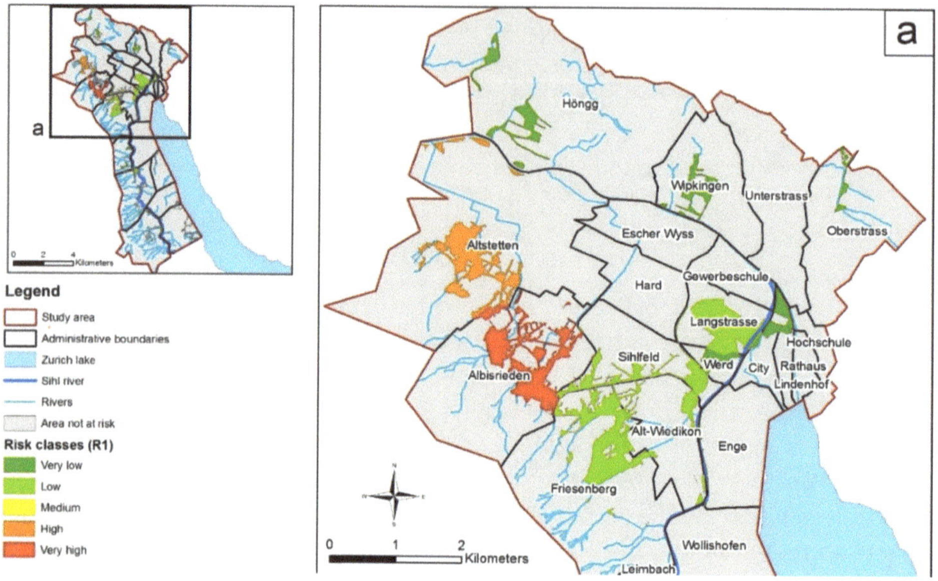

Figure 19.2: Risk map for people for the entire case study area and in particular for the city of Zurich (box a). From Ronco *et al.* 2015.

Table 19.1: Relative risk classes and range of values for injured people.

Risk Classes (R1)	Number of injuries
Very low	1–50
Low	50–100
Medium	100–150
High	150–200
Very high	>200

19.4.3.2 Economic Activities

19.4.3.2.1 Buildings

For the sake of simplicity, it has been assumed that all the buildings present in the study area (19,430 items; total buildings coverage: 10.67 km^2) are characterized by the same structure (i.e. susceptibility constant), therefore it is possible to evaluate risks (R_3) by considering the relationship between the hazard classes and the potential structural damages, as proposed by Clausen and Clark (1990). This method provides three risk classes (inundation, partial damage, total destruction), differentiating the potential consequences in a qualitative way, based on thresholds determined by flow velocity values and by the product of water depth and flow velocity.

The GIS-based risk map (Figure 19.3) shows the spatial risk distribution to buildings along the studied area where all the buildings affected by the flood event would be inundated (Table 19.2) with no dramatic structural damages (Ronco *et al.* 2015). The total number of buildings at risk is 3,267 and the related surface area is 2.2 km^2. The percentage of flooded buildings is around 17% while the percentage of flooded areas is almost 20%.

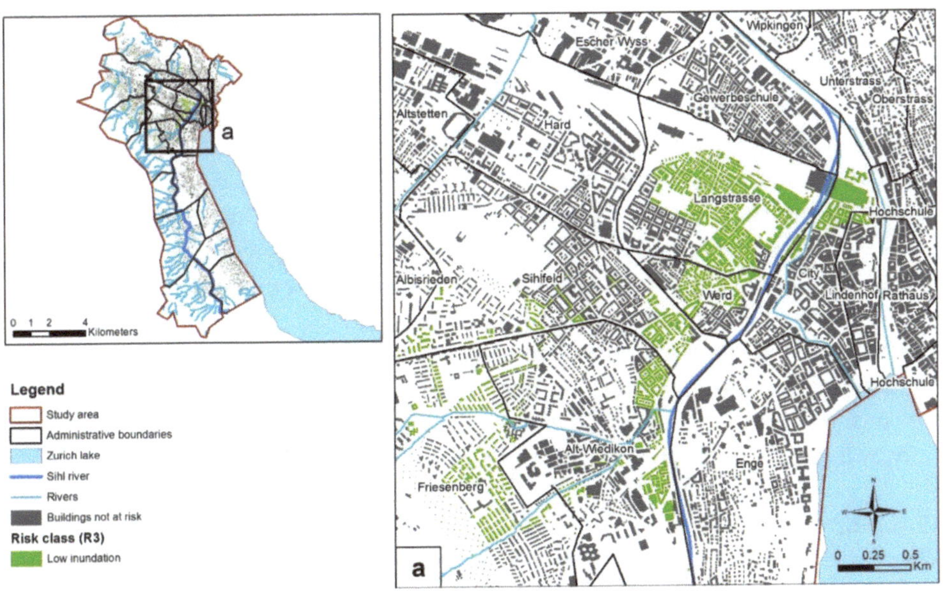

Figure 19.3: Risk map for buildings for the entire case study area and in particular for the Zurich city centre (box a). From Ronco *et al.* 2015.

Table 19.2: Relative risk classes and range of values for buildings.

Risk Classes (R3)	Description	# of inundated buildings
Not at risk	Not inundated	16,163
Low	Inundation	3267
Medium	Partial damage	0
High	Total destruction	0

19.4.3.2.2 Infrastructures

The flood hazard assessment step has considered the flood coverage related to the 300 year return period scenario as representative for the spatial characterization of the hazard to the target infrastructures. No other flood metrics have been considered since the analysis is not oriented to the evaluation of direct structural damages for infrastructures, but rather to the loss of service. For this reason, susceptibility has been considered to be constant. The exposure assessment has addressed the localization of the infrastructures (including railways, road and pathways present in the study area) using the Roads (Strasse_CH_line) and Railways (Eisenbahn_CH_line) TLM3D shapefiles. Therefore, the infrastructure-related risk (R_4) has been calculated from the intersection between the flood coverage and the infrastructures map in order to identify the length (in km) and the percentage of infrastructures inundated by a flood event of that particular magnitude. This results in around 209 km out of 1,540 km of infrastructures that currently rely on the study area, i.e. less than 14% of infrastructures are at risk. In particular, around 54 km refers to a railways network and 155 km to roads and pathways. Langstrasse and Albisrieden are the districts most affected by the flood event, belonging to the very high class and high class of risk, with 32 km and 24 km of inundated transport infrastructures respectively.

The infrastructures receptor is one of the most relevant for this case study, if we consider that the Sihl river flows underneath Zurich main station and that many railway lines are located just beside of the river.

19.4.3.2.3 Agriculture

When assessing the risk to agriculture at the meso-scale, the aim of the KR-RRA methodology is to define the percentage of the harvest loss due to a flood event. The hazard assessment requires the identification of water depth and flow velocity as relevant flood metrics, pointing out different thresholds for selected agricultural crops characterized by a different response (susceptibility) to flood events (namely: vegetables, vineyards, fruit trees and olive groves). Within the study area, the agricultural receptor is defined by CLC classes of non-irrigated arable land and pastures. Since none of these agricultural typologies are actually present in the Sihl valley, it has been assumed that arable lands and pastures should be classified as vegetables, with similar thresholds.

Moreover, for the sake of simplification and according to the overall scope of the analysis, it has been assumed that the crops in the Sihl river valley have similar growing patterns (low growing plants) and, therefore, the same susceptibility score. The agriculture-related risk (R_5) has been calculated by considering that above that crop-specific thresholds the harvest is lost, while below that level a non-destructive inundation is reported.

As a result, the risk for the agricultural cluster is very limited: the destroyed agricultural area amounts to 0.59 km^2 (around 8% of the total agricultural coverage). Out of this, 0.53 km^2 belongs to the non-irrigated arable land class and 0.07 km^2 to the pasture class. The most affected districts are Albisrieden and Leimbach. It should be noted that the total surface at risk is probably underestimated because the exposure classification has been performed according to the CLC resolution, i.e. some small agricultural areas, relevant for selected cash crop cultivation, could have been neglected. The area of the Sihl river valley is mainly devoted to residential and commercial purposes, therefore agriculture can be considered less important than other targets such as people, buildings and infrastructures.

19.4.3.3 Natural Systems

The flood hazard assessment to natural and semi-natural systems considers the coverage of the flooding area as relevant metric. Moreover, in order to perform the susceptibility assessment and therefore to evaluate the degree to which the receptor could be affected by the 300 years flood event scenario, the environmental pattern of the area should be classified. The valley is characterized by two different kinds of forest systems: coniferous (0.21 km^2), and mixed forest (19.98 km^2), which occupy most of the natural environment along the Zurich lake and valley.

The risk for natural and semi-natural systems has been calculated by considering vegetation cover, slope and soil permeability as susceptibility factors (no wetlands are present along the river valley). Each susceptibility indicator has been classified and scored using expert judgment. The susceptibility indicators have been aggregated by means of a Multi-Criteria Decision Analysis (MCDA) function named “Probabilistic Or” (Kalbfleisch 1985), in order to provide a single normalized score for homogeneous areas (Ronco *et al.* 2014). Finally, the hazard and the susceptibility scores have been aggregated into a relative risk score that supports the identification and prioritization of natural and semi-natural systems at risk. The application of the KR-RRA methodology to the Sihl river valley identified that only a limited portion of forest is at risk of inundation (0.29 km^2, 1.4 % of total forest areas), with two classes of risk being identified: a very small part (625 m^2) belongs to the high risk class, while the rest (around 289,000 m^2) belongs to the very high risk class.

Based on these results, the flood risk for natural and semi-natural areas along the Sihl river valley can be considered as not relevant. This conclusion is justified by the particular pattern of some key susceptibility factors, in particular the impermeable ground of the area and the degree of slope. In fact, forests are generally stable and resilient ecosystems. Growing along rivers, they are usually very well adapted to occasional and seasonal flooding. In this sense, the ecological, recreational and economic functionality of the Sihl valley forest ecosystem is not seriously compromised by a flood event of such magnitude (Ronco *et al.* 2015).

19.4.3.4 Cultural Heritage

The flood hazard assessment identifies the flood coverage for the 300 year return period scenario as relevant metric. The exposure assessment requires the localization of the cultural heritages in the case study area. In the Sihl valley, 416 cultural assets are currently present, mainly classified as ancient buildings. They include different historical churches, such as Fraumuster and Grossmunster in the Zurich city centre, the Swiss National Museum, the central library of Zurich, the Rathaus (the municipal building), the Synagogue, the Operahaus, several ancient residential buildings and villas in the centre as well as along the Zurich lake.

The susceptibility assessment assumes a score equal to 1 for all the targets (this means that all the cultural heritages are impacted in the same way). Accordingly, the cultural heritage-related risk (R_7) has been calculated from the intersection between the flood coverage map and the spatial pattern of the selected cultural assets. As a result, 40 items are at risk, corresponding to 9.13% of the total within the area (416 items). These assets belong to different cultural protection levels (regional and cantonal). In this framework scenario, the Swiss National Museum is at risk of inundation, while the districts belonging to a higher risk class (with 10 to 15 inundated items) are Langstrasse (close to the Zurich city centre) and Langnau am Albis (along the lower Sihl valley).

19.4.4 Total Risk

The total risk index has been calculated by aggregating the different receptor-related risks, by means of the MCDA weighted average equation for identifying and classifying areas and hotspots at risk across the studied area. The weighting process has been implemented during a roundtable meeting organized with several experts involved in the project, and the assigned weights are as follows: Infrastructures (0.8), buildings (0.6), people (0.4), agriculture (0.2), cultural heritage (0.1), and natural and semi-natural systems (0). It is noteworthy that the higher importance (and weights) have been assigned to those targets considered to be most relevant for the Sihl river valley, namely infrastructures, buildings and people (Ronco *et al.* 2015).

The total risk map shows the spatial pattern of flood risk within the studied area (Figure 19.4). The total area at risk extends for 7.98 km^2 and the total risk index ranges between $0.6 \bullet 10^{-5}$ and 0.24, which mainly represents the lower class of risk (highest risk score is 1).

In order to better visualize the spatial distribution of risk belonging to these classes, the green to red colour classification, normally tuned within the 0–1 range, has been re-tuned according to the calculated range (Table 19.3). The total risk map points out that Langstrasse district and part of the city of Zurich are subject of the highest risk; as well, areas within the districts of Werd, Sihlfeld, Alt-Wiedikon and Friesenberg, that are located next to the Sihl river course, also show relatively high risk levels. Despite being very dependent on the weights assigned to each receptor, the results are plausible because they demonstrate for the study area that the overall risk of flooding is higher around the main station of Zurich, where the most important hub for infrastructures of communication and transport (railway lines and buildings) is located.

19.5 Discussion

The total risk index allows for the identification and ranking of hotspots and areas at risk to support an appropriate science- and evidence-based land planning and decision making process. However, users must be aware of the simplified approach (participatory weighting process) that has been followed to produce this spatially distributed index. Indeed, the set of weights selected for the various targets could be subject to criticism. However, the expert judgements are based on their own experience during past events and are thus a valuable and important source of information. For the Zurich case study, in fact, the forensic approach to analyse past events and examine the complex and underlying causes, dynamics and consequences of disasters (e.g. Burton, 2010) has been performed in close collaboration with local stakeholders, authorities, and infrastructure providers by means of a participatory process (bottom-up approach) to come up with a site-specific configuration of weights, thereby improving the adaptation to the local situation. Moreover, interdependencies and criticalities needed to determine potential cascade effects and to better identify vulnerabilities and synergies among the different targets have been analysed and taken into account for a realistic and relevant setting using a robust ranking process. Therefore, when discussing the importance of different factors, it was agreed that infrastructure might be a major source of risk relative to human life mainly because of the configuration of the site, the legislative context, the level of protection due to structural and non-structural measures, and finally the history of past events (see Ronco *et al.* 2015). In addition, it is worth noticing that the ultimate decision-making process should also consider the factors that contributed to determining that particular final risk index value (i.e. susceptibility indicators, hazard metrics). A correct

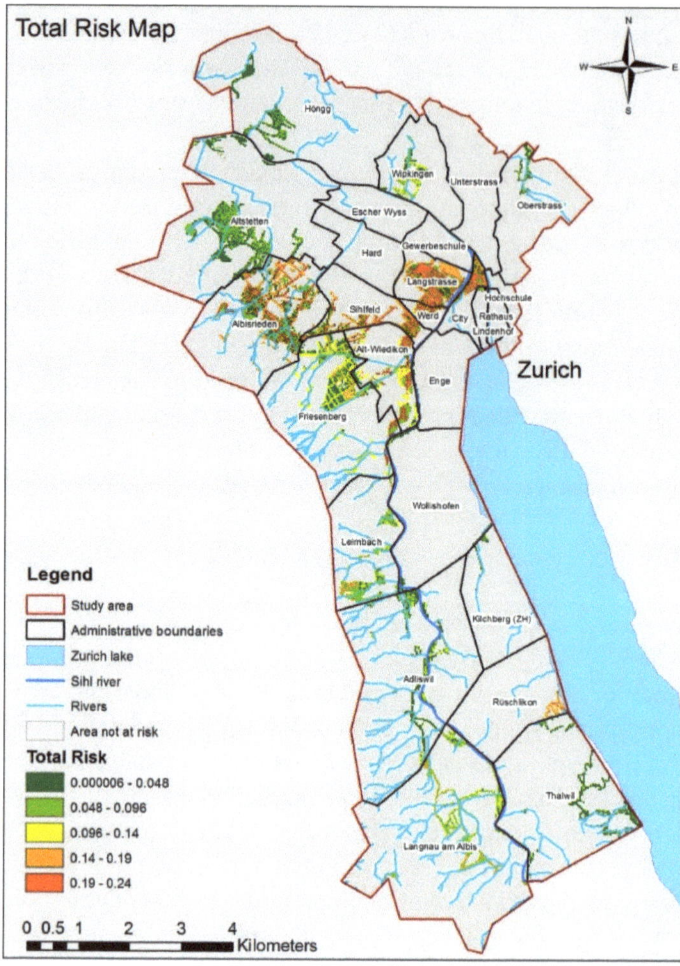

Figure 19.4: Total risk map for the Sihl river valley, considering the 300 year return period scenario, from Ronco *et al.* 2015.

Table 19.3: Total risk index classification and range of values.

Total Risk Classes	Score
Very low	0–0.048
Low	0.048–0.96
Medium	0.096–0.14
High	0.14–0.19
Very high	0.19–0.24

interpretation of these factors is particularly relevant for the analysis of the potential prevention measures that could be suitable for reducing the risk for current hot spot areas (Torresan *et al.* 2012).

19.6 Conclusions

This study addresses the application of the KR-RRA methodology to a site-specific case, the flood risk for the Sihl river valley including the city of Zurich, in Switzerland. The methodology allows for the identification and ranking of areas at risk in an urbanized and semi-urbanized area, not attempting to provide absolute predictions about flood impact. Rather, it supports the classification of the sub-areas (hot spots) as well as selected targets (among people, economic activities, natural and semi-natural systems and cultural heritage) that are more vulnerable and possibly more dramatically affected by the flood risk within the investigated region.

The proposed methodology is applicable within different contexts, case studies and spatial scales, with the aim of providing a benchmark for the implementation of the Floods Directive at the European level. The application of the KR-RRA at the meso-scale provides a screening analysis to support the assessment and prioritization of targets and, therefore, the selection of appropriate structural and/or non-structural measures and systems at the regional level, according to different (baseline and alternative) scenarios. However, a more detailed analysis (at the micro-scale) could be required in the areas at higher risk or where more specific information is available. GIS-based maps could be used to communicate the potential implications of floods in non-monetary terms to stakeholders and administrations. In the framework of spatially explicit and knowledge-based management of flood risk, the KR-RRA facilitates the quantification, in physical terms, of the risk avoidance related to the implementation of the (proposed) prevention measures, and considered by the different scenarios and settings. In fact, the methodology can provide information about the number of inhabitants, the type of economic activities, natural systems and cultural heritages potentially affected by flooding (Ronco *et al.* 2014) and, therefore, it demonstrates that prevention can be accountable and the benefits measureable.

Moreover, it is worth noticing that the final total risk index aggregates different scores coming from multiple heterogeneous parameters. The final decision-making process should therefore consider also the factors that contributed to determining that value (i.e. susceptibility indicators, hazard metrics). A correct interpretation of these factors is particularly relevant for the analysis of the potential prevention measures that could be suitable for reducing the risk for current hot spot areas.

19.7 Acknowledgements

This work was found by the Seventh Framework Programme (FP7) of the European Commission within the collaborative project "Knowledge-based approach to develop a culture of risk prevention (KULTURisk)", FP7-ENV-2010, Project 265280; www.kulturisk.eu. We thank Tessa Hegetschweiler (WSL) for assistance with data collection.

References

Addor, N., S. Jaun, F. Fundel, and M. Zappa. 2011. An operational hydrological ensemble prediction system for the city of Zurich (Switzerland): skill, case studies and scenarios. *Hydrology and Earth System Sciences* 15: 2327–2347.

Annamo, E., and G. Kristiansen. 2012. Challenges in flood risk management planning. An example of a flood risk management plan for the Finnish-Norwegian river Tana. Olso: Norwegian water resources and energy directorate. Report 16. Issn:1501–2832.

Apostolakis, G.E. 2004. How useful is quantitative risk assessment? *Risk Analysis* 24(3): 515–520.

Buchecker, M., M. Ferri, M. Mukolwe, J. Neal, and S. Rusjan. 2011. Risk prevention policy framework in the considered case studies. KULTURisk project deliverable 1.5.

Bründl, M., H.E. Romang, N. Bischof, and C.M. Rheinberger. 2009. The risk concept and its application in natural hazard risk management in Switzerland. *Natural Hazards Earth System Sciences* 9, 801–813.

Burton, I. 2010. Forensic disaster investigations in depth: A new case study model. *Environment Magazine* 52(5): 5.

Cirella, G.T., E. Semenzin, A. Critto, and A. Marcomini. 2014. Natural hazard risk assessment and management methodologies review: Europe. In *Sustainable cities and military installations: climate change impacts on energy and environmental security*, ed. I. Linkov, 329–358. Springer.

Clausen, L., and P.B. Clark. 1990. The development of criteria for predicting dam break flood damages using modelling of historical dam failures. In: *International conference on river flood hydraulics*, ed. W.R. White, 369–380. Wallingford: John Wiley & Sons Ltd.

DEFRA. 2006. Flood risk to people phase 2, FD2321/TR2 Guidance document. London: Department for environment, food and rural affairs.

EEA-European Environment Agency. 2007. CLC2006 technical guidelines. Luxembourg: Publications office of the European Union.

EP-European Parliament. 2007. Directive 2007/60/EC on the assessment and management of flood risks. Brussels.

Forte, F., L. Pennetta, and R.O. Strobl. 2005. Historic records and GIS applications for flood risk analysis in the Salento peninsula (southern Italy). *Natural Hazards and Earth System Sciences* 5: 833–844.

Forster, S., B. Kuhlmann B., K.-E. Lindenschmidt, and A. Bronstert. 2008. Assessing flood risk for a rural detention area, *Natural Hazards Earth Syst. Sci.* 8: 311–322.

Giupponi, C., V. Mojtahed, A.K. Gain, C. Biscaro, and S. Balbi. 2014. Integrated risk assessment of water related disasters. In *Hydro-meteorological hazards, risks, and disasters*, ed. P. Paron, and G. Di Baldassarre, 163–200. Elsevier.

Hall, J.W., R.J. Dawson, P.B. Sayers, C. Rosu, J.B. Chatterton, and R.A. Deakin. 2003. Methodology for national-scale flood risk assessment. *Proceedings of the ICE-Water and Maritime Engineering* 156(3): 235–247.

Hilker, N., A. Badoux, and C. Hegg. 2009. The Swiss flood and landslide damage database 1972–2007. *Natural Hazards and Earth System Science* 9: 913–925.

Hirabayashi, Y., R. Mahendran, S. Koirala, L. Konoshima, D. Yamazaki, S. Watanabe, H. Kim, and S. Kanae. 2013. Global flood risk under climate, *Nature Climate Change* 3: 816–831.

IPCC-Intergovernmental Panel on Climate Change. 2012. *Managing the risks of extreme events and disasters to advance climate change adaptation. A special report of working groups I and II of the IPCC*. Cambridge and New York: Cambridge University Press.

Jonkman, S.N., J.K. Vrijling, and A.C.W.M. Vrouwenvelder. 2008. Methods for the estimation of loss of life due to floods: a literature review and a proposal for a new method. *Natural Hazards* 46: 353–389.

Kalbfleisch, J.G. 1985. *Probability and statistical inference. Vol. 1-Probability*. Springer.

Kubal C., D. Haase, V. Meyer, and S. Scheuer. 2009. Integrated urban flood risk assessment – adapting a multicriteria approach to a city. *Natural Hazards Earth System Science* 9, 1881–1895.

Landis, W.G. ed. 2005. *Regional scale ecological risk assessment. Using the relative risk model*. Boca Raton, FL: CRC Press.

Mazzorana, B., L. Levaggi, M. Keiler, and S. Fucha. 2012. Towards dynamics in flood risk assessment. *Natural Hazards Earth System Sciences* 12, 3571–3587.

Meyer, V., S. Scheuer, and D. Haase. 2009. A multicriteria approach for flood risk mapping exemplified at the Mulde river, Germany. *Nat. Hazards* 48:17–39.

Pro Sihltal: Lebensraum Sihl, Jahrheft Nr. 58/2008, Druckerei Studer AG, Horgen, Adliswil, 2008.

Ronco, P., V. Gallina, S. Torresan, A. Zabeo, E. Semenzin, A. Critto, A. Marcomini. 2014. The KULTURisk regional risk assessment methodology for water-related natural hazards - Part 1: Physical-environmental assessment. *Hydrol. Earth Syst. Sci.* 18: 5399–5414. doi: 10.5194/hess-18-5399-2014.

Ronco, P., M. Bullo, S. Torresan, A. Critto, R. Olschewski, M. Zappa, and A. Marcomini. 2014b. The KULTURisk regional risk assessment methodology for water-related natural hazards - Part 2: Application to the Zurich case study. *Hydrol. Earth Syst. Sci. Discuss.* 11: 7875–7933. doi: 10.5194/hessd-11-7875-2014.

Santato, S., S. Bender, and M. Schaller. 2013. The european floods directive and opportunities offered by land use planning. Hamburg: CSC report 12, Climate service center. http://www.climate-service-center.de/imperia/md/content/csc/csc-report_12.pdf. Accessed 3 Oct. 2015.

Schwanbeck, J., D. Viviroli, R. Weingartner, I. Röser, and J. Trösch. 2010. A statistical model derived from process-based modelling to assist flood risk minimising reservoir operation in a pre-alpine flash flood prone catchment. EGU General Assembly, Vienna, 2–7 May 2010, EGU2010-4432-1.

Torresan S., A. Critto, J. Rizzi, and A. Marcomini. 2012. Assessment of coastal vulnerability to climate change hazards at the regional scale: the case study of the North Adriatic sea. *Natural Hazards Earth System Sciences* 12: 2347–2368.

Luca Franzi[54], Gennaro Bianco[55], Alessandro Bruno[56], Sara Foglino[56]

20 Flood Risk Assessment and Quantification in the Piedmont Region, Italy

Abstract: A major shift in flood management strategies is currently underway in many countries throughout the world. After the approval of the 2007/60 directive on floods, the Italian government empowered the Po Basin District Authority to carry out certain activities that correspond to the actions required by the 2007/60 EU directive on flood risk management. The quantitative flood IRP risk assessment methodology has been proposed by Regional Piedmont Administration, in collaboration with the Politecnico of Turin, with the aim of developing a system to quantify flood risk throughout the entire Piedmont region. The IRP methodology is described in this chapter. Some study cases on quantitative risk assessment regarding the Po basin territory, in the Turin, Susa and Arona municipalities, and in the Orco watershed basin are discussed. The proposed IRP methodology can be considered a first step toward a quantitative analysis of risk and a valuable means of supporting decision making. It is currently being developed to help decision makers compare different political strategy options and improve risk mapping in preparation of the next review foreseen by the EU directive.

Keywords: Hazard, Exposure, Vulnerability, Risk index, Piedmont

20.1 Introduction

A major shift in flood management strategies is currently underway in many countries throughout the world. The notion of flood risk management (FRM) is well established, as:

> "The systematic application of management policies, procedures and practices to the tasks of identifying, analysing, assessing, mitigating and monitoring risk" (WMO 2009).

The notion involves many elements, from integration with the management of water to emergency flood management, and from local hazard assessment to the adoption of the best mix of strategies at a basin scale (Klinke *et al.* 2002).

The effective implementation of *FRM* requires an "enabling environment", in terms of policy, legislation and information.

54 Regione Piemonte, Italy, luca.franzi@regione.piemonte.it
55 DIATI-Politecnico of Turin, gennaro.bianco@polito.it
56 Eng. Professional consultant

From an organizational point of view, the definition of clear institutional roles and functions is a central point in taking the necessary steps for FRM implementation. As is well known by regional and local administrations, the nature of floods and the definition of FRM strategies generally create a complex situation of competing requests, and, particularly after major floods, poses some difficult questions about the kinds and the hierarchy of interventions necessary for risk management pertaining to the choice of the most appropriate strategy, the targets and priorities, as well as balancing costs and benefits.

In this context, the Po Basin District Authority (PBDA), in collaboration with governmental agencies (AIPO Agency) and the regional authorities, plays the most important role in decision making in the Po basin.

The legal framework consists of legislation act no. 183/1989 (Italian Parliament, 1989), which enables the basin district authorities to implement PAI – the Hydrogeological Asset Plan throughout the entire Italian territory. After the approval of the 2007/60 directive on floods, Italian government act no. 49/2010 (Italian Government 2010) empowered the PBDA to carry out certain activities that correspond to the actions required by the 2007/60 EU directive on flood risk management.

The first activity is related to the preliminary assessment of flood hazards and risks. The assessment includes maps of river basin districts, a description of the floods that had occurred in the past and an assessment of the potential adverse consequences of future floods on health, economy and culture. The second activity consists of the identification of the high-risk flood-prone areas. This activity has concluded in December 2013, with the publication of flood risk maps. The third activity is focused on the implementation of risk management maps, and concluded in 2015.

Some issues related to the second activity are discussed in detail in the following sections.

The Piedmont Region Administration (RPA) has already published qualitative risk assessment maps pertaining to the entire territory. The dissemination and confrontation phase, according to the flood directive roadmap, is underway, with the local administrations and the general public.

A quantitative flood risk assessment methodology has been proposed by RPA, in collaboration with the Politecnico of Turin, with the aim of developing a system to quantify flood risks throughout the entire Piedmont region (which covers about 25,400 km^2), on the basis of the available knowledge and resources and within the frame of the aims and purposes of the 2007/60 EU directive, according to which

> "In order to have an effective tool for information available, as well as a valuable basis for priority setting and further technical, financial and political decisions regarding flood risk management, it is necessary to provide for the establishing of flood hazard maps and flood risk maps".

Quantitative risk assessment is of particular interest, considering that the total number of people living in flood prone areas (about 900 km^2 for the 500 return period inundation) is about 700 thousand, that is, about the 16% of the total population.

The most recent relevant floods in 1993, 1994 and 2000 have consequently drawn the attention of administrations to soil use and consumption, and have led to the revision of municipality territorial regulator plans throughout the region.

The research results proved to be useful for practical application by technical practitioners (mainly engineers and geologists) and by public administrations, as a valuable basis for decision making. The different roles played by the factors that influence risk assessment, including climate change, will be discussed thereafter. Some study cases on quantitative risk assessment regarding the Po basin territory, in the Turin, Susa and Arona municipalities, and in the Orco watershed basin will also be discussed (Figure 20.1).

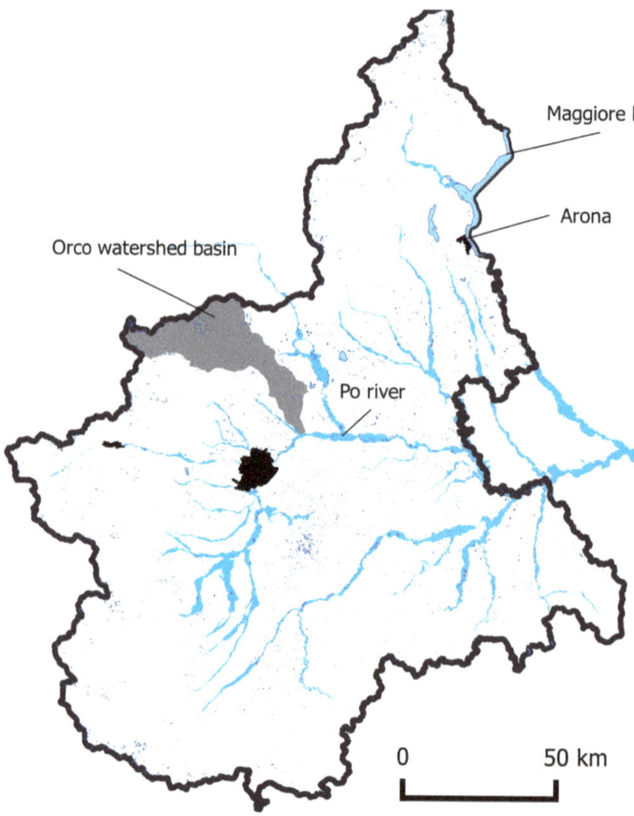

Figure 20.1: Map of the Piedmont region, showing the study case areas discussed in the chapter, i.e. the Turin, Susa and Arona municipalities and the Orco watershed basin.

20.2 Risk Assessment Quantification

From a technical viewpoint, the following definition is here assumed (WMO 2009; Franzi 2012) for flood risk:

> "potential losses associated with a hazard or an extreme event in a given place within a given period of time, which can be defined in terms of the adverse consequences (damage/losses) and the probability of occurrence".

Therefore, the concept of risk necessarily implies the concept of loss, the probability of flood occurrence, the intensity of the phenomena, the damage produced and the vulnerability of the anthropogenic context. Risk can be considered the superimposition of three factors, which are generally indicated as Hazard (H), Exposure (E) and Vulnerability (V) (Figure 20.2).

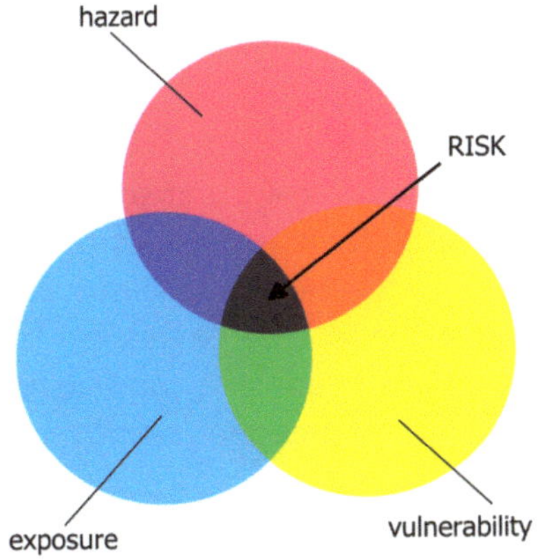

Figure 20.2: Risk considered as the superimposition of three constitutive factors, H (hazard), E (exposure), V (vulnerability) (Franzi, 2012).

As shown in Figure 20.2, risk assessment should entail the assessment of each constitutive factor, its variability in time and space, the complexity of natural/social/organizational factors and the social resilience against floods and natural disasters.

On one hand, this would allow a wide and comprehensive risk assessment to be made and thus provide a good framework for territorial planning and civil protection countermeasures, and this would also allow the best management strategies to be detected.

On the other hand, in risk assessment quantification and mapping, each national/regional/local Administration, including the district authorities, faces the problem of constraints that substantially condition the decision makers in their choice of the most appropriate methodologies, as well as of the scale and of the extension of the territories where the flood risk is assessed, mapped and evaluated. The main constraints that condition public administration depend on the context (geographic, administrative) in which the directive is implemented, that is:

- the available time to fulfil legislation deadlines: the 2007/60 EU directive imposed production of flood risk maps by members by December 2013;
- the available information and datasets, such as the total amount of information and data available at the time the administration starts to work; when planning its own activities, a public administration generally assumes that the discovery of new relevant resources (or deposits of known resources) should be excluded or is improbable;
- the available resources, i.e. at the time the work starts; resources should be intended from an extensive point of view from human resources (i.e. the number of workers that can work in the implementation of the directive, their technical abilities and formation, such as engineers, geologists etc.) to software resources (availability of free software, computing time, CPU computing capacity, total number of personal computers) and to economic resources (with the possibility of outsourcing some work).

Therefore, the context extension (i.e. the total extension of the areas where the risk should be assessed and mapped) and the detail at which it can be mapped or assessed is a consequence of a careful balance of requirements. Several scientific aspects can have different relevance as far as organizational and time factors are concerned (Figure 20.3). Rough estimations over a short time can sometimes prove to be preferable to detailed investigations, especially when they are time and resource consuming.

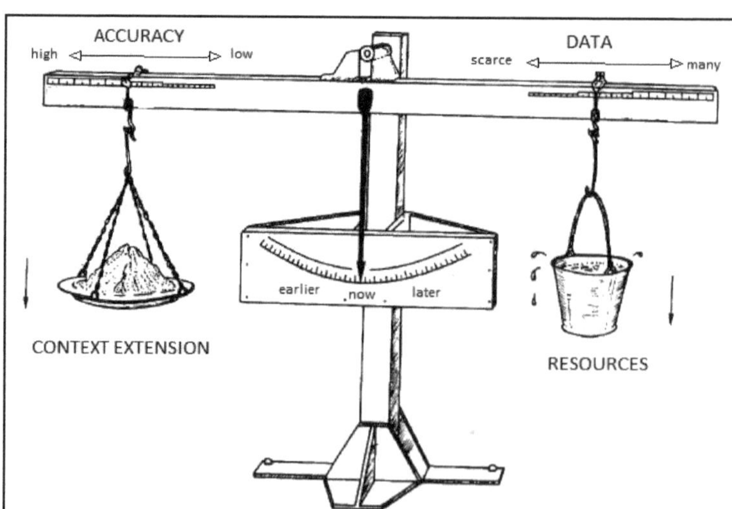

Figure 20.3: The so-called Lane balance (Lane 1955), revised according to risk assessment mapping.

As will be shown in the following sections, scenarios concerning the impact of climate change on precipitation, water runoff, flood frequency and flood intensity are probably affected most by great uncertainties, and often the results obtained by means of climatic simulation models should be confirmed by additional investigations over time (STRADA 2013; ARPA 2007).

An example of the necessity of simplification in order to fulfil the requirements of the 2007/60EU Directive is that of the complexity of concepts, in particular the concept of damage and losses. Theoretically, as far as damage is concerned, it should be pointed out that direct and indirect losses should be considered equally. Direct losses concern human life, structural losses and the loss of functionality of the anthropogenic structures, such as houses, bridges, levees, roads, dams, etc. Similarly, indirect losses such as:

1. the psycho-social impact, that is, the psychological effects on people affected directly or indirectly by the flood due to the loss of property or of livelihoods, the displacement from one's home, the disruption of economic, family and social affairs;
2. the functioning disruption, that is, the interruption of interconnections between people, services and webs, such that people and economic activities that are far from the place where the flood event occurs also suffer from the effects of the break in interconnections; this is the case, for example, for oil pipelines, water pipelines, railways;
3. the economic impact, that is, the hindering of economic growth and development, due to the high cost of relief and recovery, which may have an adverse impact on investments in infrastructures and the development activities in the area; both private and public sectors are generally discouraged from investing in recurrent high flooding conditions;
4. the economic cost of the emergency countermeasures taken for civil protection purposes and other actions taken to prevent flood damage and other losses;

should all be considered, but their evaluation and quantification may take time and this may not be compatible with the deadlines given in the 2007/60 EU directive.

Therefore, a robust (Marijolein *et al.* 2011) risk assessment would be preferable for public administrations in charge of complex and long-term projects.

This research, in collaboration with the Politecnico of Turin, started with the definition of the constitutive elements of the risk concept. A discussion about the definitions of the H, E, V terms is provided in the following sections.

20.2.1 Hazard

Hazard (H) can be defined as the probability of the occurrence, within a specific period of time in a given area, of a potentially damaging natural process (UNDRO 1980), of a specific intensity.

The regional public administration in Piedmont has extensively debated flood hazards with local administrations at a provincial and municipal scale since 2001, with the implementation of PAI (PBDA 2001), basing the hydraulic and hydrological analysis on the different available data.

By applying hydraulic and geomorphologic models, the PBDA has proposed a classification of hazards that is mainly based on a river corridor concept (Figure 20.4), a concept that has been implemented since 2001 with the PAI, and which was modified after the implementation of the 2007/60 flood directive; the river corridors are as follows:

- corridors with a low flooding probability (L): this corresponds to areas that can be flooded by a design discharge of 500 years, Q_{500};
- corridors with medium flooding probability (M): this corresponds to areas that can be flooded by a design discharge of 200 years, Q_{200};
- corridors with high flooding probability (H): this corresponds to areas that can be flooded by a design discharge of 50 years, Q_{50}.

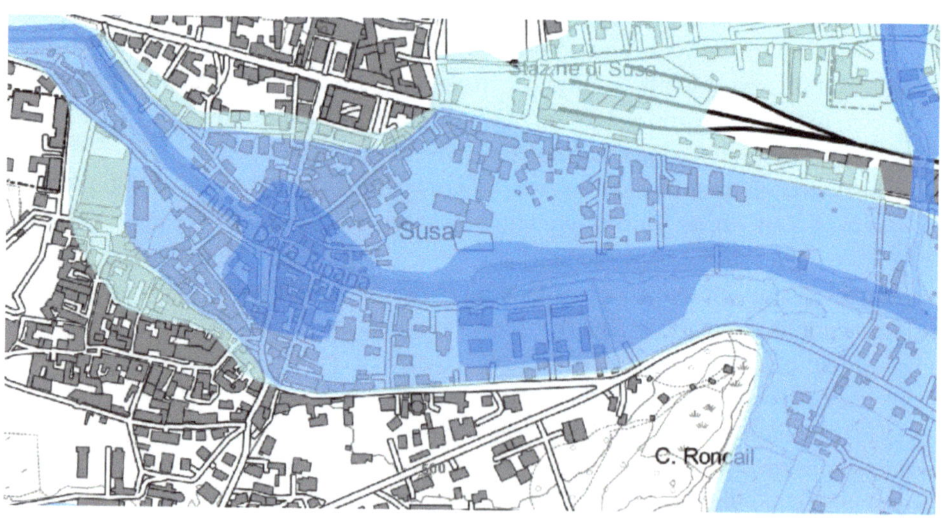

Figure 20.4: River corridor approach, developed by the PBDA, referring to the Susa municipality. Legend for hazard: (High)=dark blue; (Medium)=blue; (Low)=light blue (free online at http://osgis2.csi.it/webgisAtlante/qgiswebclient.html?map=qgis_cloud/direttiva_alluvioni).

According to the most recent evaluations, published online by RPA (hazard maps) and available for public confrontation, the H, M, L corridors extend over an area of about 400 km^2, 580 km^2 and 900 km^2, respectively.

According to (Maione *et al.* 1976), the mean occurrence probability per year of a given Q_{RP} flood is given by:

$$P = 1/RP \tag{1}$$

All the information available on the corridor mapping (available online at Regione Piemonte website), has been taken into account, especially that already contained in PAI, which has been in force since 2001. In this way the hazard maps, in the implementation of 2007/60 EU directive, are based on the hydrological evaluations and hydraulic models that are already available at a basin and local scale, and which are currently in force, amended in time and shared between public administrations.

An example of a hazard mapping methodology has been proposed by Franzi and Rinaldi who work at RPA (PBDA 2012). This methodology is based on the interpolation of 1D hydraulic simulation results, in order to obtain flooding inundation maps in a reasonable time and on the optimization of the information available on flood inundations (PBDA 2012). The approach is similar to those proposed in FEMA and IACWD (IACWD 1982; Noman *et al.* 2001; FEMA 2003; Merwade 2008) for flood inundation mapping. It involves the following procedure:

1. The design flow Q_{200} (i.e. the 200 year return period – RP -flow) is estimated using a calibrated hydrologic model and precipitation input, or through statistical analysis; as far as Piedmont rivers are concerned, this means the estimates proposed by ABDPO in the Design Flood Directive (PBDA 2001) (currently in force with amendments);
2. A water surface elevation from the hydraulic model is mapped on a digital terrain model, and a water surface (usually a triangulated irregular network – TIN format) is created;
3. The digital terrain model (DTM) is subtracted from the water surface to obtain a water-depth map[57];
4. The area with positive values in the water-depth map leads to the flood inundation map.

Flooding results are compared to an available dataset and maps, and checked by means of (1) aerial photo-interpretation and (2) geomorphologic river assessments, based on recent and past river channel changes.

57 In this chapter, the digital terrain model is intended as the representation of the bare ground surface without any objects, such as plants or buildings.

Figure 20.5: Flooding map methodology pertaining to the Orco River. A water surface elevation (middle) is superimposed onto a DTM (below). Subtraction of the two layers allows the water-depths to be obtained.

The methodology allows flood maps to be obtained in a reasonable time and optimizes the best resources available at present. Applications have shown that the results are basically affected by uncertainties on the ground elevation data, which are generally obtained by means of LIDAR and the semi-automatic procedures that allow a DTM to be obtained, starting from raw DSM data (see e.g. Gerstenecker *et al.* 2005).

Other flood mapping uncertainties arise from uncertainties in hydrology and hydraulics, which are, in turn, affected by the effects of climate change on rainfall (Merz *et al.* 2009; Todini 2007; Klinke *et al.* 2002).

The results from hydrological studies conducted according to the analysis adopted by the Environmental Regional Protection Agency (ARPA 2007) on climate change, have so far been inconclusive regarding the impact of climate change on rainfall and runoff. Different studies have been performed by ARPA to draw up scenarios on the effects of possible climate changes on rainfall, runoff, temperature, sea level, glacial extension and snow coverage, based on simulation models (Tebaldi *et al.* 2006). These scenarios can be useful to suggest and implement strategies that can be used to cope with climate changes, to adapt to meteorological variations and to climate variability, and therefore can be considered a useful tool for decision makers to visualize the possible effects of climate.

ARPA's conclusions have been confirmed from the results of the recent STRADA project (FESR project; STRADA 2013), which was focused on regional hydrology and hydro-meteorological models of Piedmont and Lombardy; the project did not lead to the definition of robust conclusions:

> "The study of the temporal rainfall distributions has actually highlighted elements of non-stationarity, due to either an increase in observational data, or to variations in climate, although it is difficult to discriminate between stable trends and fluctuations in the medium term (multidecadal)".

Recent hydrological research in France (Dumas *et al.* 2013) on alpine watershed basins contiguous to the Piemonte Region has shown that the effects of climatic changes on the estimates of the 100-yr return period discharge: (1) depend to a great extent on the downscaling techniques that are used, and, for a given technique; (2) vary remarkably from region to region. In particular, the return period (RP) under climate change associated with the present 100-yr RP level, under the WT downscaling assumption (Figure 20.4 in Dumas *et al.*) can be much higher than 100-yr, implying that climate changes could cause 100-yr floods to be less frequent in the near future.

Similar conclusions have also been shown by Kundzewicz *et al.* (2010); the recurrence interval (return period) of today's 100-yr floods in most of Piedmont's watershed basins might be less than 100-yr, when the Hirabayashi *et al.* (2008) results are considered as a basis of investigation, or higher than 100-yr if the Dankers and Feyen (2008) emission scenario is considered.

Robustness of evaluation is a crucial requirement for institutions and public bodies in charge of formulating flood risk management strategies for the river Po. *Robustness* can be defined (Marjoleine *et al.* 2011) as the ability of a system to remain functioning under disturbances, where the magnitude of the disturbance is variable and uncertain. The proliferation of methods for uncertainty analysis should be placed within a coherent framework (Merz 2009). As well as estimating the amount of uncertainty associated with key decision variables, aids to decision making should identify the most influential sources of uncertainty, and the implications of uncertainty for the ordering of preference between options. Moreover, even without considering the uncertainties due to climate change, it is has been documented in literature that hydrological models suffer from uncertainties due to data incompleteness, etc.

Moreover, the uncertainties in climate change should be compared with those due to hydrological modelling obtained in steady climatic conditions.

A simple analysis of the uncertainty management due to the use of hydrological models has been proposed by Franzi and Rinaldi and implemented in act no. 2-11830 (RPA 2009). The analysis is based on hydrological estimates obtained by means of the VAPI model, which was developed by CNR-CUGRI (Villani 2001) and is based on the geospatial statistical approach. The VAPI model, which computes regional estimates of discharges Q_{200}, is free and available for practitioners and public administrations to compare the local hydrological Q_{200} estimations (obtained by means of an arbitrarily chosen hydrological model) with those of VAPI regional estimations. Two sets of estimations were considered in the analysis, for a dataset of 71 instrumented or not-instrumented watershed basins in the Piedmont region:

- the $Q_{200VAPI}$ set of estimations obtained by means of the VAPI model;
- the $Q_{200PBDA}$ set of estimations used by PBDA for flood mapping[58].

The $Q_{200PBDAi}$ and $Q_{200VAPIi}$ rates were compared for each basin in the dataset, in the following way:

$$\left(\frac{Q_{200\,VAPI}}{Q_{200\,PBDA}}\right)_i = f\,(Area)_i \tag{3}$$

where: A is the area of the i_{th} watershed, $i = 1,2,....71$, Q_{200} is the computed flood discharge, the $_{PBDA}$ index refers to the vales used for the implementation of the directive and the $_{VAPI}$ index refers to the VAPI model. As shown in figure 6, the rate given by (3) ranges over a wide interval, from a minimum value of 0.65 to a maximum value of 2.05.

By looking at Figure 20.6 and considering the deviance between the different estimates, different questions can arise: is the Q_{200} estimation uncertainty, associated with climate changes, higher than that associated with modelling? Under which hypothesis? For which watersheds? No conclusive answers have formally been given for the watershed basin in the Piedmont region. The method given in (3) shows that, heuristically, for the dataset regarding 71 basins and for a given hydrological model (VAPI), the official Q_{200} estimates can vary over wide range.

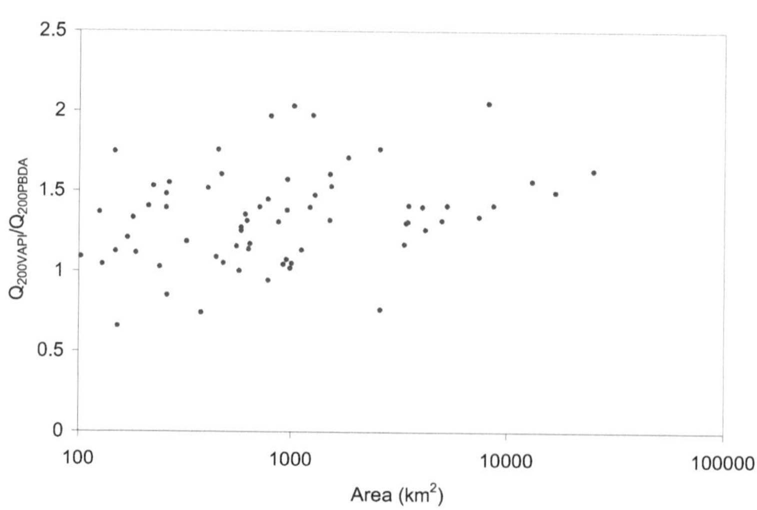

Figure 20.6: Comparison of the Q_{200} estimates, by means of two hydrological models.

58 The Q_{200} values were published in 2001 and have been updated over time.

Nevertheless, research conducted within the PBDA is underway to detect the most appropriate strategies to adapt flood risk management to possible climate changes (PBDA 2014). This has the aim of collecting recent hydrological data, evaluating possible trends in hydrology and possibly revising the 200-yr return discharge.

Evidence on soil use has induced decision makers to focus on the most relevant causes that have substantially and unquestionably increased risks in the Po watershed basin, i.e. the increasing number of receptors in the flood-prone areas, a process that started after the WWII and is presently ongoing. Quantification of the exposure (E) and of the vulnerability (V) of receptors is therefore a crucial task, as will be shown in the next section.

20.2.2 Exposure

The definition of exposure in literature generally converges to "elements at risk, or receptors, that is, people, properties and goods that can be lost, injured or damaged during an event" (UNDRO 1980). According to this definition, exposure varies according to the hazard level.

The RPA activities, which have been carried out in collaboration with the Politecnico of Turin, have focused on the exposure of structural elements at risk, which are generally referred to as receptors.

At present, the total number of receptors in the Piedmont region is systematically updated through an analysis of aerial pictures (photo-interpretation). Identification of the use of receptors, such as commercial/industrial/residential ones, is not straightforward, as the same receptor can refer to different uses with different economic relevance. The analysis has shown an increase in the total number of receptors over the last few decades (Figure 20.7) in their economic value. The latter is constantly updated by the Agenzia delle entrate (AdE).

The AdE website (OMI, observatory of the housing market) allows the available information of the economic values e_i of residential/commercial/industrial categories of receptors, expressed per square meter (m^2) to be downloaded free of charge. Each Italian municipality has been divided into zones, which are considered to be homogeneous from an economic point of view. Subcategories have been created for a given zone and for a given category (residential/commercial/industrial), with minimum and maximum economic e_i values. The OMI e_i values are constantly updated and can obviously be used to estimate a medium housing market value. Therefore the available data should obviously be used in a proper manner. Moreover, the OMI datasets may only be representative of the property values, and disregard the content value. Other damage (indirect or direct as described above) needs to be estimated by means of other parameters.

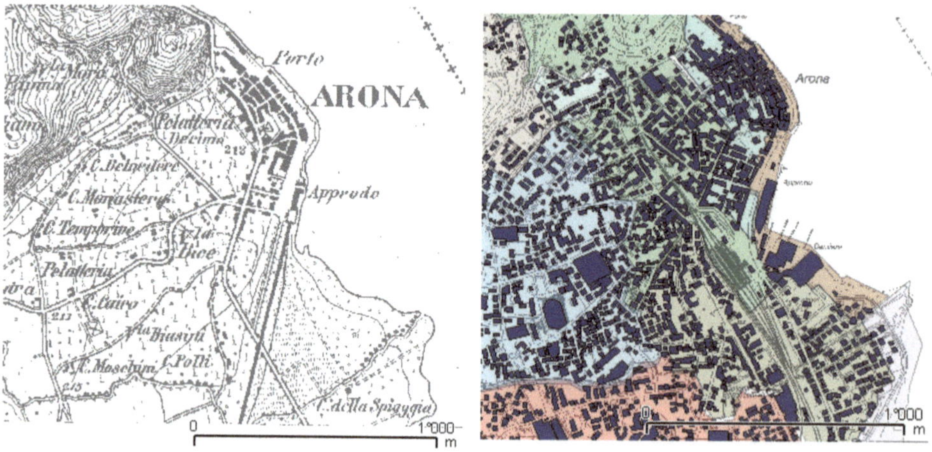

Figure 20.7: The figure shows Arona (lake Maggiore) in 1945 (left) and at present (right). The OMI zones have been highlighted with different colours (right). The receptors at the present time are shown in a dark colour (right). The inundation area is about 50 ha for TR = 500 years. A comparison of the figures shows a clear increase in the receptors.

Assuming that the total exposure of the *i-th* receptor is proportional to its area, disregarding indirect damage and considering the OMI zone datasets, a robust assessment of exposure E_i can be computed as follows:

$$E_i = e_i S_i \quad (4)$$

where e_i is the mean economic value given by the AdE per unit area (m^2) in the corresponding OMI zone and S_i is the total surface of the element at risk. The minimum and maximum values can vary over a wide range for the same homogeneous OMI zone, and the e_i values can vary from zone to zone. For example, e_i values vary significantly for civil housing in Arona in B2 OMI zone, from a minimum of 1,300 €/m^2 to a maximum of 1,850 €/m^2, while, in the *B1* zone, they vary from a minimum of 1,850 €/m^2 to a maximum of 2,750 €/m^2. In spite of this variability, the methodology described above, which refers to the mean value in a given OMI zone, allows a homogeneous evaluation to be made all over the entire region.

20.2.3 Vulnerability

From a technical viewpoint vulnerability is defined as:

> "the degree of a loss to a given element at risk, or set of such elements resulting from the occurrence of a flood with a given intensity" (UNDRO 1980).

Vulnerability is a function of the hazard level. As mentioned for the risk concept, vulnerability shows the same conceptual complexity (Franzi 2012) as risk, because structural, organizational and community vulnerability should all be considered and taken into account. Moreover, as far as structural vulnerability is concerned, no vulnerability curves have been proposed in the literature for the Italian situation or derived from the data regarding the documented property losses after floods. Therefore, vulnerability quantification is affected by heavy uncertainties.

As established in the STRADA project (STRADA 2013) a simplified formula for vulnerability, derived from those available in the literature (USACE 1992) has been implemented by RPA, according to which the vulnerability of the *i-th* element at risk is a function of the water depth *h*, i.e.:

$$V_i = V_i(h) \tag{5}$$

The water depth *h* is calculated at the barycentre of the receptor. For those receptors that are only partially enclosed within the flooding area, *h* is calculated by referring to the barycentre of the flooded area of the receptor.

The adopted vulnerability curve is that proposed by USACE (1992) (Figure 20.8), that refers to FIA (1970) two or more stories, with basement depth-damage curve (USACE 1992: 89).

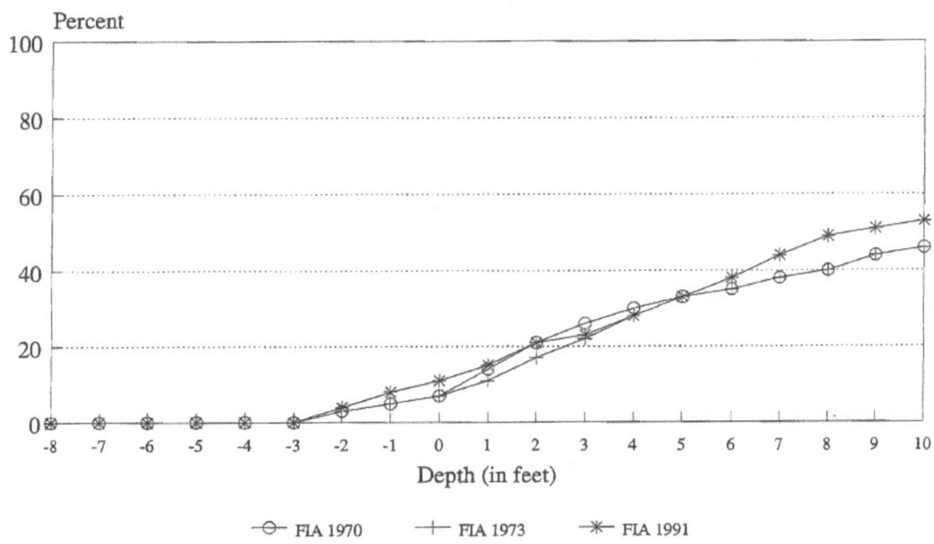

Figure 20.8: Two or more stories, with basement depth-damage curves (USACE 1992: 89). In the application that are shown in the chapter, the FIA (1970) curve has been used.

20.2.4 Application Cases

The described methodology, developed by RPA in collaboration with the Politecnico of Turin, has been systematically applied to some study cases, which were chosen from different study cases in the Piedmont region territory. Selection of the application cases was influenced by the relevance of the situation, with respect to risk, and by the availability of consistent data.

Risk has been mapped by superimposing the three risk components, described in the previous sections, which allowed an Index of Proportional Risk – IRP to be defined in the following way:

$$IRP_i = H_i E_i V_i = \frac{1}{RP} e_i D_i V(h)_i \tag{6}$$

where, together with the variables already defined and IRP is the proportional risk index, computed for the *i-th* receptor and for a flooding event which has a Q_{RP} flood discharge with return period RP.

The total risk due to the presence of *N* receptors can be computed, for a given watershed basin, or for a given geographic area, as:

$$IRP = \sum_i^N H_i E_i V_i = \sum_i^N \frac{1}{RP} e_i S_i V(h)_i \tag{7}$$

where *N* is the total number of receptors in the considered geographic area and IRP is expressed in €/year. A similar approach can be found in Hall *et al.* (2008).

The following will be shown hereafter:

- IRP quantification at the confluence of the Po river and Dora Riparia, in Turin (Bruno 2013);
- IRP quantification at the Dora Riparia river, in Susa (Foglino 2013);
- IRP quantification at lake Maggiore.

All the study cases have been conducted according to the procedure described above to obtain hazard maps and risk maps.

The risk assessment in Turin has been conducted with the aim of evaluating the total risk for a low (L) probability of occurrence due to flooding of the Dora Riparia at the confluence with the Po river, where houses, properties and even a University building are prone to flooding, and at present are not protected from inundation. The study is a good example of how risk assessment can be used as an effective tool for information and a valuable basis for priority setting and further technical, financial and political decisions regarding flood risk management.

The risk analysis refers to the 500-yr RP Dora Riparia flood discharge (catastrophic discharge), as evaluated by the PDBA and published in the Directive on design discharge. Downstream levels in the Dora Riparia exclude contemporaneous inundation by the Po river in the surrounding area.

The inundation map has been obtained (Figure 20.9), at the confluence with the Po river, by applying the same procedure described in §2.1, t. The inundation map only refers to the inundation due to the flooding of the Dora Riparia river, and does not take into account the simultaneous flooding of the Po river.

The study has proved to be useful to calibrate and modify the inundation maps obtained in 1991 and mapped in PAI (red lines in Figure 20.9), which refers to the state-of-knowledge at that time.

The inundation maps of the Dora Riparia and Po river, which have been updated and published by the PBDA in 2013, are free online.

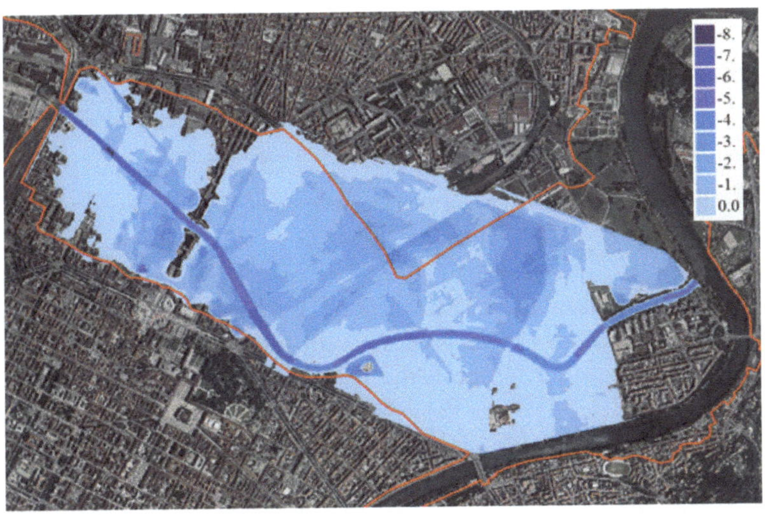

Figure 20.9: Inundation map due to Dora Riparia Q_{500} flooding. The depths are expressed in meters (Bruno 2013).

IRP has been calculated for each receptor according to the procedure described in § 3. The receptors have been divided into four risk classes (R1, R2, R3, R4) as a function of the maximum computed IRP value, according to the following table.

Table 20.1: *IRP* classes calculated on the basis of the *IRP* maximum value (taken from Foglino 2013; Bruno 2013).

IRP classes	Assignment criteria
R1	$0 < IRP_j / IRP_{max} \leq 0.25$
R2	$0.25 < IRP_j / IRP_{max} \leq 0.5$
R3	$0.5 < IRP_j / IRP_{max} \leq 0.75$
R4	$0.75 < IRP_j / IRP_{max} \leq 1$

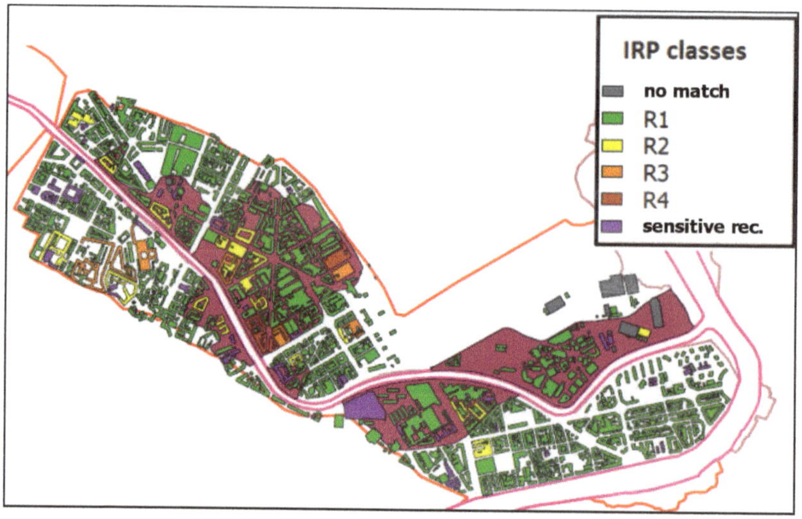

Figure 20.10: *IRP* map at the confluence of the Po and Dora Riparia rivers, RP = 500 years (Bruno 2013).

The sensitive receptors (e.g.: museums, libraries, schools, etc.) are not included in the risk class list in Table 20.1, because the analysis was only focused on economic and residential receptors. However, some receptors did not show a direct match with the receptor typologies in OMI.

The second case study deals with the Dora Riparia and in particular with the reach of the confluence of the Cenischia torrent that crosses the city of Susa, where some different protection alternatives have been considered and compared. In particular, a reduction in flooding (Foglino 2013) has been evaluated for each different option, on the basis of technical feasibility. The risk, after the implementation of countermeasures, has been mapped and compared with the existing risk.

Figures 20.11 and 20.12 represent the current state and the design state risk maps and the limit of the Q_{200} flooded areas. The design state maps include the reconstruction of a bridge, the superelevation of river banks and the reconstruction of a check-dam. The effects due to the designed structural option have been mapped, in terms of IRP, and compared with the current state (Figure 20.12).

The effects of the designed countermeasures are (Figure 20.12):

- a scale effect, due to the reduction in the total number of receptors affected by floods;
- a shift in the IRP distribution mode, which demonstrates that the most exposed receptors will draw benefits from the proposed design projects.

As far as Lake Maggiore is concerned, the study was principally motivated by the necessity of mapping hazards over a short period (three months). No official hazard maps were ever adopted for that region, up to 2013, and, in order to respect the

Figure 20.11: Map of *IRP* referring to the current state of the Dora Riparia in Susa (Foglino 2013).

Figure 20.12: Map of *IRP* referring to the design state of the Dora Riparia in Susa (Foglino 2013).

roadmap and deadlines established by PBDA, it was necessary to start a close collaboration between the public regional administration (Piedmont and Lombardy regional administrations) and Canton Ticino (Switzerland), which borders the lake. In this case, it was necessary to reach a political and technical agreement about the lake levels the hazard maps refer to, in order to publish hazard maps before the end of December 2013.

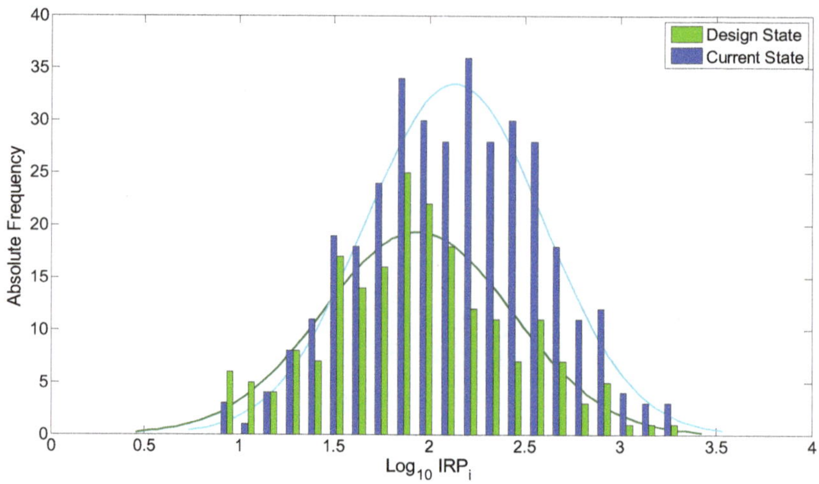

Figure 20.13: *IRP* comparison between the current and the design states. The lines represent Gaussian distributions (Foglino 2013).

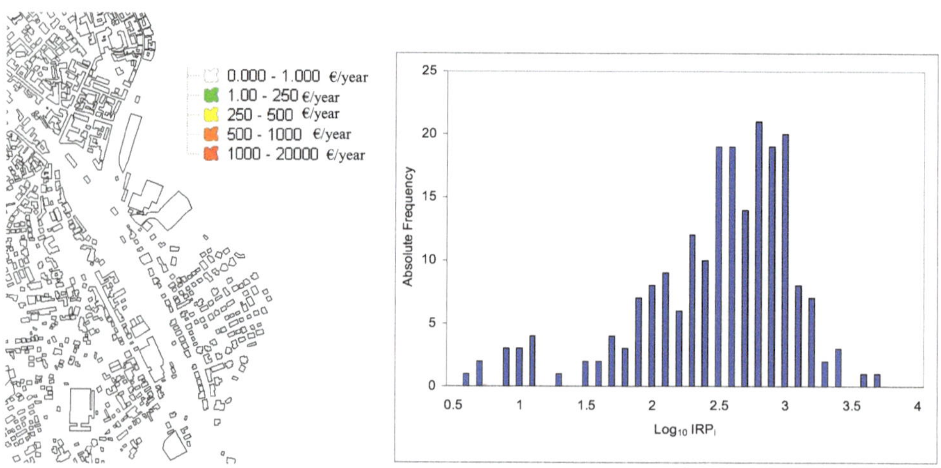

Figure 20.14: IRP Map and frequency distribution (expressed in log scale) referring to the receptors flooded in the Arona municipality. The values are expressed in €/year. The blue lines (left) indicate the inundation extension. Receptors partially enclosed in the inundation areas have been mapped.

Inundation simulations were made in static conditions, that is, assuming that the water gradually flows into the surrounding areas. Only the areas hydraulically connected to the lake were considered in the risk evaluation.

The IRP results for Q_{500} can easily be mapped using GIS software (Figure 20.13). The frequency distribution of the receptors versus IRP can be considered a useful tool to describe and understand the risk of flooding.

20.3 Conclusions

Risk assessment and mapping is a very complex topic and activity. It involves different expertise and makes it necessary for regional governments to set priorities, to define the scale and details of the investigations and to optimize the resources.

As far as the collaboration between RPA and the Politecnico of Turin is concerned, the studies that have been carried out, and completed, are a first step towards a comprehensive management of the risk of flooding. The proposed quantitative IRP methodology for risk assessment has proven to be useful for decision making and for the description of the flooding conditions, and has so far been extended to about 40 municipalities throughout the entire regional territory.

Implementation of the IRP methodology requires free software, a GIS-based operative approach, and the availability of databases, especially as far as soil use is concerned. Some of these databases can be found online (e.g. the OMI database), while others are not available directly for the public, such as DTMs, river topography, flood depths and receptor uses. Moreover, a technical/engineering approach is required, especially in order to obtain consistent inundation hazard mapping.

The results show that simplification is necessary to obtain reliable and *robust* estimations of hazards and risks. In this respect, the main uncertainties that can affect hazard, exposure and vulnerability are represented by the quality of the datasets that are available at the present state-of-the-art, especially as far as vulnerability and exposure are concerned. In particular, proper vulnerability curves or vulnerability estimators should be proposed in the scientific literature for practical use.

Hazard mapping is influenced by several uncertainties (Hall *et al.* 2008), most of which depend on the availability of updated information as far as DTMs are concerned.

Even without considering the effects of climate changes, the uncertainties in hydrological estimates are an important topic of discussion (Merz 2009). It is well known that deviance of the hydrological models can significantly affect flood estimation and therefore risk assessment. As already mentioned, the hazard maps available online are based on hydrological estimates that are contained in official documents approved by the PBDA, and this topic has therefore not been systematically approached up to now. However, this topic could be discussed in the next updating of the maps (the 2007/60 EU directive compels the member states to update maps every six years).

Uncertainty analysis, including the analysis of climate change uncertainties, should also lead to a qualitative and quantitative treatment of the available information. In particular, a quantified approach is required to understand the magnitude of the uncertainties and to focus on the resources necessary to reduce uncertainty. In the real world, where resources are finite, the available time is short and the risk protection requirements are pressing, public administration needs to carefully balance requirements and actions, in order to at least mitigate the actual and documented risks.

At present, hazard maps and qualitative risk maps are available in the Piedmont Region and are online for dissemination and confrontation with the public, politicians and local administrations.

The proposed IRP methodology can be considered a first step towards a quantitative analysis of risk and a valuable means of supporting decision making. It is currently being developed to help decision makers compare different political strategy options and to improve risk mapping in preparation for the next review foreseen by the EU directive.

References

Agenzia delle entrate. OMI dataset. http://www.agenziaentrate.gov.it/wps/content/nsilib/nsi/documentazione/omi. Accessed 5 Dec. 2015.

ARPA-Environmental Protection Agency. 2007. Il Piemonte nel cambiamento climatico. Report. Isbn: 978-88-7479-066-1.

Bruno, A. 2013. Valutazioni quantitative del rischio alluvionale. Thesis (Italian).

Dankers, R., L. Feyen. 2008. Climate change impact on flood hazard in Europe: An assessment based on high resolution climate simulations, *Journal of Geophysical Research* 113, D19105. doi :10.1029/2007JD009719.

Dumas, P., S. Hallegatte, P. Quintana-Seguì, and E. Martin. 2013. The influence of climate change on flood risks in France – first estimates and uncertainty analysis. *Nat. Hazards Earth Syst. Sci.* 13: 809–821. doi:10.5194/nhess-13-809-2013, 2013.

FEMA-Federal Emergency Management Agency. 2003. Guidelines and specifications for flood hazard mapping partners. Appendix C: Guidance for riverine flooding analysis and mapping. http://www.fema.gov/pdf/fhm/frm_gsac.pdf_.

Foglino, S. 2013. Valutazione della pericolosità e rischio idraulico nei territori fortemente antropizzati: il caso della città di Susa attraversata dalla Dora Riparia. Thesis (Italian).

Franzi, L. 2012. Flood risk management in rivers and torrents. In *Risk management for the future - Theory and cases*, ed. J. Emblemsvåg. doi: 10.5772/16448. http://www.intechopen.com/books/risk-management-for-the-future-theory-and-cases/rivers-and-torrents-flood-risk-assessment.

Gerstenecker, C., G. Laufer, D. Steineck, C. Tiede, and B. Wrobel. 2005. Validation of digital elevation models around Merapi Volcano, Java, Indonesia. *Natural Hazards and Earth System Sciences* 5: 863–876. doi:10.5194/nhess-5-863-2005, 2005.

Hall, J., D. Solomatine. 2008. A framework for uncertainty analysis in flood risk management decisions. *Intl. J. River Basin Management* 6, (2): 85–98.

Hirabayashi, Y., S. Kanae, S. Emori, T. Oki, and M. Kimoto. 2008. Global projections of changing risks of floods and droughts in a changing climate. *Hydrological Sciences Journal* 53(4): 754–772.

IACWD-Interagency Advisory Committee on Waster Data. 1982. Guidelines for determining flood flow frequency. Bulletin 17B of the Hydrology Subcommittee, Office of Water Data Coordination, U.S. Geological Survey, Reston, Va.

Italian Government. 2010. Decreto Legislativo 23 febbraio 2010, n. 49. Attuazione della direttiva 2007/60/CE relativa alla valutazione e alla gestione dei rischi di alluvioni. *Gazzetta Ufficiale*, April 2, 77.

Italian Parliament. 1989. Legge 183/89. *Gazzetta Ufficiale*, May 25, 120.

Klinke, A., and O. Renn. 2002. A new approach to risk evaluation and management: Risk-based, precaution-based, and discourse-based strategies. *Risk Analysis* 22(6): 1071–1094. doi: 10.1111/1539-6924.00274.
Kundzewicz, Z.W., S. Kanae, S.I. Seneviratne, J. Handmer, N. Nicholls, P. Peduzzi, R. Mechler, L.M. Bouwer, N. Arnell, K. Mach, R. Muir-Wood, G.R. Brakenridge, W. Kron, G. Benito, Y. Honda, K. Takahashi, B. Sherstyukov. 2014. Flood risk and climate change: Global and regional perspective. *Hydrological Sciences Journal* 59(1): 1–28.
Lane, E.W. 1955. The importance of fluvial morphology in hydraulic engineering. *American Society of Civil Engineering. Proceedings* 81, paper 745: 1–17.
Maione, U., and U. Moisello. 1976. *Elementi di statistica per idrologia*. Ed. La Goliardica Pavese.
Marjolein, J.P., F. Klijn, K. M. de Bruijn, E. van Beek. 2011. The meaning of system robustness for flood risk management. *Environmental Science & Policy* 14(8): 1121–1131.
Merwade, V., F. Olivera, M. Arabi, and S. Edleman. 2008. Uncertainty in flood inundation mapping: Current Issues and future directions. *Journal of Hydrologic Engineering* 13 (7): 608–620.
Merz, B., H. Annegret Thieken. 2009. Flood risk curves and uncertainty bounds. *Nat Hazards* 51: 437–458.
Noman, N.S., E.J. Nelson, and A.K. Zundel. 2001. Review of automated flood plain delineation from digital terrain models. *J. Water Resour. Plann. Manage.* 127(6): 394–402.
PBDA. 2001. PAI. Relazione generale (design flood directive – Annex).
PBDA. 2012. Progetto di variante al PAI: mappe della pericolosità e del rischio di alluvione. www.adbpo.it/on-multi/ADBPO/Home/articolo1351.html.
PBDA. 2014. Idrologia di piena e cambiamenti climatici http://pianoalluvioni.adbpo.it/wp-content/uploads/2014/06/Allegato_3_Idrologia-Cambiamenti-Climatici.pdf
Regione Piemonte. 2009. DGR 2-11830. *Bollettino Ufficiale Regione-BUR* 34 27 Aug. 2009. http://www.regione.piemonte.it/governo/bollettino/abbonati/2009/34/siste/00000512.htm.
STRADA project. 2013. Strategia di adattamento ai cambiamenti climatici, per la gestione dei rischi naturali, nel territorio transfrontaliero. http://www.progettostrada.net/media/report_conclusivi/Report_Conclusivo_STRADA_Azione_5.pdf
Tebaldi, C., K. Hayhoe, J.M. Arblaster, and G.A. Meehl. 2006. Going to the extremes: An intercomparison of model-simulated historical and future changes in extreme events. *Clim. Change*, 79: 185–211.
Todini, E. 2007. Hydrological catchment model: present past future. *Hydrol. Earth Syst. Sci.* (11)1: 468–482.
UNDRO. 1980. Natural disasters and vulnerability analysis. Office of the United Nations
USACE. 1992. Catalog of residential depth-damage functions used by the Army Corps of Engineers. In A. Stuart, L. Davis, and L. Skaggs. *Flood damage estimation*.
Villani, P. 2001. Il rapporto sulle piene in Piemonte. CUGRI Relazione delle attività 2001.
WMO-World Meteorological Organization. 2009. Integrated flood management. Concept paper. Geneva: WMO 1047.

Maurizio Tiepolo[59] and Enrico Ponte[60]

21 Conclusion

21.1 From the State of the Art to the Book Approach

Scientific literature and manuals on climate change (CC) characterisation and planning in the urban areas is, by now, monumental, as we are reminded by the bibliography of the 5th Assessment report drawn up by the IPCC Working Group II (2014).

To what extent does this knowledge impact on CC mitigation and adaptation planning?

To find out, we chose two climate areas: subtropics and tropics. Within these, we identified 368 large cities. We then identified how many of these large cities have a climate tool, which included mitigation, adaptation and emergency plans, strategies and policies.

The analysis of 82 planning tools adopted by the subtropical and tropical large cities (Tiepolo and Cristofori) highlighted a gap between the characterisation of CC and knowledge summarised in the 5th IPCC Assessment report. With regard to planning, this seemed to be influenced by a variety of limits.

Only 24% of large cities have a climate plan. Those cities that do not have a plan are not unaffected by CC: quite the contrary. The large cities south of the Sahara (almost all of which are without a plan despite being repeatedly hit by natural disasters) are the most resounding examples.

A large part of climate planning is funded, often entirely, by donors. Aid is not aimed at those countries that are completely lacking resources, nor those where CC has the most influence. So much so that until today (December 2014), none of the 20 large cities of the Least Developed Countries has a climate plan, despite these being human settlements among the most hit by CC.

Having clarified that climate plans barely cover a quarter of the large cities and not those most in need, let us summarize here how the existing plans are made.

Forty percent of the tools are emergency plans. A quarter of the tools are mitigation and/or adaptation plans. The rest are master plans, sustainability plans, or local development plans.

Few emergency plans characterize CC, identify the areas at greatest risk, practice scenario planning or are well linked with national emergency planning (Ponte).

59 Maurizio Tiepolo is an associate professor of urban and regional planning at DIST-Politecnico and University of Turin, he is author of paragraphs 21.1 and 21.3, maurizio.tiepolo@polito.it

60 Enrico Ponte is a planner PhD holder of a post doc grant at DIST-Politecnico and University of Turin. He is author of paragraph 21.2, enrico.ponte@polito.it

Usually these plans focus mainly on short-term measures. We can understand how this indispensable form of planning is nevertheless an insufficient solution to CC.

Climate planning usually only considers the main capital municipality, despite the fact that most large cities in the subtropics and, particularly in the tropics, consist of many administrative jurisdictions. Coastal areas are rarely planned.

Emergency or adaptation plans are mainly aimed at protecting large cities from flooding. Drought, heat waves, and dust storms are rarely characterised and subject to specific adaptation measures. Furthermore, the hot spots on which to focus the adaptation measures are not identified because there is a lack of risk mapping.

With the mere exception of some members of the Organization for Economic Co-operation and Development (OECD), the implementation of large cities' climate plans remains imprecise, in costs and programming. These aspects have to do with the operational quality of plans. We have investigated this important character of climate planning with three simple indicators regarding the cost of planning, the measures and the origin of the resources necessary to fund the latter. As these three aspects almost always remain imprecise, we can conclude that the operative quality of the plans is rather low. So low that in the majority of cases, climatic planning seems more like a politically correct compliance to the requests of the international community, the donors, and the international conventions signed by the individual countries than a set of measures driven from the local stakeholders and which find a place in the municipal budget.

Having found these discrepancies, half of the case studies included in the book regard Least Developed Countries (LDCs): Dar es Salaam (Tanzania), Méké (Senegal), Nawalparasi and Pragatinegar (Nepal), Niamey and its hinterland (Niger), Tabarre (Haiti), Gaza province (Mozambique).

The book also provides CC analysis and tools for decision making in contexts characterised by scant information on the climate and hydrology.

21.2 Lessons Learned from Case Studies

The book aimed to increase the knowledge of four aspects that have been insufficiently investigated so far: comparative analysis of similar cities, joint analysis of city and hinterland, climate characterization and adaptation planning, and methodologies adapted to the needs of local governments.

Similar cities. At least a quarter of subtropical and tropical large cities (Tiepolo and Cristofori) have climate plans. Despite the impulse in the adoption of climate plans over the last three years, the large cities of LDCs continue to lack climate planning tools. Exceptions aside, the quality of the climate analysis and, more generally, of planning remains low, even in the case of specific human settlements such as coastal cities, where demographic density increases the amount of people and goods prone to natural hazards (Ponte).

City and hinterland. Since 2008, Niamey has been flooded every year, as have many other rural municipalities several kilometres away (Tiepolo and Braccio). These have also been affected by drought (Bacci and Tarchiani). In such cases, CC is manifested with a delay in the onset of rainy season (which often causes a false start in sowing by farmers) as well as an early end. Temporary exodus by those peasants who have lost their harvest due to drought is an extreme adaptation measure. The nearest destination is Niamey. Once in city, places where it is cheap to settle are often those most at risk of flooding. Similarly, the North coastal region of the State of São Paulo, which comprises the municipalities of Caraguatatuba, São Sebastião, Ilhabela and Ubatuba, is one of the Brazilian areas most prone to flooding and debris flow deposition, owing to hydrological extreme rainfall events usually coupled with extreme tidal levels. Events such as the catastrophic scenario of Caraguatatuba on 18th March 1967, which resulted in one of the most serious natural disasters in Brazil, fosters discussions about probabilities of heavy rainfall and causal events, as well as the rise in the sea level in coastal areas (Sakai *et al.*).

Integration of climate analysis and planning. Many chapters contribute to knowledge on this topic. What causes flooding? It depends on the type of water course (flash flood, riverine flood). In the absence of daily river discharge and daily rainfall records from meteorological stations, planners can rely on other, specific information. Three-hourly rainfall (TRMM) is a useful source in these cases. Furthermore, they show that the majority of rainfall that causes flooding happens at night, when people are less ready to react to the event (Bacci).

Sometimes, cities are exposed to a contextual hazard: river flooding and sea level rise (Sakai *et al.*) or pluvial and river flooding (Tiepolo and Braccio).

Early warning is one of the first measures to adapt to CC. This is always provided with climate plans.

The identification of a correlation between local drought and global phenomena, such as the El Niño Southern Oscillation cycle, could permit early warning on a local scale, such as in the case of the Paraguayan Chaco, where impacts differs according El Niño or La Niña (Pezzoli and Ponte).

In large countries such as Mozambique, early warning in the event of heavy rains would not generally be launched, except in those areas most at risk: along the river and where population is most dense (Cristofori *et al.*).

Reducing groundwater salinization in Dar es Salaam leads to reducing the demand for water and to increased infiltration (Sappa and Luciani). This would require measures of water conservation, rainwater harvesting and awareness on a local scale (Shemdoe *et al.*).

The alteration of the micro climate is due also to the settlements' physical configuration and to the distribution of large infrastructures, such as in the case of sea-spray dynamics along the shoreline (Piazzola and Tedeschi).

Methodologies adapted to local government needs. The mainstreaming of natural resources local monitoring, local committees and pilot projects may be adapted to various contexts, such as in the case of Dar es Salaam (Macchi and Ricci).

In Nepal, we can ascertain which basic administrations (village development committees) are most vulnerable to drought under both financial and human aspects (Giri).

In Senegal, vulnerability to drought must be managed on the local community scale (Biconne).

Examples are also given directly from local administrations, such as is the case of Catalonia (Prohom and Puig), Piedmont (Franzi *et al.*) or Zurich (Ronco *et al.*), where CC is monitored and the risk of flooding is mapped by application of the European directive FD 2007/60/EC. Mapping the flood risk and measuring it through indicators allows us to focus the adaptation in specific sectors and hot spots.

21.3 Recommendations to Make Climate Planning More Incisive

Recommendations are aimed at ensuring that large cities are less prone to natural disasters through the adoption of plans managed by competent local structures. Cities need to act quickly, using snapshot as well as incremental geographical information systems that start from preliminary risk mapping, to allow decision makers to identify (and implement) measures for hot spots. We have identified six recommendations for donors, local administrators and planners:

1. Co-funding climate plans in LDCs and large cities most exposed to CC. It is good to combine the drawing up of plans and local planning capacities strengthening. This recommendation may seem banal, but how often are climate plans the product of temporary international consortiums that do not really strengthen local capacities, but rather supply a "turn-key" plan that is then not applied locally?
2. Characterising high temperatures, dust storms and droughts. Also in this case, it is good practice to mobilise and strengthen local capacities, especially those of the national weather services.
3. Consider metropolitan core, belts and rural hinterland of LDCs. The outcome is two-fold: (i) having better knowledge of the impact of CC on urban areas, especially in the terms of climatic migrations from the hinterland towards those urban areas exposed to the risk, and (ii) identifying the jurisdictions requiring adaptations that are not necessarily those in which floods, drought and heat waves occur.

4. Establishing or strengthening the local GIS data on CC. Each case study collected in this book demonstrates the importance of local databases on climate, hydrology, flooded areas, damage, and vulnerability over the longest possible period of time, without which it is impossible to characterise the hazard, identify hot spots and provide accurate adaptation solutions.
5. Producing preliminary risk maps through high resolution remote sensing, and investigate poverty and adaptation. The aim is to identify the hot spots requiring priority measures.
6. Increase the implementation of mitigation and adaptation measures. Usually in the subtropics (OECD countries), a majority of the mitigation and adaptation is carried out by individual citizens when asking for permission to build. In the tropics, this is not possible due to the cities' unregulated physical growth. Donors may address the infrastructure support (e.g. Millennium Development Goals 7c) on adaptation hot spots. However, it's time to improve local taxation, as the large cities can have great problems, but also have great resources, starting with their vacant lands (Tiepolo and Braccio 2014).

FSC
www.fsc.org
MIX
Papier | Fördert
gute Waldnutzung
FSC® C083411

Kolibri 360 GmbH
Ferdinand-Jühlke-Straße 7
99095 Erfurt, Deutschland
produktsicherheit@kolibri360.de